"The scope of this book is nothing less than the
scope of man."
Newsweek

"An endlessly enchanting, stimulating work of
high literary value."
Modern Medicine

"Stimulating . . . profound . . . vivid."
Wall Street Journal

"A modern classic."
Columbus Dispatch

"As lively as life itself, as surprising in its twists and turns
as a roller-coaster ride, as humane and lucidly down to earth as
a book can be."
Saturday Review

"This is the gamiest treatise ever written on the body . . . the reader
moves through the pages continually alert, both enlightened
and delighted."
Kirkus Reviews

Gustav Eckstein was a medical doctor, scientist, writer, teacher and philosopher. He taught mainly at University of Cincinnati, but also visited the Soviet Union several times and studied with Ivan Pavlov, the pioneer in the field of conditioned reflexes. Eckstein was the author of 10 books ranging from fiction to scientific, including *Canary*, and the bestselling *The Body Has a Head*, first published in 1969. He died in 1981.

THE BODY HAS A HEAD

The inspirational introduction to
how the body works

GUSTAV ECKSTEIN

PRELUDE

This edition published in 2017 by Prelude, an imprint
of Prelude Books Ltd
13 Carrington Road, Richmond, TW10 5AA, United Kingdom

www.preludebooks.co.uk

First published by Harper & Row in 1969
Condensed edition published by Bantam in 1980

ISBN: 978-1-911440-57-4

1 3 5 7 9 10 8 6 4 2

Publisher's Foreword

Despite being published in the same decade as *The Selfish Gene*, Richard Dawkins' exploration of genetics, *The Body Has a Head* belongs in a different stylistic universe. Dawkins is approachable, but writes in a clinically precise style – by contrast, Gustav Eckstein dazzles the reader with a literary flurry. Although he rarely uses one word when he can get away with ten, he still manages to produce a sense of tearaway pace. The first chapter takes us in just 20 pages from early Ancient Greek sages Homer and Thales to Descartes' 17th-century description of the body as a machine.

Eckstein achieves his energetic delivery with a unique, staccato style that owes as much to poetry as to traditional non-fiction writing. Where conventional prose gets its narrative flow from the use of linking words, here the often sparse sentences are pulled together by their rhythm. Take, for instance, his opening description of the male sexual function: 'Into the town comes the swashbuckler, has something to sell. That is his role, or his illusion. Assault is his physiology.' To understand the approach, it helps to see something of Eckstein, the man.

Though a medical doctor, Gustav Eckstein was anything but a conventional physician. Biographical notes describe him as a scientist, a writer, a philosopher and more. He even wrote a play called 'Christmas Eve' that appeared briefly on Broadway, though the response of the *New York Times* critic was not entirely positive: 'Dr. Eckstein has struck a bold note in anti-entertainment'. Eckstein's writing style seems to have strongly reflected his personality. A fellow medic, Stanley Block, described Eckstein's reaction to discovering something new: 'he would become so energized that he was literally unable to sit still. He would leap from his chair (usually it was a high lab stool), and pacing around the room gesticulating wildly, his voice rising an octave, he would allow the new insights to tumble out …'

Block also describes the experience of visiting Eckstein's lab: 'There was a doorbell beside the screen door with a large sign reading "Do Not Enter. Ring The Bell!!". When you rang it, and waited, Dr. Gus would eventually appear wearing a large, straw farmer's hat. The reason for all this was the fact that a small flock of free-flying canaries lived in his lab. He did not want them loose in the college building; hence the screen, the sign, and the doorbell. And when they became excited, he wanted to spare himself their droppings; hence the straw hat, as well as the newspapers that were spread over all the surfaces in the lab including the piano which he kept there to play when the spirit moved him.'

Inevitably some of the science has dated. Our detailed understanding of the human body has moved on, particularly in brain science and genetics. *The Body Has a Head* spends a lot of time on, for example, the impact of lobotomy, the work of Pavlov (who Eckstein studied under on visits to Russia) and the theories of Freud, where now these might be given less weight or (in the case of Freud) ignored altogether. Bearing in mind the relative proximity to the Dawkins book, genes get all of three lines of coverage. And there are some oddities, such as a weak understanding of British history (we are told Newton was 'Shakespeare's contemporary', despite the playwright dying 26 years before the scientist was born) and bizarrely, though somehow typically, when writing about chlorophyll, Eckstein strays off to Mars where we are told that 'Mars has a moss-like green arriving and departing through the Martian year'. This was a fictional view of Mars from a different age. However, the vast bulk of the physiological information has not changed significantly since Eckstein's day.

It's perhaps best to think of each chapter here as a collection of reflective essays that deliver plenty of physiological facts – but even more philosophical musing. With Eckstein as a guide we don't so much explore the workings of the body, as ponder their contribution to the whole that makes a human. Because always there is a link back to the brain and the mind. In Eckstein's hands, the body, very firmly, has a head.

To Martha Keegan,
who put her spirit into my writing

Contents

III. ENVIRONMENT
Up from the Sea

IV. MUSCLE
Ivory Until It Moves

V. HEART
Regular Day and Night Deliveries

VI. BLOOD
Five Quarts

VII. LUNG
Rock of Earth in Raiment of Silk

VIII. DIGESTION
It All Works like a Hungry Dog

XIII. NERVE
Carrying the Torch

XIV. BRAIN
Cold Earth of Life

Preface to the 1980 Edition

Almost ten years ago, when I was working at the galleys and then at the page proofs for the first edition of THE BODY HAS A HEAD, I realized that the book was what I had hoped for but I also suspected that it was too long. The, then 21, divisions each seemed successful in itself, and they fell in the right sequence, but the total meaning of the book seemed almost lost in the mass. After it was published, I learned that I was not alone in thinking that it might be improved by judicious pruning.

When an author has written a book, people who read it talk to him about it, reviewers too, and though no one as I remember told me outright that there was too much, that is what I felt they meant. They said they skipped as they read. They found something in the index that they were personally concerned with, then found something else. In that zigzag manner they might end by reading the whole book. There were, of course, the resolute who kept going from beginning to end.

Think, if you will, of Tolstoy's *War and Peace*. It is one of the greatest achievements of the creative mind. Yet, several times when people have talked with me about it, I have come to the conclusion that they have not read all of it. I once remarked on this to Vivien Leigh, who was in Philadelphia for a pre-Broadway tryout. She listened to me thoughtfully. Then she said, with beautiful innocence, that she had read *War and Peace*, but "not the war parts, of course."

It was a satisfaction therefore when I learned that the publisher wanted me to do this later version. I know it to be a better book now, that it will be read in all its parts because, besides holding tighter to its intention, it is more readable. I also know that, for the purpose of the book, the ten years that have intervened since the first edition have required some changes, but really only a negligible amount.

I decided very early in the revising to drop two of the 21 divisions. I can easily recognize now that those two had been written not with an eye to the reader, but to the physiologist in me. They were adequate, were clear, but they got in the way of what I was trying to say. From this gross cutting I saw the principle I should use throughout the abridging: everywhere the nonessential could and must be pruned. This I then applied to pages, paragraphs, lines, phrases.

I believe my text has lost none of the sense of wonder in the human body that I feel. And I believe that I did not lose the poetry. Especially, I believe I did not diminish the sense of excitement about what science speaks of as brain-mind, and what the person in the street is apt to call simply mind.

Gustav Eckstein
September 1978

Introduction

The intent of this book is to make the human body more familiar to anyone who owns one.

During a summer vacation years ago, I read back into history to find where the idea of the body as a machine began. That idea—our body a machine—has been a main stitch in the fabric of thought for centuries. To contemporary man, his body *is* a machine. He accepts this early in his life, takes it literally, not as analogy, and feels no irony. I did not read that summer as the scholar reads but as I liked, much or little.

That was how this book got started. The first short chapters, the history, are flashbacks, scant biographical sketches, each of a person who more than others was responsible for placing one milestone on the long road that led to our present-day understanding of the body, our experimenting with it, our resolve to wrench from it its secrets.

After that the book keeps to a line, *Growth, Environment*, a special environment inside us and curious; then the organs, *Muscle, Heart, Lung*, roughly how each works, roughly how the parts of the total work together, how the body is provided with a shrewd *Defense*, and so on; short chapters, conceptions as much as facts, pictures as much as dynamics, the hope being that there will rise before the reader a better of himself.

A literary tone fell of itself over the writing, was allowed to, because the wonder of the body, met at every turning, inspires moods of wonder, of philosophy, of far distances, thoughts of universe or universes, of the majestic vague, frequently of our smallness in immensity, but also the reverse, of our private immensity in the presence of those particles of which there are always more and more, and of which finally we are constructed.

The human mind is the body's master and the book's destination. It is a mind that sometimes seems half-ashamed of its extraordinariness, that forgets its unparalleled story, which began in dimness so-and-so many millions of years ago to arrive finally at the power of a single mind to write *Hamlet*.

I

HISTORY

Better than Any Machine
of Human Invention

HOMER:
World Before Naturalists

If Homer was one man, which I believe, he lived in the ninth century B.C., or the eighth, or the seventh.

To Homer all natural forces—the flaming sun, the shipwrecking wine-red sea—were either gods or the contrivances of gods; and so too what went on in the houses and the streets, so too the hewing of the timbers for the building of the ships, the blowing of the bellows for the forging of the shields, the maidens down at the river doing their washing; so too chicanery, treachery, badinage, passion, victory in battle, or defeat. Cause and effect were personal. Under their heroic masks those gods were human beings, but they did not die, and like human beings were by turns mysterious and not at all mysterious. Bedroom and drawing room and courtroom proceeded on Olympus much as they proceeded here below, and since things up there did mirror things down here, man took on an added stature. He was not merely of earth but of sky.

The human body to Homer stood forth now in striding beauty and now in a heap of hacked-off limbs. Guts spilled onto the plain. A spear was thrust at one side of the skull of a hero, cracked it open, came out the other. Eyes were popped from sockets. A warrior was cut through at the neck from behind, the blade reappearing at his chin, and his body wilted. It was known to Homer that a blow on the temple or on the orbits of the eyes might wipe out life. Much like that was then known, doubtless more than is said or suggested in the *Iliad* or the *Odyssey*. Homer's panorama was one of poetic grace and rough force, neither accounted for physiologically.

THALES:
Who Measures Is a Scientist

Two hundred years passed by, or three hundred. Now there were naturalists in our sense. Trying with slaughter of sheep and heifers to flatter the gods while they were making administrative decisions on Mount Olympus was giving way to watching the storm advance in the valley. The gods did not settle all things; some went on rules of their own.

Among the early naturalists, the name Thales stands high. Thales thought the magnet had a soul, whence its power to move iron—soul and movement, cause and effect. Thales was a Greek. He traveled to Egypt. There the Nile regularly rose over its banks, receded again, and the Egyptians knew how to measure and reestablish rapidly the boundaries of each man's land. Thales was interested in the measuring but also in the principles of measuring. Up to then no one in the Greek world had been much concerned with that.

Thales is credited with bisecting a circle and proving the halves equal. Is credited with calculating the height of the Pyramid of Cheops from the length of its shadow, did it by a geometry of triangles simple today, not then. Calculated the distance of ships at sea as a modern surveyor might. Was a clever man. Accurately predicted an eclipse. Which suggests he was also a lucky man. Hundreds of years after his death people would say that he led thinking from theology to physiology, and though physiology could not have meant what it means to us, the reasons for the skill of the body had begun to excite man's analyzing imagination.

DEMOCRITUS:
Atom

Two hundred years again passed by. The Greek was restless. The Greek was wanting to find the bottom of things, wanted the ultimate, the final finality, wanted that illusion.

Here are some words of Democritus: "By convention sweet is sweet, by convention bitter is bitter, by convention hot is hot, by convention cold is cold, by convention color is color. But in truth there are atoms and the void." *Atomos* meant uncut. Of the atomists the most distinguished was Democritus. To him atoms were small, in rapid motion, met at random, differed in shape, and their shape determined their action. To Democritus soured wines and acids that pricked the tongue were composed of atoms with sharp points. On the contrary, if atoms slipped easily in and out and nudged a passage between other atoms, they must be round. Flame had round atoms and so did the soul. It followed that reality was atoms and space and atoms and space and atoms and space, and was discontinuous. The search for the ultimate, the irreducible, never again would cease. It would go where Democritus' wildest imagining could not have imagined. He lived four centuries, probably, after Homer, fifth century B.C. By the twentieth A.D. the ultimate would have become a relation of energy and mass definable to the physicist, no doubt lucid to him, somewhat murky to most of us. An electron microscope would attempt to photograph the atom and display it before us, not succeed, quite. The unseeable nevertheless would be split. The splitting would result in a bombed Hiroshima, result in nation after nation adding itself to the proud list of those where science is paramount and bombs are built, with the probability of more Hiroshimas yet to come, and end of war, if achieved, would not be because thinking man had delivered himself beyond bombs but because he did not want to die. His instinct to escape destruction would have maneuvered him into a bargain with others who had the same instinct. Trapped man would be, body and mind, by atoms.

SOCRATES:
Seminar

Socrates asked. Socrates answered. Usually he did both, asked, did not wait for an answer, answered. He was a great intellectual, who talked. The manner of the talk would be remembered through millennia, the artificiality, the annoying precision, the flashes of wisdom, the genuine feeling. Socrates was serious. Socrates was moral. Socrates was not easily turned aside, was both patiently and impatiently insistent. It was always said that his wife Xanthippe nagged, but he also nagged, and we do not know much about her and we do know about him. Indeed, he bludgeoned men to precision, to stating exactly what they meant. They must define each term. They must listen to each question. His own questions he slanted so that the questioned would look into himself and discover the truth that had been there all of the time. This was the Socratic method. It would survive. The relentlessness, the coolness, the vague or less vague fretfulness, would like a ghost from Attic streets be standing somewhere behind each chair at each seminar in the mid-twentieth century. His method was his milestone.

The bright Athenian youth flocked to him, the foolish youth, forsook felicity to learn, found felicity in this seminar of the streets. It was a situation sure to enrage someone. Furthermore, Socrates was fearless as to what he asked or whom it galled, turned on respected citizens in public places and asked them questions they could not answer, which was bad. He joined no party, which was bad. After a while he managed to offend all parties. The outcome? He was accused of impiety, brought to trial, found guilty. What surprised him was that the vote of the jurors was close. "Had thirty votes gone over to the other side, I would have been acquitted." Instead of regarding his trial as a blemish on Athens, we should probably regard it as a tribute, consider that no later people might so far have suppressed their distaste toward an unpleasant person in favor of justice as did those Greeks. Even during his trial and after his conviction he remained his usual gadfly self, inquiring, pronouncing, persecuting words. He conducted this seminar like the others, unruffled, lucid, sometimes as a man not concerned, as if the issues debated actually were more important than whether he lived or died, as if death actually were a life-to-afterlife passage-way. Even on the long day of his execution he asked and he answered.

PLATO:
The University

Plato reported those disputatious seminars of the Athenian streets, later brought seminars within walls, brought learning and philosophy within the confines of the Academy, which has correctly been described as the first university in the world.

Plato revered Socrates, but seemed himself to have been different in genius and in tact, seemed not one to have gotten himself into such a quandary that it would be necessary to test his readiness to die for his thoughts, lived with success until he was eighty.

Plato theorized about the earthly—how it reached toward the unearthly. Theorized about government—how it could be made to bring to the greatest number of citizens the fullest realization, not of their rights, but of their capacities. To theorize with talent in any age, one must be born with the temperament, and Plato was. He had one of the strongest and yet light-spirited intellects of all time. University dialogue would, because of him, though diluted by the centuries, have more give and take, more ostensible tolerance, less humor, more urbanity, more of the conscious, less of the unconscious.

Plato theorized that beyond the apparent was the ideal. Reality merely approximated the universal, mostly distorted the universal. He would not have wasted his time or a pupil's on, say, a particular triangle. He might use a geometric form in illustration, had curiosity about geometric form, but why dwell on a single isosceles triangle when there was existent the lovely principle, triangle? Similarly with good and evil: not a good man, but good; not an evil man, but evil. And behind good and evil? Virtue. Virtue was the intimate of knowledge, and knowledge could be learned, and taught.

When Plato theorized about the human body, he brought fantasy (possibly it was not fantasy to him, but conviction) to the concrete. He said: "The gods made what is called the lower belly to be a receptacle for the superfluous meat and drink, and formed that convolution of the bowel, so that the food might be prevented from passing quickly through, and compelling the body to require more food, thus producing insatiable gluttony, and making the entire race an enemy to philosophy and music, and rebellious against the divinest element within us." He theorized that the flesh of the neck served as an isthmus to separate the mortal below from the divine above. He theorized that the lungs served as cushions to prevent the undisciplined heart beating itself to shreds during bouts of passion.

Twenty-three hundred years afterward all of this might seem quaint as statement and quaint as physiology, but the direction that had begun to show in Thales was plainly pushing forward.

SOPHOCLES:
The First Freudian

Athens was a place half the size of Waukesha, Wisconsin. Twenty-one thousand citizens. There were also ten thousand foreigners and forty thousand slaves. In that city in its great period were living the following: Thucydides, the almost unparalleled historian. Herodotus, historian also, his mind filled with strange biological observings and stranger biological interpretings. (Biology had begun pushing forward.) Thucydides died at seventy-one. Phidias, builder of the Parthenon, died at sixty-eight. Pindar, poet, died at seventy-four. Pericles, statesman, orator, writer, died at sixty-six. Socrates drank the hemlock at seventy. Then there was Plato, and Democritus. And Aristophanes, via a pricking irony letting the air out of various windbags, now and then letting blood out, as when he helped hound Socrates to his death by jabbing at him from the stage, Socrates sitting there in the audience. Aristophanes could do this and still be a poet. That was Athens. Aeschylus, the first of the three mighty writers of tragedy, lived until 456 B.C., well into the great period. Sophocles was second. Euripides third. These all were in that city.

Sophocles was the most polished of the three. The most probing. In his quiet there was always an urgency. He never let go of his story. He began with what seemed tragedy's highest feelings, yet the feelings mounted, as an agonizing thirst mounts, pulled from within to the point of pain, each emotion transferred to each spectator as he sat waiting for thirst and pain.

Sophocles was an actor in his own plays. He was born at Colonus. It was said he was a handsome man. A general in the army. A rounded, a perfect Athenian. Lived to be ninety. Lived between the battles of Salamis and Marathon, therefore not only within the great period but at its peak. Year after year, almost, he won first prize for tragedy. He wrote one hundred and twenty plays. Seven remain. *Oedipus Rex* was his greatest. The whole afterworld agreed but not the Athenian critics. *Oedipus Rex* did not win first prize.

The argument runs so. Oedipus, the king of Thebes, in an altercation at a crossroad slew his father, not knowing who he was, then wedded and bedded his mother, not knowing who she was. Patricide and incest. Many years elapsed. Pestilence fell on Thebes. An oracle brought down from the Pythian temple pronounced pollution was within the city, and it must be swept out. The king vowed it should be. But what was this pollution? Where was it? A soothsayer, unwilling to tell what he knew, refused to say. The king, in a rage, accused the soothsayer, who in turn accused the king. The king was the pollution! For this seeming falsehood the king might have struck the soothsayer dead, instead dallied with the accusation, was lured by the dalliance, followed threads, sharpened his questions, could not halt his self-pursuit, by a marching compulsion uncovered fact on fact, until his guilt stood forth as clear as sunlight—one thought one saw the literal sunlight falling on that crowded theater of Dionysus—so clear that he ripped from his wife, who had hanged herself, the brooches of gold that fastened her robe, lifted them high, brought them down, gouged out his two eyes, blotted out earth and sky, presented to his audience his white face with its two blood-streaming sockets; theater and tragedy.

How much of this took place at the behest of the hidden? How much was chance? What was this relentlessness? What events were symbol? How much in gouging out his eyes was he gouging out some other organ? A long and penetrating probing of the normal and the abnormal, the direct and devious, the bizarre ways of our human mind must have preceded the writing of such a play. The play would be interpreted and reinterpreted. It would rise in the psychiatry of the twentieth century, sit there in the king's place, lash on the human mind to continue its perpetual harassed self-analysis.

HIPPOCRATES:
Learning from the Sick

He was Dr. Hippocrates, of Cos, Greece. "Redness of the face, eyes fixed, hands distended, grinding of the teeth. . . ." That was his economical description of a man who had had a stroke. Often of an evening this physician walked by the harbor. He reduced a fisherman's dislocated collarbone, diagnosed an intermittent fever. Watched sickness. Watched himself. Watched men. He said so. What unexpectedness he saw in a human body, heard in it, felt with his fingers, he wrote down. He had stubborn conviction as to what the practice of medicine might teach the practitioner. "I think that one cannot know anything certain respecting nature from any other quarter than from medicine." To the Coan school of medicine, which was in competition with the Cnidian school, he added instruction and fame, both schools spreading their learning and their often opposite views through the ancient world, and then through the rest of the human world.

A marble bust of him is in the British Museum, a style more than a likeness, and better so, because what do we know of him as a person? Two thousand years after his death, the Spanish realist, Velázquez, painted his portrait. Hippocrates may have looked like that portrait.

From a vast mass of medical writings gathered under his name, one pieces together how this physician probably thought.

There is an anecdote that when he was young he had to flee from his town because he set fire to the library so that no one might know what he knew. He got to be quite old. The years 460 B.C. to 377 B.C. are accepted. Granting that he recognized he was unusual, which is probable, he could hardly have anticipated that men at this far point of time would still be saying that he was the Father of Medicine.

The Hippocratic oath—repeated by medical students about to become physicians—believably attaches to him. It bespeaks the rules of an honorable profession, and it bespeaks some of the mundane realities of the medicine of that time. Hippocrates had done what he could to rip that medicine from the supernatural. He objected to anyone's saying disease was a punishment from the gods. It had natural causes. The physician must pay attention "to the patient's habits, regimen, and pursuits; to his conversation, manners, taciturnity, thoughts, sleep, or absence of sleep, and sometimes his dreams, what and when they occur; to his picking and scratching, to his tears; to the alvine discharges, urine, sputa, vomitings. . .."

Epilepsy, he insisted, ought not to be called divine. It was a disturbed brain. He noted how the epileptic wanted to be alone, and this had nothing to do with any fear of the divine but was the epileptic's shame at having his fit in public. Different if the epileptic was a child. "Little children at first fall down wherever they may happen to be, from inexperience, but when they have been often seized, and feel the approach beforehand, they flee to their mother, or to any other person they are acquainted with, from terror and dread of the affection, because being still infants they do not yet know what it is to be ashamed." He learned from all the ages of the human being. He drew truth from the disease itself, from the physician himself, from the man in the sick man.

Something like that was Hippocrates' milestone.

ARISTOTLE:
Animals Added

Here is what Aristotle said: "If there is anyone who holds that the study of the animal is an unworthy pursuit, he ought to go farther and hold the same opinion about the study of himself." With his clear mind in that later, still-golden age of Greece he went from animal to man, man to animal. Sharks. Rays. Sponges. Octopi. Cuttlefishes. Sea urchins. Deer. Weasels. Aristotle dragged the Mediterranean for specimens. Raked the land. The comparing of animal with animal—comparative biology—begins with him. After him the comparative would increasingly include man. To read him is like a fresh cup of morning coffee. "The elephant's nose is unique owing to its enormous size and its extraordinary character. By means of his nose, as if it were a hand, the elephant carries his food, both solid and fluid, to his mouth; by means of it he tears up trees, by winding it around them. In fact, he uses it for all purposes as if it were a hand. . .. It is the opinion of Anaxagoras that the possession of hands is the cause of man being of all animals the most intelligent." So, already in the fourth century B.C. someone was not only taking for granted that man was an animal but was speaking freely of it, and someone was struck by the intimacy of hand and mind. "But, it is more rational to suppose that this possession of hands is the consequence not the cause. . . . The invariable plan of nature in distributing the organs is to give each to such an animal as can make use of it . . . . For, it is a better plan to take a person who is already a flute-player and give him a flute, than to take one who possesses a flute and teach him the art of flute-playing." "Quadrupeds find it no trouble to remain standing, and do not get tired if they remain continually on their feet—the time is as well spent as lying down, because they have four supports underneath them. But the human being cannot. . . . His body needs rest; it must be seated. That, then, is why man has buttocks and fleshy legs, and for the same cause has no tail; the nourishment gets used up for the benefit of the buttocks and legs before it can get as far as the place for the tail." Though his reasoning here may be upside-down, though we know that in obvious respects it was wrong, though we always have realized that we might not be reading exactly what he wrote, none of that is important. What is important is how in his observation and his speculation he went from animal to man, man to animal. In the sciences of the human body, this comparative approach was henceforth a part of all thinking of any size.

He concerned himself with everything—birds, talk, poetry. There is his realistic theory of tragedy, with which all subsequent writing on tragedy was apt to begin, even late psychoanalytic writing, the theory so trenchant in Greek ideas of art that one has needed sometimes to remind one-self that Aristotle did not write the great Greek tragedies, only confidently explained them, left us his theory of the tragic.

The gods on Olympus, throughout, were like shadows fading before the accumulating light. Aristotle, with his force, his intellect, his statement, was supporting the new facts, or the new illusions that would grow in power and produce eventually naturalistic science and mechanistic logic. Around him was still the youth, the gaiety, the lightness of Athens, but these were lessening, and there might in the future be left only the logic, and logic without youth, gaiety, lightness, and creation becomes mathematics, analysis, presumption, and peremptoriness.

LUKE:
Made in His Image

Greece declined. The myth and the reality were in the glory of the sunset, where they would stay. The Hebraic world had begun earlier and was where it was. Rome had grown to power. Julius Caesar was born and died. Jesus Christ was born and died. That last gave a twist to history. Christianity's strengths, its weaknesses, its sentiments, its sentimentality, its tendernesses, but especially its way of regarding the sick—the stress put upon suffering and dying—would affect all later thinking about body and mind.

Luke, the beloved physician, what do we know of him? Turn to the Bible. It will surprise you. You will be sure there was more and that you have forgotten. Humility he had; we may say this confidently because of the Christmas story, but if fact is looked for, there is not much. Nevertheless, we were not wrong in what we thought we remembered, because in Luke the physician we saw Christ the physician. Luke had taken from Christ the compassion that we thought was Luke's alone. Another meaning had come to the physician. The halt. The lame. The blind. The healing of the leper. Made whole by faith. The resurrection. There were a hundred new phrases, with many implications. The idea and the ideal of the human body were changing. It never again would be what it had been to the Greeks. That shine was off it. It had become a garment. A mystery touched it, gave it importance, reminded it of its brevity, then left it. There had been similar thinking in different corners of the world, but this now was sewn with such insinuation into the cloth that it would have taken a fervor equal to that which sewed it in to pluck it out, and there were no Greeks to do it, and they alone would have had the intensity and the cunning; probably not even they.

GALEN:
The Clinician

Rome in turn declined. The Western Empire—what we mostly think of as Rome—passed its peak.

Dr. Claudius Galen, of Rome.

For four years Galen had been surgeon to the gladiators in Pergamum, where he was born, must often have looked across the sea to Rome. "That is the size of city for me." Six years, then, this Greek physician practiced in Rome with a wild vitality. After that, we do not know much of his whereabouts, but what he was at we know. He was writing the most voluminous physiological and clinical work of all antiquity, volumes of blunt fact and free interpretation; with this was welding an authority that would reach through twelve centuries. "Galen and Aristotle," men said. "In the name of Galen." "Galen did." "Galen spoke." "Galen wrote." The writing reveals his range.

Animal Experimentation. He probed the living animal—watched its organs while they were in action. Was saturated with the temperament

of the vivisector. Cut into swine and oxen and the Barbary ape without benefit of anesthesia. Cut open the living skull, looked in, saw
the pulsations. Cut the nerves to the larynx, paralyzed the voice, concluded that the control of the voice must be from somewhere above,
where those nerves came out of the brain. Associated what he saw in
the vivisected with what he saw in the sick human being. Before death
a patient had had difficulty breathing and at postmortem, Galen recollected, a tumor had been found in the brain. Reported the facts
with cool clinical-pathological-conference acumen. Later, the part of
the brain where the tumor was pressing or gouging would be known
as the respiratory center. A slave, Galen recollected, had committed
suicide by holding his breath.

Anatomy. Probed the dead human body—was saturated with the
temperature of the dissector. Of the twelve nerves that go from brain
to head, trunk, chest, abdomen, he knew seven. (This was the second
century A.D.) Left rules for dissecting a brain. Gave permanent names
to parts of it. Knew two of the three membranes that wrap it.

Physiology. Probed the living human body—was saturated with the
temperament of the medical investigator. Was interested always in
the work of the kidney. "Now, the amount of urine passed every day
shows clearly that it is the whole of the liquid drunk which becomes
urine, except for that which passes away with defecation or passes off
as sweat. This is most easily recognized in winter in those who are
doing no work but are carousing, especially if the wine be thin and
diffusible. These people rapidly pass almost the same quantity as they
drink." An informal medical observation but important. Later, physiologists would be measuring the quantities exactly in hundreds of
circumstances, would be continuing a line that can be said to join the
measuring of Thales to the measuring of Galen to the measuring of
any modern anyone cares to mention.

Obstetrics. Probed the parturient—was saturated with the temperament of the obstetrician. "The midwife does not make the parturient woman get up at once and sit down on the chair, but the midwife
begins by palpating the os as it gradually dilates, and the first thing
she says is that it has dilated enough to admit the little finger." Obstetricians would be saying that eighteen hundred years afterward,
tonight.

The sick are always with us, but the talent of a Galen occurs only sporadically through centuries.

His rationalized explanations went in all directions. He was vivisector, dissector, physiologist, obstetrician, neurologist, psychiatrist, every species of clinician, businessman, and writer.

His milestone was conspicuous. He sank it with assurance. He seemed by intellect and temperament to anticipate the modern multiple approach to the understanding of body and head.

CHAULIAC:
Plague

A man lies sick. A doctor has been sent for. This now is a thousand years after the fall of Rome. The doctor arrives on horse. He does not dismount. He is as groomed as the horse. A member of the family comes to the door. From the horse the doctor makes some formal inquiry, hardly deigns to look at the member of the family, nods to his assistant, who stands at a respectful distance. The assistant acknowledges the nod, disappears into the house, takes the sick man's pulse, asks him to stick out his tongue, reappears at the door with a sample of the sick man's dejecta. He on horse bestows a glance on the dejecta. Knows the date of the sick man's birth, the hour, the astrological conjunctions, the magical and alchemical possibilities, if the man was rich has already sent his ring for blessing to the king, finally goes into a séance with himself, comes out with a prescription for rhubarb. There were exceptions, but the common run of doctors conversed of Galen and Aristotle, and they knew less and less of both. As for the living body, they rarely laid hand on it, and as for the dead, the dissection of two corpses in a lifetime might give a man eminence, at least add eminence to one who already had it. This was true for Guy de Chauliac, who practiced a brilliant surgery in Avignon, France, between 1342 and 1368.

The Dark Ages were not dark, not literally. The sun did rise every morning. Birds sang. There was a hot bun at supper. Babies laughed, fell asleep, the moon came out, its magic drove lovers into each other's arms. And doctors continued to converse of Galen and

Aristotle, and to dispute of the status and the rights of barbers versus physicians.

300 A.D. 400 A.D. 1100 A.D. 1300 A.D. Then—plague struck Europe. Bubonic plague. Black Death. The Great Dying. It struck again and again. The first definite European date was 1347. Earlier it had ravaged Asia and Africa. Apparently it was brought to Europe by ship late that year. The next three years, 1348, 1349, 1350, were the frightful years. A quarter of the population of Europe was wiped out. And plague kept striking. It burned out in Sicily. It struck in Italy. In Spain. In Oxford, England. Eighty thousand dead in Danzig. Forty thousand dead in Paris in the one year 1466. Before the entire toll had been taken, sixty million were dead. We of this century have not yet been able to match this even with a war. The hearts and the eyes were on the ground. "The Lord gave, and the Lord hath taken away; blessed by the name of the Lord." Read Defoe's *Journal of the Plague Year*. Recall Boccaccio's *Decameron*. See again Raphael's agonized painting *La Peste*. The sick were deserted and their houses burned. A starving wretch touched the stone of a house where plague broke out in the night, and he was hanged next day for witchcraft. Another was hanged for pollution of the wells. Elizabeth I shut herself up in Windsor Castle, a gallows outside for anyone who dared come there from stricken London. Whole towns were evacuated. Transportation stopped. Speech all but stopped. The doctor, poor pompous ass with an engraved voice, stood revealed. The bankruptcy of his calling stood revealed. He saw nothing, knew nothing, was nothing. The human body had become a shroud. What the doctor believed to be the unchallengeable truths of Galen and Aristotle was a heap of broken type. This fellow could not perform the simplest palliative acts. He died and was himself buried with the priest. The medicine of the Middle Ages was largely buried with him. The disillusion of human beings in their doctor swept medicine toward the modern era where all the sciences one after another would be drawn into the study of the living machine. Not plague alone wrought this change, because the revival of learning touched medicine as all else, but plague put the granite hand into the doctor's silk glove.

Guy de Chauliac, the surgeon, did not run away from Avignon, as did others of his profession when plague struck their cities. He stayed in Avignon and worked. People called him Guy.

PARACELSUS:
Metals in Medicine

Dr. Philippus Aureolus Theophrastus, of Basel, Switzerland.

This one was crazy, a little. And so sane. We are in the second quarter of the sixteenth century. Milestones fly—in the pages of history. It was St. John's Day. At the celebration of the students at the medical school, by way of launching his first course of lectures, Theophrastus Bombastus von Hohenheim (Paracelsus) threw into the fire the books of Galen. In the announcement of his lectures he stated that they would be not from Galen but from Paracelsus, would be delivered not in Latin but in the vernacular, German.

This was the new world. This was the end of the millennium. This was air let into the palaces, the houses, the churches, the polluted wells, the cellars.

The birth of Paracelsus had occurred in Einsiedeln, a small place. It was the year after Columbus discovered America. Paracelsus' father was the village doctor. Later the family moved to a mining town where there was a mining school, and Paracelsus learned chemistry, especially of metals, worked in the mines and in the mine laboratory. Then began his *Wanderjahre*. All his life he enjoyed travel, and traveled.

In 1526 Paracelsus was preparing to settle down as a doctor in Strasbourg, but was invited to the medical school in Basel, accepted, and that was when he threw the books into the fire. The faculty at Basel did not enjoy him. The great humanist Erasmus did. The faculty wanted him dropped. He petitioned city council. It would not have looked well for city council to have supported the faculty immediately. An event mediated. A burgher of Basel had been ill and no doctor had helped him, so he promised a reward to any doctor who would, one hundred guldens. Paracelsus cured him. The burgher, cured, paid ten guldens. Paracelsus took the matter to court. The judges decided in favor of the burgher. Paracelsus told the judges what he thought of their justice, and therefore had to steal out of Basel under the cover of night. He had been in that city two years, and faculty and council were rid of him without the slightest self-sullying act.

In spite of all the energy he flung around him, Paracelsus had energy left for experimental investigation. He turned to profit what he had learned in the mines, started chemical pharmacology, introduced metals into the pharmacopoeia. No one yet knew that a tiny amount of some metal might be essential for life, the body that lacked it not surviving, but we cannot be sure of what Paracelsus did not know, or guess. Galen had prescribed mainly herbs; Paracelsus prescribed antimony. Paracelsus was the first to employ the word *zinc*. He argued for the healing value of salts, sulfur, mercury.

One knows that if one had been ill in Basel and free to call whomever one liked, it would have been Dr. Theophrastus Bombastus; in Rome, Dr. Galen; in Athens, Dr. Hippocrates.

Professional modern opinion passing on Paracelsus' skeleton has held that he did not die of a blow on the head. (The blow was purported to have been received in a tavern brawl.) We are not sure, but in any case the hour came that always seems to him to whom it comes to come too soon. His death took place at Salzburg. His last will and testament he had dictated to a notary three days earlier. The words on his tomb are: "Here is buried Philippus Theophrastus, distinguished Doctor of Medicine, who with wonderful art cured dire wounds, leprosy, gout, dropsy, and other contagious diseases of the body, and who gave to the poor the goods which he obtained and accumulated. In the year of our Lord 1541, the 24th day of September, he exchanged life for death."

LEONARDO:
The Human Body in the New World

The Renaissance arrived early in Italy. From imperceptible sources suddenly dawn was everywhere. In that soft light all human beings must have been observing the human body for the artists to have observed it with such fervor. Art had become again a force in the advancing history. The *Mona Lisa* took four years to paint. Vasari said of the portrait: "To look closely at her throat you might imagine that the pulse was beating." This would not have been said two centuries earlier, the human eye not having then that quickly shifting vision, for art, for science. In the course of the years of his life, Leonardo designed mills, invented instruments, drew plans for a war tank, a steam gun. He thought like an artist, and thought like a scientist, employed both capacities as they could be employed in the Italian Renaissance, had violence, had intelligence, had peace of mind. He constructed clay models, sketched with color on cambric and linen, talked, talked, talked. How convincing that talk must have been. He persuaded a man that a certain temple might be lifted from its foundations and moved to another place, and the man did not realize how grossly impossible it was until Leonardo and his voice had vanished. He slid open the door of a vendor's bird cage, let the birds escape, paid the vendor, vanished. He drew plans, every schoolboy knows, for a flying machine. Chauliac had dissected two human bodies. Leonardo dissected thirty. He brought the knowledge learned from the dead to the living, his paintings and the drawings of the living having always under them the firm structural framework that he had found in the dead. He has been blamed—and when he was coming down to his own end it seems he blamed himself—for starting so much and leaving so much unfinished. However, had he finished what he started, and one doubts he ever meant to, the loss would have been ours, because we would have been trading for completed works that consumed his strength the wonder of beholding the human body in the many guises he beheld it in, at the peak of inspiration, unbelabored, unchilled. The shapes that the mouth takes in battle. The unseeing eye of a thoughtful child. Heads rotated in passion. Torsos twisted. Arms forcibly flexed. Abdomens bent. Figures kneeling in adoration. These are as perfect often as those of the Greeks, never as young, more actual, more touched by something that we do not miss in the Greeks, a darker psychology, and this not in the fact alone but falling like a shadow over the whole. Christ as he sits at the middle of the table in *The Last Supper* definitely has a body. Above it is a human head, a human mind, a human sadness, and this does not subtract but adds to that body's divinity.

VESALIUS:
Dissection in the Morning

It would have been one of the sights of this world—Andreas Vesalius, the Belgian, dissecting. A legend has it that a robber was hoisted by a chain and slowly burned to death, a roast for wild birds that picked and picked till finally there was left the skeleton. Vesalius never could resist a skeleton. At night he let himself get locked outside the city walls, tore off single bones, brought these one after another into the city as there was opportunity, assembled them, remounted them, then, on being questioned by the authorities, presented them with the skeleton. It was sin besides outrage to hash the image of God.

Andreas Vesalius dissecting—in Paris. In Paris labored the famous anatomist Sylvius. Sylvius read aloud in the gradually discredited texts of Galen while a demonstrator cut at a corpse to demonstrate what Sylvius read. If a part mentioned in the texts could not be found in the corpse, and if Sylvius was forced by the headstrong Vesalius to admit an error in Galen, he commented that man's body had changed since Galen, and not for the better. Sometimes Galen had dissected a dog's body, and the demonstrator was dissecting a human body.

Andreas Vesalius dissecting—in Padua. Italy was the center of the world in anatomy, Padua the center of Italy, Vesalius the center of Padua. He was twenty-three. Corpses, corpses, corpses, he wanted corpses. A legend has it that he stole them when he could get them in no other way. The errors in Galen mounted. There were hundreds. By 1543 Vesalius' immortal work, his atlas of anatomy, *De Humani Corporis Fabrica*, was published. His fame reached over the earth. He was twenty-nine.

Back in Paris, Sylvius was continuing to read Galen and continuing to denounce Vesalius. Sylvius hated Vesalius. Many anatomists may have hated him—and why not? He withdrew to Spain, to Madrid, had been summoned there as physician to the court, had become less the anatomist and more the fashionable doctor. He lamented that in Madrid he could not "get hold of so much as a dried skull, let alone the chance of making a dissection." Back in Belgium, Maximilian d'Egmont was ill. Vesalius not only predicted that Maximilian would die but when, the day, the hour. "The dread anticipation occupied the Count's mind. He called his relatives and friends to a feast, distributed gifts, declared his last wishes, took formal leave, waited with suppressed emotion, and at the hour predicted, died." All doctors ought to be reminded of the power in a prognosis.

Throughout these latter years a pupil of Vesalius, Fallopius, quietly dissected. Fallopius published his own atlas of anatomy. Vesalius began to hate Fallopius. A final legend has it that Vesalius got permission to dissect a corpse in order to discover the nature of the fatal malady, and when the corpse was opened the bystanders saw the heart beat. Unlikely, but Vesalius had been an envied man and the luster was off him, and the youth. The family of the deceased appealed to the Inquisition. The emperor intervened in Vesalius' behalf, ordered him to make a pilgrimage to the Holy Land, and it was on his return that either he was stricken with fever or he was shipwrecked, swam ashore, died of exhaustion, this on the Mediterranean island of Zante, and a ship companion, a stranger, a foreigner, placed a stone by the dissector's body. *Requiescat in pace.*

HARVEY:
Quantitative Experiment

It would have been one of the sights of this world—William Harvey vivisecting. Frogs. Pigeons. Eels. Crabs. Oxen. Sheep. Shrimps. Serpents. Slugs. Wasps. Hornets. Flies. Chicks. Snails. Mice. All of them have blood in them, and William Harvey's sharp dark eyes looked into them all. Harvey was using all of them to quench his curiosity, to find every proof he could that the flow of the blood was continuous, from the heart back to the heart, an irregular circling but a circling.

This was the seventeenth century—Shakespeare's great decade near the beginning of it, Newton's near the end.

Harvey spoke of the difficulty his eyes had to follow the beat of an animal's heart. Today, any freshman medical student follows with ease, sees—because he knows beforehand what he ought to see. To see first, first in the world, anything, is psychologically a completely other process than to see second.

Harvey measured in a human corpse the quantity of blood the heart can hold, then in a sheep the quantity its heart can hold. The two quantities were comparable. Next he bled a sheep. The sheep's entire blood was not more than four pounds. At each beat the heart forced out some. The total forced out in half an hour must be greater than all there was in the sheep or—Harvey stated this in different ways—must be greater than the sheep's body needed for nutriment, greater than could be produced by the nutriment eaten by the sheep. Therefore, the blood pumped out by the heart could do nothing else than return to the heart. It was ejected, went in a circle, returned.

It circulated. Harvey was not the first to use that word *circulate*, and he was not the first to conceive the blood as departing from a point and again arriving at that point, but he was the first to prove it. And he established, we rightly say, the quantitative experiment. This was Harvey's milestone. He brought number to the sciences of the body. Hereafter the living would be investigated with mathematics.

Add to this brief portrait the detail that completes all our human brevities, the signs and symptoms of an apoplectic stroke. He discovered that he had a "palsey," "and he saw what was to become of him, he knew there was then no hopes of his recovery. . . ." The apothecary was sent for "to lett him blood in the tongue, which did little or no good; and so ended his dayes." It was the year 1657. Harvey was born in 1578. So, he had reached seventy-nine, a good age even in our day. Shakespeare was fourteen the year Harvey was born. One wonders what the boys and girls in their respective English streets thought of those two, William Shakespeare, William Harvey.

DESCARTES:
"Give Me Matter and Motion"

I said in the Introduction that I had read back into history to find where the idea of the body as a machine began. I had thought it must have aggregated during centuries. But no. At least, a first statement came full-blown from the mouth of one man. "The body is a machine made by the hand of God, infinitely better than any machine of human invention." Those were the words of René Descartes.

A geometrician, Descartes invented analytical geometry. A physiologist, he wrote the original description of a reflex. A philosopher, he arrived at the conclusion *Cogito, ergo sum*, three Latin words known to many who know no Latin. "I think, therefore I am." For a time Descartes had been a soldier. Then he seemed to experience the need to learn everything afresh and from the foundations, a need some men experience, and in order to accomplish it, settled away from his countrymen, far away, out of Paris, out of France, in Holland. His mechanistic thoughts he expressed neatly. "Give me matter and motion and I will make a world." Expressed them too neatly, possibly, but that was his temperament, Gallic, thrifty, terse.

The seat of the soul to René Descartes was in the brain. He knew where. The pineal gland. The pineal was not paired as were other parts of the brain, was located in the center, and he mistakenly believed it to dip into the fluid-filled cavities that Galen had described. To a mathematical mind like Descartes's the fact of this one structure in the midst of the twos of the brain seemed proof. Descartes's reasoning went as follows. The soul was in the pineal. He did not say the soul was the pineal; it resided in the pineal, temporarily; he was a Catholic and his faith was firm. The soul stirred. The pineal moved. It acted like a valve, guided the flow of the vital spirits through those cavities, the ventricles. Everybody in that day believed in vital spirits. These traveled out the hollow tubes of the nerves—everybody believed in hollow tubes—and at the other end of the tubes the muscles contracted. As simple as that. As machinelike. The human machine to Descartes was a mechanical machine, whereas to us it is in addition chemical, thermal, electromagnetic.

Descartes lay on his couch. He did not doze as he lay. He kept thinking. He wrote. During those twenty years in Holland he went back to Paris only three times.

Letter writing was the chief means of spreading knowledge in that busy century. Anyone wrote freely to anyone. Many all over Europe corresponded with Descartes. Christina, queen of Sweden, was wanting to know how to live happily and still not annoy God. Who has not wanted to know that? The French ambassador advised her to write to Descartes. She did. Soon Christina was inviting him to Sweden. It proved a long journey, and when he arrived in the autumn of the year he regretted he had come, found Sweden cold, especially the hearts. A Frenchman, and fifty-three, old for that day, he was accustomed to rising late and having breakfast in bed. Christina, a Swede, and twenty-three, preferred her lessons by candlelight before sunup. At that wretched icy hour Descartes had to walk to the palace, or ride in a sleigh, caught a cold which became pneumonia. The queen— we are all so foolish—sent to this Frenchman a Dutch doctor. That was nearly to send a German. The doctor concluded he must bleed the Frenchman. Moribund, Descartes rose from his bed. "You shall not sacrifice one drop of French blood." He said that over and over, lingered between delirium and lucidity, commended his soul to God, and died. This was 1650. The queen—we are all so foolish—wanted to bury this Catholic in a Lutheran graveyard. To that the French ambassador forcibly objected, and the corpse went to the graveyard for those children who had not been baptized, the foundlings, the little bastards. Years later, Descartes's greatness evident in the world and in France, his body was exhumed, taken to Paris, buried again, and, as if that were not enough, exhumed again, buried again, in St. Germain des Prés, lies there now, what is left of it. Impossible to escape the irony that this corpse, which during its lifetime had been the shining new philosopher who proclaimed across the centuries and down into our day that the body is a machine, should have been thus bumped from junkyard to junkyard.

II

GROWTH
I Becomes 1,000,000,000,000,000

FEMALE:
Lamps Trimmed and Burning

She has plenty of them. They have been counted. They may not fall short of half a million. Only one needs to go forth each lunar month, so a simple arithmetic makes clear that the total used in a female life is only one tenth of one percent of her inheritance. In a modern marriage two may be expected to get to college. Experts formerly believed her allotment was in her when she was born, but later evidence suggests she goes on making them. Eggs. Like enough to a hen's, containing yolk enough to keep a new life from starving during its first inexperienced days. For greater dignity our species calls ours the human ovum largest cell of the human body.

A female whale has an ovary weighs three pounds. Odd to think of an ovum in an ovary in a whale who is keeping her lamps trimmed and burning.

A tube—fallopian—points in the direction of the ovary, is wide-open there, like a funnel, its fringed border in ceaseless motion. Funnel and tube are the open roadway when that egg sallies forth. Careful nature is almost careless here, because an egg if it happened not to slip into the funnel might into the abdomen, get lost, does occasionally, is fertilized there, rarely. For any routine lunar month, funnel and tube are sufficiently near. Inside the *tube* are microscopic lashes that have a rhythmic beat, create a current, paddle along that small important object, its escape impossible once it is caught in the millstream. At the other end of the tube is the brood-chamber—the egg's destination. If fertile, it will dwell there a time. That brood-chamber, the uterus, has the shape of a pear, its cavity in the intervals between pregnancies a mere chink, tremendous at term, its muscle cells grown seventeen to forty times their interim size, and their myriad number myriadly increased. The pregnant uterus is one single-minded muscle when it needs to be.

At the tip of the pear an opening leads into still another passage, three inches long, which joins the brood-chamber to the outside world. At its entrance the caller is received.

Everything except the ovaries are the female's accessory reproductive organs. Everything is enclosed within the bony framework of the pelvis. Everything is ready. Later, at that entrance a small blurry transient will be ushered out.

Far and wide over the female body is a feeling of welcome. Far and wide the lamps are kept trimmed and burning. By those lamps one sees in a special way, separates the sexes, distinguishes the female abdomen, chest, waist, hip, small of the back, hair on the head meeting the brow in a bowlike curve. Large breasts. High-pitched voice. A temperament that varies female to female but is female, a quality of mind, a manner of thinking, of sighing, complaining, understanding, misunderstanding, talents all its own, old often though the body still is young, and when modified by life modifies in its female direction. That temperament especially keeps the lamps trimmed and burning, waits.

MALE:
Into the Town

Into the town comes the swashbuckler, has something to sell. That is his role, or his illusion. Assault is his physiology.

He is the owner of two testicles. Carries them outside in front of him, a unique idea that even the male mind never quite gets used to. They are in a hanging pouch, skin mostly, no fat, sweat glands to help cool them. The reason for the air conditioning could be that the male sex cell, the spermatozoon, manufactured by the testicles, dies if subjected to the high temperatures inside. A fever kills spermatozoa in catastrophic numbers.

Loops of tubes and loops of tubes and loops of tubes, hundreds, closely packed, lead to a delta of channels draining into a dozen large tubes—that is the testicle. The dozen drain then into a single tube, twenty feet long, *feet*, microscopically thin, coiled, also closely packed—that is the epididymis. A small organ, the epididymis, tacked to the top and back of the testicle, the total of its channels, its corridors, leading in one direction, out. That seems forever the spermatozoon's ambition, out. Either it is on the road on its travels, or it is gathering itself together back at the start, where in layers of cells the production of the spermatozoon proceeds stepwise from the most external layer to the most internal, and there the finished marvel emerges, a no-nonsense construction. Whereas the ovum is the largest cell of the human body, the spermatozoon is the smallest. As slim as an iron filing and lured a magnet, small spermatozoon, sad in its self-important way.

A duct continues the travels, runs under the skin of the groin, up over the top of the pubic bone, down beneath the bladder, approaches its companion of the other side, and, as if this were not too much already, receives there its own offshoot, which delivers a glistening fluid. Everything suggests a rich concoction brewed in stages. The final lap is via a tunnel, the urethra, that traverses eventually the frank instrument, a common passageway for urine and spermatozoa, insulting, as if Nature even at this late date did not take seriously our place in her plan. All the accessory male parts are supplying ingredients to the fluid that is the liquid world in which the spermatozoon first floats as a physical object, then races on its own propulsive power. Once it has struck its stride, it may attain a speed of more than two inches an hour.

The swashbuckler when he entered the town carried with him his secondary sex characteristics—angular look, visible muscularity, low voice, a mind that touches with maleness everything that body touches, and a confidence born of the knowledge that he is possessed of the talent to keep life alive by fertilizing an egg.

MATING:
The Presentation

They meet. They partly know, partly blunder, partly learn. It states a fundamental fact about us—some learning attaches even to the most instinctual performance. In the scale of animals, the higher the creature the greater is the capacity to learn, and the more manipulated the instinct. A young male chimpanzee does not just know how to mate, but fumbles (a distinguished primate student describes to us), and a female chimpanzee, this is more amazing, has to learn how to take care of her infant, is clumsy with her firstborn.

The male's duty is to deliver the spermatozoa deep inside the female. Nature has great interest that this delivery should be guaranteed to the full, 60,000,000 to 200,000,000 spermatozoa in a cubic centimeter, and two to four of these at each encounter. If fewer than 60,000,000, the male is apt to be sterile, though he may be sterile even if he has the 60,000,000, and fertile sometimes when he has not.

A combination of senses, sight, touch, smell, stimulate the male, the act advancing until at a moment with a burst those spermatozoa are released, after which the entire male body relaxes. In the female the mating is apt to be quieter. In her as in him there are serial contractions, their mechanical purpose being to help the male complete the release and to assure the spermatozoa arriving duly where they should, in the uterus. At the end of the act her body relaxes. The mating is complete.

CONCEPTION:
Toast to the Future

In the warm wet dark an egg slips out of an ovary, once in twenty-eight days, but only on the average, as every woman knows. One egg in an ovary packed with eggs sidles over to the wall of the ovary. The wall thins at that point. A blister forms. The blister bursts. The egg is free. It enters the tube, starts toward the brood-chamber. If the blessed event is to occur this month, it must soon, and within the tube, because the journey down the tube, four inches, takes three days, and an egg is dead in hours. That journey is rapid at the start, rapid at the finish, slow in the middle, where the egg appears to loiter expectantly. Meanwhile the appointed spermatozoon has been coming the other way, if the mating has been properly timed, or is lucky. Or unlucky, should this not be what the lady and gentleman meant. Spermatozoon and siblings enter the necklike entrance to the uterus, hustle up, thence on out into the tube. Millions are hustling. This is not merely nature's usual making sure. The spermatozoon besides being a living cell is a chemical able to dissolve the jelly that embeds the egg. Millions do the work of dissolving, one small spermatozoon sneaks in, and that woman has conceived. Here is the moment to toast the future. The Japanese reasonably consider us a year old when we are born, but even a Japanese waits till the actual birth to go to the front of his house and hang verses on the trees.

ZYGOTE:
Genius or Blockhead or Plain Citizen

Every cell in the human body has forty-six chromosomes, as thirty days hath November. For years it was thought forty-eight, also forty-seven, but now forty-six. The chromosomes have our genes, our genes have our heredity, and of the study of genes there is no end.

Only the sex cells have half as many chromosomes, the reduction to half, to twenty-three, occurring in the female and male primary organs.

At conception a twenty-three male goes up to a twenty-three female, tips his hat, and without further fuss they are united in a forty-six zygote. That cell—zygote—is the first cell of the impending new human life. It is the fertilized ovum. It has in it the contribution of both parents, and their parents, and the parents back to Jehovah. It possesses the entire stupendousness.

The zygote's sex also has already been decided. There are two kinds of male cells, and whichever kind won that illustrious race up the uterus and out the tube decided. Ova have sex chromosomes of one kind, X. Spermatozoa have two kinds, X and Y, the Y smaller, but important.

If in the tube Y met X, boy. If X met X, girl. Ergo, the zygote was not only the first cell of the new life but had upon it the vestment of man or of woman. What a moment! The clock stops in Tristram Shandy's kitchen, or it should.

Such fascination was in what so far has taken place, one might forget the something alongside: namely, this zygote had besides the potentiality of a human body, the potentiality of a human mind different from every other. It is conventional nowadays to emphasize the influences from without that play upon the zygote. Many evidently do, and upon what ensues, but as evidently, and this fact is disturbing sometimes and satisfying sometimes, that zygote was from the start touched with destiny. Where it concerns a genius one feels knocked down. Edmund Baehr complained one day to God, that He should sit up there in the sky with His pailful of the water of genius beside Him, dip in His finger, splash a drop on this one, on that, on a third, get bored, pick up the pail, pour the whole of it over the zygote of Beethoven.

PREGNANCY:
Ten Lunar or Nine Calendar Months

The zygote divides. Two daughter cells result. These divide. Those divide. The chromosomes are heralding the dividing by their own dividing. This multiplication beggars description, truly, can be observed directly only in a mouse or other nonhuman species, but we may safely assume that in those earliest hours a family man and his house mouse are brothers under the skin. The dividing goes on. Soon there is a lump of cells, called mulberry mass, that hollows, fills with liquid, and to one side of the hollow is growing a future president of the United States, or someone greater, or someone less, but whoever he is, he has henceforth the rights and privileges of a still more amazing title, embryo.

About one week after conception the entire product, hollow and that growing to the side, Pullman and passenger, attaches itself to the inner lining of the uterus, and here is the beginning of that famous connection between mother and offspring, the placenta.

The dividing goes on. Besides increase in number of cells, a separation into kinds is occurring. Each kind contributes to the building of special tissues and organs of the growing body. Try to keep always in mind that that small zygote had in it these unnumbered dynamic possibilities. Embryologists have pieced together the steps that simultaneously are taking place in different directions at different rates.

While you have been reading, the embryo has been bulging farther and farther into what was originally that chink of cavity in the uterus. Embryo is growing. Uterus is growing. Placenta is growing. Umbilical cord, which makes the deliveries between mother and fetus, is growing. Nature spares (or is it deprives?) the mother from most of this. She is just another pregnant woman. On her side of the placenta, blood is moving slowly in what has become a lake or bayou into which the ends of the blood vessels of the embryo dip. The two bloods, mother's and embryo's, do not mix, but nourishment goes across from mother to embryo; wastes go the other way, the mother adding these to her own and disposing of both through routine channels. The great source feeds and cleanses the small spring. In the meantime the developing parasite lolls there, quite lordly, hands in his pockets.

That woman carries about inside her an expanding universe. Whatever way one looks back over the course—ovum, spermatozoon, zygote, embryo, perfect small foot of a perfect small fetus—the word *marvelous*, the word *miracle*, is not enough. Her disposition too, the mother's, has been changing. She is annoyed at one special kick from that small foot directed straight at her stomach. Once the young lord stands on her backbone and delivers an oration. Her appearance is changing, oh yes. She may be annoyed at that too. How can she be? How can she be fussing with her gaudy tent of dress designed to hide the most astonishing truth in the whole of our astonishing world? She may be more beautiful in her face now. That's an old wives' tale, she says. But if that is not an old wives' tale—if she does have it—she will lose this particular beauty. It is not hers to keep, just loaned, and the loan passed from woman to woman to woman.

FETAL BRAIN:
Rake's Progress

When the zygote became the mulberry mass, along its back was a line of cells more active than others. These were the ground-breaking. Upon these would rise this most special part of this new human being, the base for his thinking, dreaming, gaiety, disillusion, swindling, lechery, tenderness. Four weeks after conception, the head end of the line had become so sculptured that any tyro could see the dawn of the later divisions of the adult brain. Everywhere the fetus was changing. (For the first three months in the womb a human creature is an embryo, after that a fetus.) Six weeks after conception the dawn brain had become the gross daylight brain. Three months, the same tyro could see that this was not only brain but undeniably human brain, shape of a lima bean. Four months, shape of a globe, unnaturally smooth on the surface. Five months, back of the globe rounded out. Seven months, all the main grooves on the outside had appeared but the surface still too smooth, still suggesting brain in the butcher's pan, edible, animal. Eight months, one no longer found oneself shrinking with some un-formulated feeling from an object that should be our kind and was not quite. Nine months, all as it should be, what one was familiar with from drawings and photographs, a grooved baby brain waiting in a womb. The reason for the grooving, we are told, is that the outermost surface has been enlarging more rapidly than the underneath, and it could not do that and still keep itself within the skull unless it folded. So it folded. So, more surface. At maturity only a third of that out-ermost surface would touch the skull; the rest would be down in the grooves. Because of that "more surface," the mind associated with this brain could go on conquering chimpanzee and gorilla and dolphin and everything except death.

BIRTH:
"Into This World and Why Not Knowing"

Tristram Shandy's father in the kitchen was worried about Tristram, who was to arrive on earth that night. Nothing would be surer cure for the neurosis of expectant fathers than to read *Tristram Shandy* in their own kitchens. They would learn, for instance, that there had been fa-thers before them. They would learn that they were not required to be as calm as the obstetrician. Of course, if they fully pictured to them-selves what was going on, what pushings, what twistings, they might never in their lives, not the fathers and not the obstetrician either, have a completely calm hour again.

Why this birth occurs when it does, why not earlier, why not lat-er, what starts it—experts give reasons, nobody clearly knows. If it occurs on schedule, it occurs ten cycles after conception, ten moons. The uterus at that time gets irritable, but why? Of course, the fetus is stretching, poking the uterus, and would it not be amusing if the whole cause were one final savage satisfying stretch? The placenta at the ten-cycle period is having its blood source squeezed upon, some of it cut off by the pressure, spots of tissues dying from lack of oxygen, and the irritation of the dead tissue might be the cause. Whatever the cause, it can act later, act earlier, even three months earlier, "preemies" being common and most of them surviving and growing to sturdy boys and girls and grandmothers and grandfathers and obstetricians.

Childbirth—called also parturition, labor, travail—is divided into three stages. All through the pregnancy the uterus indulged in mild contractions, and toward the end these were more frequent. One result was the steady angling of the package into a position mechanically right for delivery. In the first stage, small contractions became large contractions, each beginning feebly, rising to a crest, staying awhile, fading out, attended by pangs, the episodes five to ten minutes apart. Pressure was thus built up inside the brood-chamber. If the fetus in spite of nature's long experience was still not in a right position, the obstetrician is sure to be a fellow with clever fingers able to turn a passenger around without too much disturbing his composure. However, even an obstetrician can have trouble.

Usually it is the head that is most in a hurry, anxious to be out and get things started, but the obstetrician shoves it back, the head shoves the other way, a fight between obstetrician and head, the total creature meanwhile moving downward, the uterus behind gathering up the slack, the membranes bulging, those membranes that through ten lunar months were the thin inner wall of the fluid-filled Pullman that provided this passenger with the finest of possible accommodations in this finest of possible worlds. At last the membranes burst and from then on the passenger is constrained to take the bumps of our world direct. Body follows head. A small red or a small pale guest steps out, or is dragged out. Instead of a head, it may have been only the back of a head, or one shoulder may have got stuck, or a face may have appeared as if it wanted to see how things are, or a breech have presented itself, or one arm, and a spectacle that is, an arm reaching through; it adds to the difficulties. During most of this the woman will have given voluntary help, as Galen assured the Romans she would.

A half hour or so later, the contractions that had stopped begin again, to expel placenta and membranes, with which expulsion this female body is freed of its months of responsibility. Soon after the guest's arrival, the obstetrician tied the umbilical cord and cut it with scissors, with confidence. A cat does the same, with confidence, without scissors. The tying and cutting were delayed until the cord no longer pulsed, the mother's heart no longer able to pump blood through the narrowing narrows of that cord. Finally, the obstetrician touched the raw flesh with an antiseptic and fastened a sterile pad over the top. Maybe that tying and cutting should have been the occasion for giving this fetus his ticket of admission to infancy. He was a member of the human race henceforth. Maybe, also, some small formal mention should have been made of the famous first cry, the new citizen's first proclamation, excited by a slap on the backside in the barbarous days, his first swallow of free air on this not very free planet. Tired obstetrician is permitted now to go home to his own family.

Every human being every so often in his bathtub looks down with thoughtfulness at his navel, certificate that he likewise crossed the Rubicon, though he does not believe it really.

GROWTH CURVE:
An Uneven Slope

Ovum, spermatozoon, zygote, embryo, fetus, infant, child, bride, middle-aged matron, old lacy lady—growth curves could be plotted for each. Each is growing in its or his or her way.

During the first months in the womb there was the growth from mulberry mass to creature complete though small, not yet able to survive outside his Pullman. After those first months his chances of survival were better and better, and the nearer to birth the more the Pullman could be regarded as a luxury, he in there rolling along, not worrying about who was to meet him at the station, able to jump out practically when he liked, a type of luxury reserved for the human species and only a few others, most dumped out earlier.

For three months at the start of the pregnancy the gain in weight was slow (a slowly rising curve), sped during the last three months (a steeply rising curve), the peak reached at birth or thereabouts, with, immediately after birth, a brief loss in weight because the newborn had not begun to eat (a briefly falling curve), in ten days this loss made good (the curve back where it was), then for months and years, in general, a gain (the curve rising). Preschool. School. Postschool. In inches, the fetus made its greatest advance during the middle three months of the pregnancy, then a decline.

If the weight was eight pounds at birth, it would be roughly twenty at one year, twenty-five at two, a gradual increase from then on with a tendency to loss at four or five, followed by a methodic rise to puberty. From about eleven the girl would run ahead, weight and height, but the boy would overtake her, pass her, might pass her also in some powers of mind. After full height the young man would continue to accumulate weight, till he was thirty. This is all average. Exceptions are on every side. In the elderly there is loss both in weight and height, but "elderly" comes at different times in different persons, fast in any case, many at sixty already earnestly sunning themselves on the sands of Miami, apparently not having read in the newspapers that a human life might today reach one hundred and ten.

POSTNATAL BRAIN:
Rake's Further Progress

What, meanwhile, has happened to that brain that was to conquer gorilla and chimpanzee and everything else? At birth, fetal brain became postnatal brain. This grew rapidly. Enormous is the topmost part of the nervous system at birth, enormous all that has to do with seeing and hearing. Big head, big eyes, small nose, small everything else, is what impresses anyone when first he sees an infant daughter lying unworried in the bend of her father's elbow. Once out of the womb this daughter every minute is getting into trouble, is in danger of her life, requires that big head, requires the defending nervous system, though at birth this topmost part is still not ready to do much defending. It is about twelve percent of our body then, two percent when we are full-grown. One-fiftieth of us is the scene of our torments and delights, one five-hundredth of the lion, wherefore the lion must have its great muscular strength and we can get along with our comparative weakness. The weight of an average human brain at birth—the brain lifted out of the skull of a stillborn and put onto a scale—is 350 grams. At one month, 420 grams. At one year, half the adult weight. At seven years, nine-tenths. That last fact—his daughter at seven years having nine-tenths of all the brain she ever will have—could possibly help the mystified father somewhat to understand why she, now in her second year at school, is already as smart as he, lacks only experience. While the growing brain has been brought almost to a standstill, the rest of the body is steadily gaining. As for the skull, it also has been brought almost to a standstill, the stimulus for the growing skull having long been thought to be the growing brain. From the beginning, nerve cells have been dying, and since about as many as there ever will be are present at birth, that infant daughter in this respect is already on the decline.

> And so, from hour to hour, we ripe and ripe
> And then, from hour to hour, we rot and rot.

GROWTH THEMES:
How Far By How Soon

Growth could be discussed from a dozen points of view—geometry, heredity, health, dietetics, others. Each would bring in its knowledge, its sophistication, its dogmas.

Geometry could be the theme—a body to become a body must fall in with the forces that give it its lines and angles. There would be a geometry for the "living" wall and a geometry finally for every "living" molecule. As with the single cell, so with the single organ, the pushes, the resistances, the organ's size and shape thus eventually explained. And as with the single cell and organ, so with the entire body, scooping its spot from time and space.

Heredity could be the theme. Dumpy mothers have dumpy daughters often, small mothers small daughters. True, a ten-pound four-ounce baby boy did literally leap from the womb of his seventy-eight-pound mother into the amazed arms of a green medical intern, but it is too soon to write that baby's story, and he may yet, when his growing is finished, be a father as small as was his mother, have a body like hers. Single features may be race-begotten, also family-begotten. Heredity is stern. Doctors used to warn the expectant mother to curb her appetite, but obstetrical experience teaches that, thought she ought for excellent reasons not to overeat, if she nevertheless does the baby will probably be born no bigger. The reverse is true also, except in grimmest famine; in ordinary reduced intake the fetus may be somewhat reduced inside at birth; it is throughout quietly consuming its deprived mother. To be sure, the fetus might be deprived of a single item which its mother because she was starving was unable to supply, and that can be bad for a fetus. Most of the time the body attains the size and shape that was its fate, that was its heredity, that was written into the gene script.

Both excitants of growth and checks on growth might take orders from the script. A check might operate indirectly. The growth of the heart is limited by the amount of blood circulating through its substance, and the growth of the body by the amount of blood the heart can deliver, and so would an indirect check have prevented a human body from growing to an inhuman one. The kidney is big, not bigger. An eyelash is long, not longer. That gene script is employing its chemical agents, of course, chemical drudges that do blind labor for it, supervise the rates of growth.

Health could be the theme. A body to grow must have health, all the molecules operating harmoniously, proper order in time, proper order in space. Verily, that health may be a minimum. Should a children's hospital summon all its medical acumen to keep one of the inharmonious little ones alive, nature might not insist that this one be too happy, might give it a useless body and a sharp mind, let it be wheeled through the corridors with its thin legs dangling while it intelligently watched the other children at their play.

Food could be the theme. A body to grow must have food. Growth of ovum and spermatozoon was owing to molecules that were eaten by the mother; growth of zygote, to molecules eaten by the mother; growth of embryo, fetus, nursing infant, still to molecules eaten by the mother. Each fed on her; the nursing infant merely more conventionally. Later the young master rudely shoved his mother aside, drew his chair to the supper table, brought his own spoon into the foray.

Food factor could be the theme. Each body, and indeed each cell of each body, takes from the food that comes to it what it requires to manufacture itself. A bacterium may get all the carbon it needs from a sugar. If that bacterium is changed, made to mutate by, say, X ray, it and its descendants require something besides that sugar to get their carbon. That something the geneticist calls a food factor.

MISGROWTH:
Giant and Dwarf

Giants do not grow to specification. Dwarfs do not. Many human be-ings do not, in some detail. Many animals. Giants exceed specification, may keep growing, upwardly, until they are more than eight feet tall, long arms, long legs, long hands, long feet, bones of the torso huge. Dwarfs stop short. Though not growing to size, a dwarf may grow to fame, be right pleasant to look on, become the playmate of the infan-ta, have his portrait painted by Velázquez, a well-proportioned small man with an immense head who doubtless experiences the satisfaction of having sons and daughters of socially acceptable proportions. The infanta sometimes considered her dwarf not a playmate but a play-thing, and that was surely nice of her. Giants of the fairy tales and of the irony of Jonathan Swift have giant natures, are offspring of a fictitious race, as are also the giants of Richard Wagner, but the giants of the circus are not fiction, are offspring of chemical abnormality, molecule abnormality, have genes with an atom displaced. Now and then a prizefighter has been such a heavyweight, who through all the months of training before the moment he stepped into the ring was frightened. But sometimes a giant will be muscularly feeble, sexually feeble, speak with a high-pitched voice.

The bones of our arms and legs have long shafts with marrow in them, and expanded ends. For a time after birth the ends are not unit-ed with the shaft. If there is one kind of chemical abnormality, the union is delayed, the shaft grows excessively, the result is a giant, not a distasteful type. If the abnormality occurs after the union, the shaft does not grow excessively, but other parts of the skeleton do, also the soft tissues; the result is not a giant. This body is not tall. This mis-growth is of an abominable character. The person may for a time not know he is changing, seeing himself as he does every day in his mirror, but one morning he must admit, with perplexity, then with horror, then with disgust and defeat, that last winter's gloves are too small because his hands are too large, his shoes pinch because of his feet, his total body lacks symmetry in a manner befitting the damned. He stares at his jaw. The ugly thing pokes out. His teeth stand apart. His chewing is so difficult he could prefer to starve. His cheekbones jut. The arches above his eyes project. Nose and ears bulge. The vertebrae have gotten porous, the spine has given way, collapsed, poor hunch-back, his hands reach his knees. "Acromegalic" he has been classified. He is careless of his clothes because he is careless of his life. Some say Punch was one of them. Some say the village blacksmith.

Then, there is that dwarf unpleasant to look on, limbs too short, belly protruding, navel protruding, preternaturally old, skin that hangs and is cold and with a peculiar soft thick pad under it, bones of the face insufficient, especially of the nose and the orbit of the eyes, bones of the cranium oversufficient, a swollen tongue between thick lips that do not close, hair in patches, thinned eyebrows, half-calcified decayed teeth, stupid to idiocy in appearance and in fact. He sleeps. Dickens knew him. In him too a chemistry has gone wrong.

Other chemistries can go wrong, genetically wrong, a gene dropped in or out or substituted. May be merely an emphasis. Some such emphasis is in each of us, a private chemical change bequeathed to us from somewhere in the past.

We are born what we are, and if that was not lucky, we can make it worse by adding to it our thoughts. How the giant or the dwarf thinks of what he sees in the shop windows when he is pretending to study the new spring styles is a force in his life, also what others let him know they think. A human being is not simply cells. There is a mind attached. This may wish often it had been born a tree.

OBESITY:
Size

Obesity is not growth. It is increase. There can be a tenfold difference in the amount of fat in two human bodies. Fat is a poor conductor of heat, so the fat man sighs in summer and gasps and has the air conditioner going full blast, but he is mighty comfortable in winter. Undoubtedly a body can have too much fat, past all common sense in some, cart it around day and night. Those persons are the obese. To give them a name takes a kind of care of them, puts them in a class, provides them a communal defense so they do not need to fight it through alone. It might be more accurate simply to write them off as sick. The greatest quantity of their fat, half of it, is under their skin. Their weight increases, their size increases, but growth is always either increase in the number of normal cells or in the mass of each single cell, and not fat cells only.

Why ever is there excess fat? We could of course, should perhaps, blame the dark lady, heredity. A fat disposition foreordained in the zygote and come to fruition in a fat carcass runs through families, through generations, unhappy wretches haunted to consume more than is required to heat them and to get them their exercise. "For God's sake, I tell you I am hungry!" He is. Stop nagging at him who has been nagged at enough. He has given up. He has taken a job as a cook and does not think cooking respectable. He just eats, adds too much and subtracts too little, tastes, as a cook must, finds it good, eats, and on his summer holiday consumes doughnuts with a self-conscious nonfattening drink while resting in a hammock. Has waddled his way into an imbalance between the consuming and the exercise, which two must balance if fat is not to settle in grotesque amounts in grotesque places.

When the dark lady wrecks a different control, she leaves a body bloated, and that is not growth either, a dragged-out-of-the-river look, this one suffering not from excess fat, obesity, and not from excess water, edema, but he is the victim of another misgrowth, a dilute glue everywhere under his skin. When she wrecks a still different control, the muscles themselves get big; not powerful, big.

Human beings laugh in this world when they should not. We laugh at fat. The wine god was fat. Bacchus laughs, so as to be laughing before he is laughed at. Fat men like to play Santa Claus. Do they play Santa Claus to conceal their bulk in a costume?

The yearning to eat excessively goes by an excessive name, hyperphagia. *Phagia* means hunger and *hyper* means hyper, and there is an experimental procedure that induces such successful hyperphagia in a rat that the experimenter stands in revulsion before his success, a rat that eats, eats, eats, gets bigger, bigger, bigger, poor fat rat.

The top end of the obese, as of the skin-and-bones, is the head. The trouble, the explanation, is in his head, or her head, we say lightly, make the head responsible for the whole fattish situation even while we are unsure whether the head is responsible for itself. In a man's or a woman's life or her thinking about her life there can be that which lets her friends literally see her enlarge. One young woman fell out with her mother, ate determinedly for a week, gained thirty pounds.

Disappointment, frustration, chagrin—those words from the lexicon of the damned stand for drives that are weighable in pounds. The bored gain. They gain in spite of dreaming of a svelte social career. Is that heredity? Is the boredom? The dreaming? The wife who was snared? The husband who was the son of so-and-so and looked such a bargain? The brat who was unwanted? The other who was wanted too much and was overfed with coddling and food? Is nothing just chance? We say of a fat man that he is doing this thing to himself. We say he is eating to avoid facing his misery. He eats at each crisis, at each disappointment, like any addict. He tries to forget in mashed potatoes. Apparently that is easy to do. She eats and drinks and fattens and he eats and drinks and fattens and both do it to take attention from the gnawing at their hearts. But she was born fat! Yes, she was born fat. Poor woman, she whines that she is eating no more than the happier members of her family and maybe is telling the truth. She says her sister eats the same food she eats, but she gains while her sister loses, and even this might begin to be understood if the minds, the top ends, could have been eliminated before one started on the biochemical calculation. But how to eliminate the top ends? And if they were born with their top ends?

OLD AGE:
The Downslope

This gentleman who comes to mind should by the rules have been dead years before, at least should have been on the downslope, but never a sign. Then one day there was a sign. He did not tie his tie. He put on a clean white shirt but did not tie his tie. Also, instead of taking his walk around the block and then around the shopping center, he took it in the grass around the house. His body rolled a bit. Last week he let an appointment wait, five minutes, and that would have wounded his pride three years ago. One year ago in Mariemont Square a neighbor failed to recognize him, then hurried back, burst out over-enthusiastically that he had not changed in thirty years.

Did this make him glum? Why should it? The down-slope is fact, is natural, is a stage of growth. The growth curve rises, plateaus, falls. His was falling rather rapidly of late, he pointed out. It amused him.

His body had shrunk, an inch. He made the measurement himself. It is possible to shrink an inch and be not in the least bent. He also weighed less. In ten years he had gone from 146 to 132. Always he weighed himself in the bathroom and measured his height, in the morning before breakfast, was regular about everything. He had quietly and accurately looked into the face of approaching old age, turned that corner as he had others, with interest.

He was born lucky to begin with, insisted it would have been luckier not to have been born at all, been one of his mother's three miscarriages. Aggressive by the end of the first five minutes when the midwife (this was passed down through the family) caught the athlete by one foot and just missed letting him land on his head, he fought so. "It's a boy!" She said that. The mother did not say much. The growth curve fell briefly after birth, as it should, then rose, then went its inexorable course. A boy, therefore long legs that got longer. The body as a whole was on the short side but robust. With birth began the methodic tearing off of the pages of the calendar, but who in those middle European hills would have dreamed that in this instance the tearing off would not cease for more than ninety-three years?

At eleven he emigrated from the old country. That second cutting of the umbilical cord he did himself, no more hurt than the first. Never had seen the ocean. Crossed it in winter. The crowd in the steerage was seasick, not he. A new country. Rain did not give him a snotty nose, snow freeze up his joints, fears keep him locked in his boardinghouse.

Marriage. Conception. Four conceptions. Started four growth processes around him while his own continued. Those four always knew from the time they knew anything that they had a father with a healthy body and a healthy mind. The latter kept growing luxuriantly.

Prime. A satisfying intelligence. A gay cynicism. An amused irreligion. Possibly the irreligion was overvocal. His advancing years could not have provided a life insurance company with data for supreme old age. In this living machine of René Descartes, as once there had been normal acceleration of parts, now there was normal deceleration. He would have agreed that he was a Descartes machine. It would have satisfied the intelligence and the cynicism and the irreligion—the human body a trim machine. His was. It made use of all the advantages of the planet, green things, flesh broiled, fried, stewed, and at the other extreme the advantages of human history, human learning. Then there came that morning in Mariemont Square when with a sardonic chuckle he took official note of the downslope. He had too much health, some thought, also too much cleanliness. That of the body remained as on that first day the midwife gave him the first bath. Recently the baths may have required a few minutes longer, but he came out scrubbed. The fiftieth birthday was behind him. The sixtieth. The seventieth. The eightieth. A noisy glittering party at the ninetieth. The girls of twenty and forty and sixty kissed him and he enjoyed it and slept well that night. He slept well all nights, always had, slightly fitfully of late.

It was the greatness and decline of one man's empire.

What happens in the healthiest and often even in the young is that a single system of the body is not as brisk as the others, and in old age definitely is not. His skin was soft, some flakiness here and there. The single cell was probably able to do less work. The single organ was able to do less. The stream from the bladder was lazier even though the drainpipe was not pressed on. At the ninetieth birthday party the guests were saying that his back was as straight as ever. "How he marches, that philosopher there," remarked a stranger who from a distance observed his confident locomotion. Without a doubt he spoke more deliberately, but also uniquely lucidly, had his private sense of the funny, said the downslope ends in the horizontal, where one can at last lie flat on one's back until the trumpet calls.

No doctor ever had the honor of taking his pulse. Took his own. Had for years. It may have become an odd pulse even to him. Always had been slow, was faster now, not much. His brain according to growth curve statistics would have grown smaller by so-and-so many million cells. Then, one morning, unannounced, he departed this felicity awhile. He himself had said that the best time was early in the morning. People were surprised he should look so spruce in the coffin.

CANCER:
Out of Bounds

Decorum was in all so far. Control. In the giant and the dwarf the processes of growth had a different inherited base, but upon that base the processes maintained decorum. An obese Negress will move decorously in Harlem at midnight, in Paris as she steps from restaurant to restaurant to seat herself majestically on two chairs, and no Frenchman says of her, "badly put-together American."

In the artificial growth of cells in a test tube, the growth of each cell is decorous. Control. It is built into the processes of that growth. Millions of dollars in grants to laboratories have been appropriated for the study of these cells. Study in a test tube would seem ideal, except it makes some difference whether a cell does or does not have the influence of the cells of a living body around it.

During the development of the human embryo there is perfect regard for time and space, the cells multiplying, changing type, differentiating so as to become the different parts of the body. Control. Then comes birth, and after birth we speak mostly in sociological terms. "Harry has gained five pounds again, dear boy." The cool precisions are going on inside Harry. Each cell seems individually to know when to stop dividing. Order.

Then one happy summer evening something changes. Growth becomes an ogre. The clock has been tampered with. No dependable growth rate enables a dependable scientist to construct a dependable curve.

Causes of cancer have been claimed, denied, hailed, scorned, a slippery minority established, meanwhile tremendous step-on-step worldwide work. Even that minority may be only exciting causes that give a determined nudge to one universal cause. Viruses could give the nudge. Viruses could be the cause. Viruses could play the role at one time and not at another. There are pathologists who believe viruses responsible for all cancers. The well-known chicken cancer (sarcoma) is definitely caused by a virus, proven more than fifty years ago, and in a gloomy moment one could think that in spite of all the grants nothing as crucial has been discovered since, and undoubtedly that discovery has been underneath mountains of speculation and experimentation.

But the cause might not be a virus at all, might be some altogether different dynamics, a mutation, the production "spontaneously" of a different type of molecule, this the result of any of many factors that became a chemical factor, any chemical factor that became a physical factor.

The popular explanation for cancer has often been chronic or acute irritation, or inflammation. "He had an odd bruise last year."

That bruise became something else. Tolstoy motivated his gnawingly brilliant novel *The Death of Ivan Ilych* on such an "odd bruise." He also throughout made the reader feel that that might not be the truth though Ivan was convinced of it.

Carcinogen is a summarizing term. It applies to anything that is cancer-causing. Radiations. Poisons. The friendly cigarette. The exhaust of automobiles. The temperature of our body. Something easily turned topsy-turvy in the growth of aging cells whether these be in a young person or an old. Normal growth might have in it agents that halt, that inhibit what would otherwise run to the abnormal, some inhibiting chemical among chemicals, some inhibiting molecule among molecules, and the cause of cancer then the failure of that chemical, that molecule.

The net result of the carcinogen?

A great rapid growth in the neck of a five-day-old infant or a leisurely fatal wart on the leg of an eighty-five-year-old woman, in either case a procreative cockeyed increase of cells, a piling on, cell on top of cell. Nothing in normal growth resembles this except some special tissue for some brief period in the embryo. On the contrary, a cancer of the skin, like the eighty-five-year-old woman's wart, may have grown for a decade and one scarcely saw an enlargement. A cancer may lazy along in the cervix of the uterus for nine years, fourteen years, abruptly break through, spread and scatter, invade and metastasize, that is, transfer cells to an unrelated part of the body.

Cancer cells are in some sense tougher than normal ones, abler to survive under adverse conditions, to live anywhere, need not recognize their own kind, and be associated only with their own kind, or so one might think, but can thrive among other kinds, make room for themselves in blood, brain, kidney.

Cancer cells are usually well on the march before the patient feels anything, before the physician finds a lump, a tumor. They have stolen an unsportsmanlike advantage before there are enough of them to crowd an organ or jam a tissue and produce symptoms. In some cancers abnormal cells do normal but excessive work, as a cancerous ovary secreting excessive ovarian extract, a cancerous thyroid secreting the hormone thyroxine, the results of the secretion on some activity of the body catching the patient's or the physician's attention. An old idea placed the origin of the anarchy in a single awry cell. Possibly it is.

But no pathologist would hang a diagnosis on the look of a single cell. He wants numbers. A reasonable number will convince him.

The appearance of the dividing cancer cell?

As in the early stage of any cell division, the chromosomes in the nucleus split and arrange themselves in figures, mitotic figures. When the physician or surgeon snips out a bit of questionable tissue, the biopsy specimen, the pathologist examines it under the microscope, and if it is cancer he can diagnose it by those figures. In a modern hospital the specimen may be sent over an automatic delivery system from the operating room to the pathologist and the report come back over the operating room TV, an irony in the promptness of reporting an insidiousness about which there may be nothing to do but wait for the dismal end. A rapidly growing cancer shows mitotic figures everywhere. The cells look like haste. The chromosomes are crooked or fractured. The nucleus is geographically misplaced. The design in that spot of walled life has lost that smug decorum. The cell's parentage may be unrecognizable, and the bastard multiplies. The progeny of cancer cells are cancer cells. The massing of such cells may so squeeze the neighborhood, so distort the picture, that even the specialist does not know whether this is kidney, brain, liver, lung. Blood vessels streak through at unexpected points. The scene could be thought aboriginal, undifferentiated, wrecked, as if life had not yet found its assignment, or had found it but had lost its pride and whatever intention it ever had of becoming the crisp nonagenarian. The price tags have been removed. Growth is splitting fees with the gravedigger.

Why does the stricken human being die?

The products of the cancer cells kill, in some. A small tumor may kill quite mysteriously, no explanation. The outlet to a necessary gland gets plugged by cells, the body deprived of a necessary secretion, in some. Blood vessels get plugged and blood does not flow where it must, in some. The mass of the tumor may squeeze and thus from the outside block the flow of secretion or blood. Streaking blood vessels may be eroded by the cancer, with fatal hemorrhage, in some. A vital control line may be cut across, in some. The malignant cells may and usually do sooner or later free themselves from the original lump and travel and procreate in soil elsewhere, consume the good earth, luxuriate while the normal cells suffocate and starve. A surgeon's knife can extirpate part or all of the growth. Radiation can dissolve it. In rare instances the cancer arrests spontaneously, not meaning that the arrest did not have a cause but that the cause was unknown. Often the cancer eludes knife and spontaneous arrest and cobalt and X ray, to disport at leisure. The invader has lodged upon the peaceful countryside, an unashamed illegitimate relative determined to stay, and this land will not be fallow again.

III

ENVIRONMENT
Up from the Sea

CONSTANCY:
A Fresh Idea

To René Descartes the body was a machine, a machine made by the hand of God and better than any of human invention.

In the nineteenth century there was another fresh idea, Claude Bernard's. He also was French. To him also the body was a machine, but in it he conceived what he called an *internal environment*.

By environment we mean usually something outside, but it is *external environment*, and Claude Bernard's was internal.

In the two centuries between Descartes and Bernard the living machine was more and more examined for the chemical, had become a chemical machine. Bernard recognized this. He was a genius, a physiologist, had lofty imagination, was limited by the techniques of his day, but he could see that anything living lived by a multitude of balances. He knew that body and brain were built of cells, and he emphasized, postulated, because he could not yet prove, that around the cells there was always fluid. In his time the fluid would have been mostly, but not altogether, the blood. Cells + fluid around = a living body. It would make a body a vast interconnected catacomb walled by cells, the corridors filled with fluid. A Japanese undersea spear fisherman and a high-and-dry investment broker on the fifty-seventh floor of the RCA Building were equally 1,000,000,000,000,000 cell islands plus the fluid around.

Bernard's words were: "All the vital mechanisms, however varied they may be, have only one object, that of preserving constant the conditions of life in the internal environment." He compared the body to a hothouse. The 1,000,000,000,000,000 cells were the flowers. Outside a hothouse the weather could be fickle, but for the flowers the environment did not change. They had the constancy. The higher in the scale of creatures, the more exacting was this requirement. The low could endure larger ups and downs. The frog could. We could not.

Bernard argued that experimental medicine—he said experimental medicine, not experimental physiology, suggesting that the health of human beings was at the back of his experimental thinking—ought to have its special instruments as other sciences had theirs, and ought to use those special instruments to study the internal environment. The lengths to which this would be carried, no one of his day could have foreseen, not even he.

WATER:
Essential to It

Bernard's constancy requires water. Blood cells contain it. Blood cells travel in a bloodstream that is largely water. So, water travels in water. And the stream streaks through flesh that is watery cells plus the watery fluid around. A quarter of a century ago a Canadian ventured the notion that the fluid around was essentially the ancient sea. We came up from the sea, an earlier model, an earlier earth, were complicated then and became more so. For ages we went back and forth between sea and land, finally stayed on the land.

Water is scant in the universe—none on the hot stars that are half the universe, none on our sun.

Substances that combine with water are fabulously altered. The water molecules themselves have a latticed structure. In ice they are rigidly bonded together, kicked about helter-skelter when steam. Water has such intelligent physical properties—boils at a heat suitable to coagulate our breakfast egg, freezes suitably for sherbet, dissolves sugar, dissolves table salt, allows the salt molecules to slip in between and make connections with the water molecules, or the other way. Some of the water of the body is intracellular (inside cells) and some extracellular (outside cells but inside us), this said in consequence of Bernard's thinking but not said in his day, and in our day said with mathematical descriptions but always with some new qualification.

Who cannot recall the shock when first he heard that an average-sized man has a hundred pounds of water in him, the half-worry that he himself might leak? However, a creature of water soon convinces itself that water is the best material to be made of. Bags we are, water bags primordially, bags perpetually balanced in the universe's perpetual balancings, bags that keep their shape, move in a water-laden atmosphere over the watery outside of a sphere that pirouettes with a tide-producing moon. The moon is dry. To dry a corpse, to desiccate it—it has been done—is not a courteous way to measure water. Other ways are neater. By all ways we are two-thirds water, 73 percent of our fat-free weight.

On a bad midnight when our bowels come to our rescue, flush us out, wash us clean of every chunk of rotten roast beef eaten in the honky-tonk, employ that vulgarest of our outlets, we might hear again in the air the lucid, nightmarish Samuel Taylor Coleridge's "Water, water, everywhere."

ACID-ALKALI:
Small Swings

Acid-alkali can speak for the other constancies.

Our body must not be too acid, not too alkali, only small shifts permissible. The pH says where the body stands. That hieroglyphic tells, in numbers, where the internal environment stands as to acid-alkali.

"The pH of a man's blood or his urine or the fluid around in the corridors between his cells is the negative value of the logarithm to the base 10 of his hydrogen ion concentration."

With that now thoroughly understood, we continue.

A living cell produces carbon dioxide, and since water with carbon dioxide produces carbonic acid, acid is added to the body. Whenever, wherever in the water-drenched flesh a water molecule consorts with a carbon dioxide molecule, the body receives a vanishingly small push to the acid side, the pH falls. A working muscle produces lactic acid—the body moves to the acid side. Rest after work removes lactic acid—body to the alkali side. The stomach manufactures hydrochloric acid, which mixes with the disgesting food and virtually leaves the body—to the alkali side. The pancreas manufactures alkaline juice, also mixes with the digesting food—acid side. We eat acid in pickles. Eat alkali in baking soda. In an illness like diabetes the body handles fat abnormally—acid side.

Despite all pushes, the acid-alkali constancy remains unbelievably constant. By human yardsticks only the smallest back-and-forth is permissible. Descartes's machine operates best when Bernard's internal environment is somewhat on the alkali side, around pH 7.4.

In short, each of us is an exact, cell-dotted, slightly alkaline marsh.

Consider the solution of an acid. Acid molecules disport among water molecules, and electrified particles, hydrogen ions, appear. The solution tastes sour because of them, stings, burns, damages because of them. Democritus two thousand and more years ago said: "Atoms of sour wine have sharp points that prick the tongue." Those were hydrogen ions. Weak acids have few, strong many. Water itself splits, vanishingly triflingly, and in pure water there are only one ten-millionth as many hydrogen ions as water molecules.

Pure stomach juice is far to the acid side, pancreatic juice far to the alkali side. Always difficult to believe that two quarts of concentrated hydrochloric acid have been produced by our stomach since this same hour yesterday, two pounds of carbonic acid in body and head. The mind feels scalded but the flesh is not. The shuttlecock internal environment keeps bobbing composedly around pH 7.4.

When the body has to lose acid, it is breathed out from the lungs, urinated from the kidneys, or fails to be cast out, stays, but stays harmlessly because it has disguised itself in a weak acid plus the weak acid's salt, an ingenuity called buffering, and chemists and nature employ buffering with the greatest of ease.

CLAUDE BERNARD:
He Who Conceived It

What kind of man?

The earliest fact is registered in the village registry. He was born at a usual convenient hour between midnight and morning, July 12, 1813. Later he was baptized. The middle name was his paternal grandfather's, Claude, the full name, Pierre Claude Bernard. That ceremony took place at Saint-Julien, canton Villefranche. Bernard *père* was a winegrower, and the mind of the son was from the beginning fed through eyes and ears and nose from the earth of a French countryside. He never cut his connection with that countryside, on his vacations must get his hands into anything that had to do with the making of wine, an excellent Beaujolais. After a time he was apprenticed to a druggist—folded square papers for powders, for eighteen months. At night he might slip away, not frequently of course, this being stern old Europe, to the nearby city of Lyons, to the theater, wrote a skit for the theater, *La Rose du Rhône*, which was produced and earned him sundry francs. A five-act play followed, naturally a tragedy, *Arthur de Bretagne*, which was in the baggage when the young man, twenty-one years old now, went where all good Frenchmen go, to Paris. There the external environment was healthily cruel. The play was read by a professor at the Sorbonne. Was read by an actor. By a drama critic. It had faults. Was rejected. But the dramatist was in the city he wanted to be in. The French world swirled and swished around him. He met old friends. They were studying to be doctors. That drama critic had given the dramatist incidental advice—why not?—and he enrolled in the College of Medicine.

Magendie was the great physiologist of France, recognized the quality in Bernard, and Bernard the quality in Magendie, worked for him, then with him, differed with him, left him, returned to him. Magendie, when his death was approaching, predicted for Bernard, if he could not bequeath, and conceivably may have been satisfied not to bequeath, the chair of physiology at the Collège de France.

Much of Bernard's work was finished before that appointment. Everyone in the fields of medical science knows the work. The reports, articles, eventually books, many translated, reach around the world. His public addresses are extant. His popular writing, and he did some, can be found in back issues of *Revue des Deux Mondes*. Despite all, one feels shut off, more than from most human beings.

The laboratory in which he worked would to us in rich America seem dingy and poor, a place where a worker all winter keeps a running nose and wears a muffler and a tall hat. Discoveries could nevertheless be made under the tall French hat of that day. Experimental medicine was in a triumphant mood. The dogs were tied down. No anesthetic. When the blessing of anesthesia came, late in Bernard's life, he used it. A friend related how the master, whose hands were in the open abdomen of a dog that was howling mournfully, turned with benignity to ask the friend to wait a bit, went on with the surgery. Two centuries earlier that dog would have been nailed down alive. In those old days dogs did not suffer pain at all.

He married late, thirty-two. He married unhappily. Two sons were born. Both died in infancy. Two daughters lived. When the parents eventually separated, the daughters clung to mama, later joined her in establishing an asylum for stray cats, doing as it were penance for papa. To Bernard the separation could have meant that nothing now stood in the way of the last chapter of some of the most far-reaching experimentation and meditation any biological scientist ever achieved. He led his own field, physiology, into several of its modern directions. He defined carbon monoxide poisoning. He proved that the liver stores sugar in the form of a starch, glucose in the form of glycogen, converts the glycogen back to glucose as the body needs it (the understanding of the steps awaiting a subsequent evolution of chemistry), secretes the glucose directly into the blood, internally into the blood, this being the discovery of the first internal secretion, therefore the beginning of today's spreading knowledge of the endocrines. He made probings of the entire digestive tract beyond the stomach. Studied and characterized the similarities and dissimilarities of plants and animals.

It was Claude Bernard who baptized the laboratory frog, gave it the name of the Old Testament personality Job.

Not anyone ever gave to the blood more total meaning. He proved the highly important fact that the size of blood vessels can vary and that nerves can control this, accordingly would have been in a position to allot to two signs of emotion, blushing and blanching, a simple mechanism: vessels open, vessels close, blood flows in, blood flows out. When, late in his life, Pasteur wrote to him eulogizing his achievements, pointing to the logic in his physiological manner of thinking, as in the discovery of the glycogenic function of the liver, Bernard, reading that eulogy, said he had had a paralysis of control of certain of his blood vessels, that he was blushing all the way up to his eyes.

In the course of many concrete discoveries there came that generalizing one, that idea of a constant internal environment maintained by a myriadfold balancing and rebalancing. It was his crown. It had put into the long road one of those rare turnings that display suddenly a new country.

HOMEOSTASIS:
A Restatement

Years later, at Harvard, Walter Cannon, American physiologist—he died in 1945—restated Bernard's idea. He reshaped it. He brought it into line with his own experimentation, his developing concepts of pain, hunger, fear, rage, and in so doing he brought the idea into line with the physiology of the time, helped thus to bridge Bernard from the nineteenth century to the twentieth.

Cannon coined the term *homeostasis*. Bernard's constancy became Cannon's homeostasis. Bernard's fluid around became Cannon's fluid matrix. Both men, of different quality, different intellect, different intensity, believed that all systems of the body contributed to the body's constancy, but it was especially the involuntary nervous system, the autonomic, that Cannon thought controlled the constancy of the fluid matrix.

His term *homeostasis* flooded the literature. At any meeting of physiologists, biochemists, pharmacologists, pathologists, every roomful might hear the term several times an hour. All categories of medical scientists every year for at least a quarter of a century would be finding new places to use the term, new experiments to use the idea. A psychiatrist might describe all of his work as a guiding of the disequilibrated mind back to homeostasis.

Even politicians and sociologists rolled Cannon's term on their tongues. This would have pleased Cannon, this drawing in of the social sciences. He always welcomed social implication. He wrote: "It is pertinent to observe that these swings from right to left and vice versa, which produce in society as in well-developed organisms a trend toward a middle course, are possible only in a democratic form of governmental organization." Democracy = homeostasis.

Another term began sweeping the literature—*mechanism*. The two met and wooed and wedded—homeostatic mechanism. The study of the single constancies, like body temperature, like body pH, like blood sugar, other components of the blood, Claude Bernard's internal secretion (he named only the one), drew investigators this direction, then that. Cannon too, but in his experiments he kept returning to the emotions, their signs, their nervous-system relations, nervous-system controls, chemical controls. He was not a chemist. He was a physiologist who throughout his career sought to fit into Claude Bernard's idea his own branching homeostasis.

DISEASE:
Large Swings

Imagine Cannon's swings swinging too wide. Imagine a healthy body trying to keep in harmony with an unhealthy external environment. That environment may be minute, bacteria, viruses; may be huge, a sick nation, a sick era, or a sick body part that has become essentially external to the workings of the whole.

Imagine what was a healthy heart now beating with decreasing efficiency against increasing friction of blood moving through narrowing blood vessels. It was a balanced pump, but it has too long been pumping blood into out-of-balance conduits, and fluid is beginning to collect in places it should not, in amounts it should not. The swings are swinging wider. Breathing is too fast or too slow. Blood pressure soars or tends to collapse. Extremes are becoming both effect and cause, and still the swings are swinging wider. The physiology is everywhere fighting excessively to make up for not fighting efficiently.

Textbooks do not express this quite this way. Tradition is strong. Disease continues to seem so-and-so many signs plus so-and-so many symptoms, with the corrective dynamics definitely not in the foreground. Claude Bernard and Walter Cannon could be considered as still around and goading physician and experimental worker to the not necessarily easy task of searching ceaselessly for the rebalancing factors in an imbalanced body and head.

The pattern of imbalancings and rebalancings is the sickness. If the strains persist, if cells change in their molecules, that is biochemical injury. If biochemical injury persists, that is pathology, may be recognized as such by the physician's fingers, seen under his microscope. Disease is a process aiming at recovery, its intent equilibrative,

slanted toward health, but, missing that, accepting the downslope toward death.

DETERMINISM:
"L'Idée Directrice"

A plan *after* the house is up, sketched by the draftsman from the finished thing, is how science understands plan. Plan for a creature that
stayed in the sea, one-celled or billion-celled. Plan for a creature that
came onto the land. Plan for a creature between sea and land. Plans,
plans, plans. Science observes what is before its eyes, wide-open eyes,
sometimes eyes that have not lost the dimension of wonder, but human eyes that see a human world, helplessly focused by the techniques,
instruments, hypotheses, fashions of a human mind in a human era.
Difficult for any of us to accept this though we know it.

The man of science is a determinist—things are as they are because
they are as they are. Everything can be brought within that enclosure.
Everything can be mathematically examined, related, and accounted
for. Alive body, alive brain, alive mind, and the dead, are only logical
steps in the blind grinding of the universe. Step, step, step, astronomy,
physics, chemistry, cytology, and, on the top step, man, the thinker,
chews the cud of his thought.

Science's overall determinism is strengthened by each man's individual determinism, the result of his experience. He is so sure of this
that one must sometimes wonder if he is not unsure. Sometimes he
vents himself in a burst of annoyance: "Vitalism!" Sometimes contempt: "Religion!" Sometimes: "Metaphysics!" Sometimes a seeming
observation: "Nothing is supernatural." Another observation: "No
man-type intelligence behind." Making certain of the last: "Nothing
behind." Correcting: "Nothing that is not science's oyster." Professori-
al: "All life can be genetically and environmentally explained."

Anyone would be able from these premises to go on to no freedom
of the will. Whatever was mistaken for freedom of the will would on
examination reveal that it was in fact determined. Safest for science to
allow no freedom of the will, as safest for the Church to hold tight to
it, because were science ever to let that go, that window open, anything
might fly out or in, even what might embarrass science, for though

science often has embarrassed the sentimentalist and the Church, it still might be embarrassed in its turn. Not likely. Not soon.

Claude Bernard stated his determinism coldly, but he too appears not to have been at ease, called something *l'idée directrice*, the directing idea. What did he mean? That has been haggled over. Physiology to Claude Bernard was the study of the constancy of his milieu. The body was a tender tremulous tremendous adjusting. The constancy of the internal environment can be considered the shadow of universal adjustment falling across the thin thread of the life path.

Behind physiology lies, if you like, evolution, the theory. Near to one hundred years ago, while Bernard was developing his theory, Darwin was developing his. Our planet is attractive when two such eminences push up their heads in the same brief period. The years since Darwin have added mainly detail, not conception. Lamarck, who earlier than Darwin pondered evolution but differently, entered the nineteenth-century mind, departed from that mind, has somewhat reentered the twentieth-century mind, while Darwin just stayed.

Behind physiology and evolution lies, if you like, astronomy. Again by adjustment there has come a universe incomprehensible except as we think, not of it, but of our measurements of it, a universe achieved from nothing, a universe that continues to be achieved from matter and antimatter, but however achieved is magnificent, produced the night sky which we have beheld from when we first looked up, the Great Bear and the Southern Cross and to the right someone shrouded carrying Mars as a crystal lamp, and all the visible stars and all the invisible, with mathematical infallibility each where it ought to be.

It leaves us with this: symmetry is everywhere, even behind any ostensible physical or chemical asymmetry. Order is everywhere, even in ostensible disorder. The random never is random. Plan is everywhere—plan after the house went up.

Nevertheless, there still is left in the close of the twentieth century an occasional soft temperament. This one clings to the discredited other plan—plan *before* the house went up. He requires Olympian motive. The blind grinding operated upon him too but left him incapable of imagining a primordial mass-energy from nowhere, then the particles, the waves, the atoms, the molecules, the molecules that grew larger, the molecules that replicated, the living cell, the orchid, then Shake-speare, then Beethoven, left him incapable in some hours of not asking if this sequence could ever, in spite of the inevitability of all the succes-sive evolutions, have occurred without an antecedent deliberate some-thing-or-other. He gropes, he sputters, he blurts out an incoherence, and someone nearby says with disgust: "Teleology." Teleology might be the reverse of determinism, and the meaning of both might change with the winds of time. Purpose could even be old-style hell-and-heav-en purpose. Granted that it would not be understandable purpose. To say that something is not understandable is not to say that it will nev-er be understood, or that it is supernatural, nor need the speaker be crossed off as benighted. He all-helplessly has the not-understandable as part of the inherited organization of his mind, so let him be. Or he has the not-understandable as a construct that the organization of his mind caused him to place in his mind, so let him be. Possibly he is stupid. Possibly not. Possibly he is arrogant. Possibly humble.

Should he be humble—this is to be recognized—he might perform even such an absurdity as to fall to his knees, not because of an outer commandment, but because his having been born, his having lived, his having entered into confusion, his having to die, continue so to over-whelm him that he is shoved to his knees by his own inner compulsion. It might seem that what science has taught him best is that body and mind were from the start beset with limitations, these assuredly leav-ing some, conceivably much, of reality beyond his reach. His human head is always getting in between. There is no escape. A silhouette is cut from the unknowable by the knowable. It would not follow—this too is to be recognized—that everyone's god must wear a white beard and a white robe, be made in man's image, think man's thoughts, float in a white cloud over the blue Italian source of it all, the ancient sea. Anyone's god just might be, and might be past all theological or philo-sophical or scientific reckoning, be timeless, matterless, an inscrutable for which an artist might daub in a term like luminosity, a luminosi-ty that some intelligences block, some centuries block, and others let through. Anyone's god just might be. Then God would not be dead as Nietzsche said, and would have been glimpsed through a crack forced open by a science that did not mean to force open anything, yet might seem more real than if glimpsed by faith alone. Even to a science that eschews first causes, a particular first cause just might be.

"It might? And upon what ground does a man then stand?"

"Upon uncertainty—an alert uncertainty that continues in all things to try to find certainty with no hope ever of finding it—where man has always stood."

"And be satisfied with that?"

"Be satisfied with that."

IV

MUSCLE
Ivory Until It Moves

KINDS:
The Silken Cord

By weight a man's muscles are some 40 percent of him.

There are three kinds, with three appearances under the microscope, and three responsibilities. *Smooth. Striated. Cardiac.*

The first—smooth—does its work in bowel, blood vessels, pupil of the eye. The second—striated—enabled Galatea to step down from her dead ivory, enabled laborers to raise St. Paul's from the ground of London, Notre Dame from the streets of Paris, the Sphinx from the sand of the desert. The third—cardiac—is right now quietly or noisily beating in your chest, and in Job's, the frog's.

Smooth looks smooth under the microscope, is laid down fiber next to fiber. The fine fibers look frail, but when they join their strengths, they can generate great force.

Striated looked striated under the microscope, its fibers striking one for their sturdiness and for their gay barberpole appearance. The fibers have alternate light and dark bands. With careful lighting, a film of striated muscle may strike one as a more delicate fabric than any one has ever seen. It has been thought that the more the striation the faster the performance. Striated causes the slight tension that is always over the body wall, that gives the body its feel of vibrancy, feel of life. Striated wrinkles a man's brow, turns his frown into a smile, tightens his eardrum, grinds a chop between his teeth, kicks a football.

Cardiac has striation too, but its own kind, one fiber everywhere bridging across to the next fiber, the ordinary light microscope quite sufficient to show the bridging. By the bridging the heart knits itself into a single mass, which beats. Any fragment beats. If the heart of a frog is cut out with a scissors, as has been done in a hundred colleges, then shredded, the shreds beat. The heart of a humming bird may beat eight hundred times a minute, ours seventy or eighty, quicker in the expectant debutante, slower in her no longer expectant bachelor uncle.

CONTRACTION:
Lifts the World

Muscle contracts. That is its genius. It shortens. Some muscles shorten slowly, some rapidly. After a muscle has shortened, it may immediately lengthen so that it can shorten again. There are accompanying electrical events, accompanying chemical events.

When a muscle shortens, it pulls on whatever it is hooked to. Say that is two bones. The two join to make a hinge. When the muscle shortens, the two are pulled toward each other, the hinge is closed. Then the muscle relaxes and the hinge opens. In the green frog, hinges close, open, close, open, close, open, and that is the frog hopping from stone to stone across the brook to the mud of the other bank.

The calf muscle, the large muscle at the back of the lower leg, is called gastrocnemius, frog or man. For the movement of the body of either there are fast fibers; for simply holding a position, slow fibers. That muscle puts the power into the hop that all day sends the frog across the brook toward town. In a man it makes possible, as Oscar Batson stated it, the springing step, tiptoeing, toedancing, raiding of pantry shelves.

It has been calculated that the gastrocnemius of a human sprinter is capable of lifting six times his weight. A calf muscle lifts six men! The sum of the muscle fibers in a human body has been calculated at a quarter billion, which seems many, but six men also is many. Were that quarter billion rigged to pull at one time in one direction, it could lift twenty-five tons. Muscle does thus become the silken cord that lifts the world.

MUSCLE-MOLECULE:
With Millions in Parallel

Picture again that calf muscle of the frog. With a scalpel it has been dissected free of the frog. Now take a needle and, carefully, there where the tissue splits naturally along the length of the muscle, tease it apart. The total muscle is constructed of bundles of fibers. A bundle is a fasciculus. Fasciculation is a contraction of such bundles just under the skin, one after another, spurts of contraction, twitchings, happens normally with a tired eyelid, but can be a sign of nervous-system disease.

Now, tease the fasciculus into its fibers. A fiber is still large enough to be seen with the naked eye. Tease a fiber into fibrils. A fibril requires the microscope. Tease a fibril into micelles. Tease a micelle into filaments. A filament is the smallest unit known.

In this stepwise fashion you have left behind you your naked eye, then the ordinary microscope, then the electron microscope, and have traveled that road that leads to where you must place your faith in molecules.

It has been generally accepted that there are two kinds of muscle filament, that the two kinds lie parallel and overlap, one thick, one thin. The overlapping occurs at the ends. On the instant of contraction, the overlapping is greater, the filaments sliding past each other, telescoping, which could explain the shortening, and could explain also the changing microscopic picture during the shortening.

A Hungarian scientist working with the chemistry discovered that if two muscle proteins (about 20 percent of muscle is protein and the rest mostly water) were brought together as a thin jelly, and this dropped into a juice expressed from muscle, the jelly contracted. It got stiff after it had been pliable. This was an early successful stride in the advance upon the nature of the chemistry of striated and cardiac muscles especially, and upon their phases of contraction.

The Hungarian's muscle molecule possessed a double talent; could contract, could excite the drawing of energy for the contraction from its energy source. It was a contractile chemical (had that energy) and could excite the contractile chemistry (an enzyme therefore). A link within the molecule supplied the energy (is supplying it for you now as you read), and the production of that energy costs the same as the production of any energy.

The story of muscle contraction is part of a serial written by successive authors as they appear and disappear, every now and then a particularly brilliant installment. On Times Square on New Year's Eve, far within the visible scene are thin filaments sliding past each other, and millions or billions or trillions of molecules interacting while great muscles get shorter and longer. Happy New Year!

V
HEART
Regular Day and Night Deliveries

FOUR CHAMBERS:
Pump

Not all animal hearts have four chambers. The turtle has three, the frog three, but dog and cat and monkey have four, like us. Two of our four are on top side by side, called atria, formerly called auricles, still often called so. (Auricle = little ear.). Two are on the bottom, ventricles. (Ventricle = little belly.)

The walls of the chambers being muscle, when a chamber contracts, the blood in it is squeezed, and any heart door that was ajar is slammed shut, and any that was shut is thrown open, and the blood rushes past the open door, the valve.

Thereafter the blood continues on the postman's appointed round.

From every part of the body, it is collected in the right auricle → is delivered from there to the right ventrical → thence lungs → left auricle → left ventricle → out to all the zip code areas. That is one round, one cycle.

The wall of the right ventricle is thin (if held toward the sun, the light comes through), and it can be thin because it delivers only in and out of the lungs. The left is thick, must deliver from toenails to scalp.

Behind breastbone and ribs, somewhat to the left, in a stout sac that prevents outward bulging during the beat, and with pads of fat cushioning it, hangs the heart. From no more than its shape and its hardness the student might suspect the economy that lets this pump with such mailclerk regularity meet its day and night appointments. In the laboratory an experiment will long be over, and still the dog heart be beating.

Through ages of change on the planet, through mammals, birds, reptiles, this organ has had the same seemingly simple design, the same seemingly simple action, and the least that can be said is: you have something long-thought-out pumping inside of you, brother.

Three thousand million beats beat in a man who gets old, a hundred thousand or so in one day. By heartbeats a life is not brief. Rarely is a beat missed, but if a few in succession are, the body drops like a stone. How much heart work is done in a life? By one melodramatic arithmetic, each ventricle puts out 150,000 tons of blood; by another, raises a ten-ton weight ten miles.

A Greek twenty-three hundred years ago watched an animal embryo before its heart had formed, could see, there where it would form, a point that paled then flushed. Harvey two thousand years later watched a point that paled then flushed, comprehended that the paling foreshadowed a heart emptying, the flushing its refilling. Paling, flushing, emptying, refilling, the auricles always ahead of the ventricles. The heart gives the force to the circulation, and every cell of our body depends on it.

ARTERY:
Road Out

From the pump go the pipelines, go the day and night deliveries.

A man's chest has been laid open, his heart beating, a meditative heart surgeon looking in, all of his face except his eyes masked. An intense man, the meditative surgeon, but quiet, compels himself to be quiet, and lucky it is for his patient.

The Middle Ages called the great artery the aorta, and the word means to heave. The aorta takes off from the left ventricle, sweeps up, then back, then down, makes an arch, arch of the aorta, after that drops almost straight through chest, through abdomen, the vertebral column behind. From the top of the arch, branches go to the head and brain. Below, branches go to the lungs, sac of the heart, diaphragm, liver, bowel, spleen, pancreas; right and left a stout branch goes to each kidney; branch, branch, branch, at last those distant ones that deliver to a man's toes.

The wall of the aorta is elastic. It is resilient. The walls of all the arteries are resilient, more so in the young, which is why a child can dash from her father and he can only puff after. But were there nothing but elastic tissue, were there no tough inelastic framework within that elastic tissue, the artery like a rubber tube inflated too often would blow out. The aorta has been studied for the way a viscous fluid leaping into it would stretch it, for the changing demands put on it.

Each heartbeat throws in the blood, the arteries yield, then pay back that yield while waiting for the next beat; hence the stream, despite the pump's intermittence, moves smoothly.

In the suburbs, at a point, the arteries take on a heavy circular muscle. When this contracts, the stream at the point is narrowed. Blood can by such contraction be diverted from here to there, from where it is less needed to where more needed. That contracting can be too tight and too continuous, emotion among the reasons, and the heart accordingly pumps against a useless narrowing, does useless work.

The branchings of a good-sized artery were counted. It was the mesenteric that delivers to the bowel. It has fifteen main branches, one thousand eight hundred and ninety-nine bowel branches. There are then the branchings of the branchings. Many speeds are in each branch—one speed along the walls, another at the center, and every manner of speed in between. A few moments ago, the single swift stream of the great artery, now the slower and slower streams of those smaller and smaller branches spreading out over the countryside, giving one a gay feeling, all that blood hurrying toward unnumbered and unnamed creeks and rills.

PRESSURE:
Pushing Deliveries

In every creature high enough to have a circulation there is friction between vessels and blood, also between blood and blood, so there must be pressure to keep the stream going. Five quarts must be squeezed through all the narrow passageways and brought back to the heart. Of course, the circulation adds no unnecessary resistance. If more than just enough were spent on each beat, those persons destined to wear out because of their hearts would wear out sooner. Nature does not wish us to wear out sooner. Has her reasons, we ours.

The first taking of a blood pressure was a century after Harvey, in 1732, by another Englishman, Stephen Hales, a clergyman, who fastened a mare, as he stated it, "alive on her back and exposed the large artery in her thigh about three inches below her belly." She was anyway "to have been killed as unfit for service." (Must not spoil a good horse.) The reverend put a tie around the artery, inserted into it a tall glass tube via two brass tubes, and made the plumbing connections with the windpipe of a goose. He said: "The design of using the Wind-Pipe was by its pliancy to avoid the inconveniences that might rise if the Mare struggled." So, perpendicular stood the glass tube. The reverend untied the tie. First with a big leap, then with small and smaller leaps, the blood rose and reached a height of more than eight feet. "Perpendicular above the level of the left ventricle of the heart." What it meant was that a mare's arterial pressure lifted a column of blood eight feet.

No one expects a man to be as compliant as his horse, and when a physician takes a man's blood pressure, nothing is nicked except possibly the man's plans for his winter vacation.

Each of us has had it done to him. A narrow rubber bag is fitted inside a cloth bag, the combination wrapped around the upper arm. A rubber hose connects rubber bag to a U-shaped glass tube filled in part with mercury and etched with a scale. Human pressures are in millimeters of mercury, the mare's was in feet of blood. Rubber bag, rubber hose, U-shaped glass tube, lead one into the other, so pressure in one is the same as in the other. A rubber bulb is for inflating. The physician locates the artery in the bend of the elbow, listens, hears nothing. Begins inflating. Up goes the pressure in the bag. Up the mercury in the tube. The artery begins to flatten under the pressure. The friction between moving blood and vessel wall increases. A noise. Finally, the artery is flattened altogether. No noise. Now the physician lets air leak out, and when enough has and the pressure in the bag falls below the highest pressure in the artery, some blood slips through, and there is a click. The physician reads the mercury. That is the highest pressure in the artery. More air leaks. More blood slips through and the quality of the sound of the clicks changes. Then the pressure in the bag drops below the lowest in the artery. No noise. The physician reads. He has now a high reading and a low, a systolic and a diastolic pressure. That word, *diastolic*, came from the medieval Latin, and that came from Greek, and the English of it was not used until the year Harvey was born. Systolic was used the next year, 1579.

Blood pressures are routine in our lives. Anyone not interested may be a fool, but if too interested is sure to be a fool. Descartes's machine of the body runs best when it is running without our thinking. Human beings think too much. A giraffe does not. A giraffe's systolic is somewhere about 360, a man's 120. A tough job for the giraffe's heart to lift her blood twenty feet above the ground. Lucky for a giraffe's life that she does not every hour worry that her machine is going to break down.

CAPILLARY:
Gone Underground

The thin walls of the vessels meanwhile have gotten thinner and thinner until they are thinnest of the thin, finest of the fine, one cell thin. A delicate membrane supports those cells.

And we have here arrived at the true capillary.

The name comes from the Latin *capillus*, means hair.

It is a cell or cells rolled into a tube.

Blood moves slowly in a capillary. When the second hand of a watch has leaped once, blood has entered a capillary, loiters, and when the second hand has leaped again, this blood has departed that capillary, (On the average.) Some of its fluid will in that time have leaked through the wall, the blood cells be more packed. The picture is always different. With injury the packing may be so great that any novice can see it.

The routing of the blood—in detail—has been from artery to arteriole, to precapillary, to thoroughfare capillary, to true capillary, to prevein, to venule, to vein. The length of the capillary was one-fifth of an inch. (On the average.)

It is evident that the destination of the entire system is the capillary. Heart and arteries and veins adjust their work so the capillary can get its work done. One might say that the rest of the circulation is the servant of the capillary.

August Krogh, physiologist, Nobel Prize winner, did some of the best studies in this world on those small vessels. He chose muscle capillaries. He showed how they open and close. Only a few will be open when the muscle is at rest, perhaps fifty times as many open when the muscle is at work. By-products of the work decide this. Krogh would poke the needle of a syringe into an artery heading toward muscle capillaries, inject, kill the animal, slice out a piece of the muscle, harden it, cut microscopic sections at right angles to the capillaries, the open ones visible because filled with the injected material. So he would prove, in numbers, the startling difference between rest and work. A bit of muscle the diameter of a pin might contain seven hundred muscle fibers, two hundred capillaries, and the two hundred all open during work, all closed during rest.

For that reader who likes things laid end to end, the capillaries of his body, connected, would make a hairbreadth pipe reaching several times around the earth.

Harvey knew that there must be connections between the two sides of the circulation. That fact troubled him, but he had not the instrument to let him see. He spoke of pores, as Galen had. Those pores are our capillaries.

Too bad that Harvey never saw a capillary. He would have needed to live only three more years.

VEIN:
Road Back

The same industrious worker who counted the arteries that deliver to the hills and valleys of the bowel counted the veins coming out of those hills and valleys from the smallest tributaries, the venules, to the large collecting mesenteric vein. Millions drain into thousands, thousands to hundreds, hundreds to one.

Veins are different in texture from arteries, less stiff, more yielding, abler therefore to hold variable quantities of blood. During the rhythmic increase and decrease of space inside a lung, there is an increase and decrease of space inside the lung's veins. They follow the lungs. It is because of that quality in veins that they can serve as a reservoir for blood, this blood always of course on the move from the reservoir's inlet toward its outlet. More than half of our blood, 65 percent, may be in our veins.

On the blood's return trip, no more force is wasted than on the way out, just enough left at the last to spill the blood into the right auricle.

It must be more when a heart is enfeebled and its work more difficult. Harvey was not surprised that when a vein was slit, its blood should flow out with only mildest pulsing, slightest pressure. His mechanical sense told him it must. From head and neck the blood needed only to drop to the heart, from legs and arms needed to be hoisted.

Hoisting is assisted in three ways: by valves in the veins, by muscles that contract around the veins, by the breathing chest.

The three kinds of muscle tissue labor for the circulation. At the start it is the heart—cardiac muscle. Along the course it is the muscle in the vessel walls—smooth muscle. Wrapping the veins and contracting on them it is the muscle of the skeleton—striated muscle. The more powerful the contractions, the faster the flow. Then there are the breathing muscles—striated also—that suck the blood upward into the huge collecting vein, the vena cava, from where it is dropped into the right auricle to be ready for its next round.

HEART SOUND:
Din of the City

When the valves of the heart bang shut, the flesh of the heart vibrates, the flesh of the chest wall vibrates, and from there the vibrations go out to the ear of the physician. His machinery codes them, the code travels, is decoded, and he hears a heart sound.

Those vibrations from the heart may instead be written on a moving surface, leave a readable report, a phonocardiogram.

The textbook says that the sound is *lupp-dup*. Day and night *lupp-dup*. Only *lupp-dup*. A medical student listens below the nipple, moves his shiny stethoscope, per instruction, rib by rib up toward the neck, under the arm, localities that last week he might have thought too remote.

First heart sound—easy to hear. Second—easy to hear. Third—not easy. Fourth—difficult.

Harvey compared the movement he saw when a horse took a swallow of water, and the sound that went along, to the human pulse that he felt and the heart sound that went along. Harvey was so sharp, and his statement always so vivid.

Valves—silken and sturdy and efficient—swing between auricles and ventricles and between ventricles and outgoing arteries. So there are four valves, two left and two right. As the ventricles fill with blood, the valves between them and the auricles first float, then when the ventricles contract and pressure mounts, bang shut. At the same instant, almost, the other two valves, toward the outgoing arteries, swing open and the blood rushes past. The combined tones—muscle, valves, blood—make the first heart sound. Most of it is those valves banging shut.

When the pressure of the blood columns in the outgoing arteries (they are columns now indeed) mounts above that in the relaxing ventricles, the valves that had swung open bang shut, the blood columns vibrate, and the other two valves, between auricles and ventricles, sweep open. The combined tone—muscle, valves, blood—makes the second heart sound. Most of it is those valves banging shut.

Should a valve leak, the normal sounds are altered, murmurs are added, and the murmurs occur because particles of blood are thrown in sundry directions by the leak in the valve, abnormal directions for a normal heart, and the sounds are abnormal. Time, place, quality, define the murmur.

A physician's trained ear may hear exceedingly faint sounds. It is quite routine for him to hear the heart of the fetus in the womb.

A heart built badly originally, or damaged afterward, may still allow for long survival. No physician may ever have listened to it, nor had it ever have called attention to itself, so no one ever knew it had a flaw; this was discovered only when the poor tramp came to postmortem. And his ear did perhaps hear his abnormal heart sounds toward the close when he was propped in a hospital bed and his head beat the pillow to the rhythm of his pump.

Until 1950 surgeons did not enter the heart, accomplished only what they could without entering. Today surgery in and around the heart has almost lost its venturesomeness. Birth defects are corrected in an infant boy or girl, a damaged valve in an eighty-year-old lady, the great pipelines directly patched or replaced, and heart sounds had their small part in the development of this surgery, especially back at the beginning.

All diagnostic techniques will have been brought in and all professional heads put together before that morning when surgeon and colleagues and assistants, the so-called team, are each in his proper place in the operating room. The surgeon makes the initial cut. He is the senior of the team. He is a meditative man, intense, quiet.

The cut looked simple, and one could fail to remember the long history of such cuts on the part of the long line of eminent surgeons. One might have expected blood to gush. It did not. It was not given the chance, every ready-to-bleed vessel being clipped as quickly as it opened its small red mouth. There followed several cuts, layers of cuts, finally a swift cut through the muscle right down to the ribs. Then the ribs were wedged apart, with care but with confidence, the chest laid wide open. The lung seemed impatient and intrusive. It was pushed aside. The heart was exposed. Beat. Beat. Beat.

Boldly, now, the surgeon clamped the clamp straight across the left auricle. It was a big clamp as mechanically gentle as a big brother, and the heart still appeared indifferent. One might fancy it was putting on a performance. Instead of all this it might early in the proceedings have been sidetracked from the circulation, a man-made pump substituted, oxygen bubbled through the blood as it passed along, the brain never for a moment deprived, and the heart brought back into the circulation at the end of the surgery.

In this present patient, this eighty-year-old lady, the valve between her left auricle and left ventricle has become scarred by disease, the passage narrowed, her life limited, and increasingly in danger. She and her surgeon have talked the matter over, realistically both sides, and decided on this morning.

This surgeon has now come to the point of threading a purse *string* of stitches along the line of that clamp he clamped across the left auricle, then a second line of stitches, on the chance the first might break. He slices off the flesh above the clamp. Takes away that clamp. Pushes his index finger down into that hole, into the heart. The wet tissues hug his finger. He is inside that eighty-year-old lady's beating heart. The entire operative field dances. Deeper into the ventricle goes the finger. Now he appears almost recklessly to thrash it side to side, knows how, rips scars, widens the passage. Then he coolly draws out the finger, replaces the clamp, washes away the debris, tightens both purse strings, reinforces with stitches all the cut layers of flesh, closes the hole in the chest. His finger has been down where the first heart sound is made.

Natural to think now of René Hyacinthe Laënnec, the Frenchman who in the early nineteenth century invented the stethoscope. Laënnec was possessed of extraordinary hearing. With unaided ears, and standing well away from the sickbed, it is reported that he could hear the tones of a patient's heart, as a rare ear hears falling snow.

ELECTROCARDIOGRAM:
Walkie-Talkie

The continuous electrical forces that are always coming off the heart are not easy to follow, not in one's mind either, though the total of them, the summing, the overall wave of odd shape is not difficult for anyone to see. This wave repeats with each beat. It is there to be recorded any hour, a regular irregularity. It is a slow wave, as such things go. This electricity excites the muscle tissue of the chambers of the heart, and the chambers contract, and the blood goes its rounds.

The fraction-of-a-second-to-fraction-of-a-second recording of the electrical sweep is not only possible but it is clear, we think. There are mountains of records. They heap up everywhere. Interpretation of them is not as clear, because though events in a single muscle fiber can be interpreted reasonably with principles of physics, events in the whole heart are far from as reasonably interpreted. It is consoling, though, that the clinical usefulness of the electrocardiogram, the ECG, is manifest fact. The useful research that it has inspired is fact also.

The instrument is the electrocardiograph. The record is the electrocardiogram. (Telegram. Radiogram. Cardiogram.)

An electrocardiogram is today part of the most casual medical examination. It helps a diagnosis, may clinch it. The record can show what and where in the heart the damage is, show it every five minutes in a hospital bed, or show it year to year in the regular checkups, so tell whether the damage is at a standstill, whether progressing, but always more secure as a record than as a prediction. The clinical usefulness often reaches beyond the heart to chemical upset in the body's internal environment, the so-called electrolyte state, clarifying problems there.

Conventionally the electrocardiograph is wired to the patient, at left wrist, right wrist, left ankle. Three electrical outlets. The arms and the leg that lead to the wrists and the ankle are to be regarded as flesh cables, gross electrical lines, because flesh in electrical respects is a dilute salt solution, a conductor. It is not electrically homogeneous but is sufficiently so. The heart also lies sufficiently near the center of the trunk, is itself a conductor besides a producer of electricity. From it electricity radiates in all directions; radiates to the pubis, thence to left ankle, to instrument; radiates to left wrist, to right wrist, to instrument. For electrocardiographic purposes the heart lies, then, at the center of an equilateral triangle, one side of this triangle being a line drawn between the two shoulders, the other two sides being lines between shoulders and pubis.

Toward these lines, the force and the direction of the heart's electricity is aimed; and from this emerge the three standard leads handed down by the father of electrocardiography, Einthoven. Einthoven's scheme had soon to be modified, and there were modifications of the modifications. The all-present computer is today brought in, and yet no instrument or combination of instruments is equal to the intricacies. And the Einthoven triangle itself, or any other triangle, becomes only an approximation whenever one thinks of the problem as the physicist would want to think of it.

Lead I. Lead II. Lead III.

Lead I picks up electrical differences between right and left shoulders, giving information mostly, as one would anticipate, about the top of the heart. Lead II, differences between right shoulder and left ankle, information mostly about the heart's right border. Lead III, between left shoulder and left ankle, the heart's left border. This is a loose summary. Any instant in any record in any lead is the resultant of many electrical additions and subtractions and neutralizations. Ingenious thinking has been stimulated and ingenious techniques developed.

Electricity can be tapped with two electrodes. It can be tapped from the two ends of any side of the Einthoven triangle, or variant of the Einthoven triangle. It can be tapped by one electrode. For a time it seemed that one-electrode tapping would revolutionize electrocardiography. This tapping can be from near the heart, or far, scant or much flesh intervening, from inside or outside the dog heart, from inside a single chamber of a human heart. No two chests are alike, and that has added handicaps to the game.

The electricity spreads through the heart, excites the successive chambers, passes through walls and partitions, but does no damage, and that is because it is everywhere neutralized, and because the voltages are small.

Today electrocardiograms are recorded photographically, or written on wax paper with a heated stylus, or by a moving stream of ink flowing from a tiny glass cannula, are like any taped news. One can rerun them. Recently volunteers have worn miniature electrocardiographs fastened to their chests, and the autobiography of a man's heart could be writing itself while he was writing another with a pen, writing the events of his day. He might do this while speeding thousands of miles above the earth, the records accumulating in a small station far below on the earth.

Vertical lines ruled on the paper make it possible to read the contours, the fractions of waves in fractions of seconds. Horizontal lines make it possible to read the heart's voltages in ten-thousandths of a volt. These voltages are ten times greater than those of the brain. That recorder fastened to the chest is essentially a walkie-talkie. It records heart cycle after heart cycle for as long as there is tape and for as long as the pump keeps pumping.

Any moment of any record is meanwhile rising above or dropping below a neutral line, or staying on the line.

Electrical activity begins in the auricles, as Harvey essentially knew, and the sign of the activity is a small upthrust in the electrocardiogram. Whether up or down does in fact make no difference, the physician being able to reverse the written directions by simply turning a dial. This upthrust normally is rounded, called P wave. It registers the electrical activity in the auricles; on behest of this electrical activity the auricles contract. There follows a neutrality, a level line in the electrocardiogram, a pause in the heart. During that pause in the heart the activity would have been moving from auricles to ventricles, then into the muscle masses of the ventricles, the sign being again an upthrust but a different shape, a spiking called QRS, and thereupon the ventricles contract. There may be no Q in the QRS or no S. After the S, again a neutrality. Again a level line. A pause in the heart. The electricity now would have been moving from S to T, where a rounded upthrust registers the electrical preparation of the ventricles for the next cycle. Sometimes there is a U.

All of this occurs for every beat, and a human heart may beat 150 or more times a minute.

But any telling of this story may annoy us for two reasons: that it is too simply told, that it is too complicatedly told. Both annoyances are justified. What is worth picturing is a heart that has passing through it these regular electrical sweeps.

FAINT:
Peak of the Journey

Every minute a pint and a half, depending on the cardiac output, is delivered up a man's neck toward his brain. Two carotid arteries deliver four-fifths of that blood, two vertebrals one-fifth. Carotid comes from Greek, means plunge into sleep, the arteries getting their name no doubt because squeezing them in the neck caused unconsciousness.

In the skull is the postman's top of the hill. Up there is evolution's masterpiece, has the royal right to one-seventh of his blood, gets it, a still larger fraction of his oxygen. If not, it quits, is apt to suddenly, and since one of its jobs is to maintain the man's erect posture, the man collapses. The cause of a faint is an abrupt drop of blood pressure, but there are causes of the cause referable to every part of the body and the mind. Faint, and something is revealed about you.

A man has been sick, has had enough of it, enough of bed, starts toward the bathroom, but his belly muscles because he has not been using them are flabby and do not give sufficient buttress to the blood vessels inside there, and the muscles of his legs are flabby and do not give sufficient buttress to their vessels. Too much of this man's total blood is in consequence virtually stagnant, lost to use, and if in the bathroom he adds some ordinary act of slight cost, brain blood drops, and should the man be lucky he will have eased himself back into bed, if not lucky down he sinks to Mother Earth.

A tall lady in good health is on this afternoon in a department store too public, too crowded, too stifling, the shunting of blood to her brain too slow (it needs to be only by a trifle), and down sinks a tall lady, her customary poise made a mockery of.

That cadet at West Point standing still in the ranks, that perfection-ist soldier, does not move one unnecessary muscle, his capillaries and veins given less than the necessary prop, the one-seventh of his blood not available for the top, and down goes a soldier and not a shot fired. He will have a headache tomorrow.

Blood sloshes freely in our veins, those who know veins tell us, and tell us that though veins have been studied for centuries, every year they present new problems. They were doing that in Harvey's day. He observed veins, did his neat experiments with them, reported them thoroughly in his immortal book.

A stuffy summer day made an elderly woman's skin aflame, her blood delivered in quantity to the surface to drop off heat, the rest of her body stinted, her brain stinted, and the climax was like the other climaxes. People past middle age may feel a bit giddy in June, never mention it. A medical student fainted in the operating room. Among the causes were erect posture, standing still, operating room temperature, strong emotion. At this particular operation both assisting nurses happened to be pregnant, and pregnancy affects circulation, therefore when the two nurses saw the medical student faint, it struck them they had an even better right to. That made three. Everybody spoke of it at lunch.

A pale mailman was threatened by a chained dog. He started to run and, if he had, his blood would have been hustled toward his heart by the contracting muscles of his running legs, but there was a gate that had locked itself. The mailman could not run, his blood was not helped toward heart and brain, and down went a mailman. The dog could not figure out what had happened.

Regardless of the cause of the faint, the body lies dropped from the vertical to the horizontal, brain brought down from the top of the hill to the level of the rest, making the heart do less lifting, and the feet brought up from the bottom of the hill, also making the heart do less lifting, the faint becoming a cure, as it always has been interpreted, a cure of a symptom whose causes are many and sometimes hidden.

CORONARY:
Point of First Delivery

The heart must eat and drink and breathe, rid itself of waste, keep its appointed rounds through snow, rain, heat, gloom of night, keep in repair, do this in health and not be stopped by any little sickness. Like the other blood rivers, those of the heart rush in (coronary arteries), go underground (coronary capillaries), wind their way out (coronary

veins). Galen, that ambitious Greek practicing in Rome, gave them
their name, coronary; it means the making of a circlet, a crown; and
Harvey, choleric Englishman practicing in London, recognized why
the animal has them. Galen no doubt knew too.

That word *coronary* is streaked with mortality. It is a word for which
the bell tolls. Long or short notices, usually on the next-to-last page of
the first division of the *New York Times*, with a photograph of a seventy-eight-year-old gentleman taken when he was forty-eight, dwelling
on his virtues, are every day the public proclamation of disaster in the
coronaries.

Coronary thrombosis. Coronary occlusion. Coronary infarct.

New Yorkers read those titillating terms the morning after, then
turn to the book reviews and the notice on last night's play. Possibly
the seventy-eight-year-old, if he could send back his message, would
assure us that it was an acceptable way to bow into the night. He had
laid out his dressiest evening jacket, next to his bed, laid himself on the
bed, naked as he was, anticipating the luxury of ten minutes before
cocktails, and so they found him, a cautious smile on his face—was
not sure that he was yet seeing straight all things celestial. Ladies suffer
coronaries less frequently, but they gossip brightly about those who
do—their mutual friend had had two busy days in Washington, returned Wednesday night, Thursday morning sent for his breakfast of
coffee, orange juice, toast, two slices of bacon, rang for the cook, sent
back the breakfast, did not want it really, nothing wrong with it, that
afternoon admitted he felt ill, not very, nothing alarming, but agreed
to go to the hospital for a routine checkup, rang violently for the nurse
at 3:00 A.M., she off to the telephone for the cardiologist, and while at
the telephone heard his body plump. He was a big-bodied man.

To calm anyone foolish enough to immediately need calming;
a person may survive his first attack, be felled by the second, third,
fourth, even fifth, sixth. But the event may also be as caressing as an
unexpected kiss—a bit of clot delivered straight to the mouth of a
large coronary, plugging it neatly, all flow stopped on the instant, no
oxygen, so adieu.

Leonardo's drawings of the heart place the coronaries correctly anatomically. It is possible that Harvey saw those drawings.

As to the amount of blood to the heart, it is greater than to any
organ except thyroid, kidneys, liver. As to coronary arteries, they are

pencil-sized. As to capillaries, the heart has twice more than other muscle. As to veins, 70 percent of their blood is poured into a lake at the back, most of the rest into the right auricle, some trickles into each of the other three chambers. After that it is the postman's routine round: right auricle, right ventricle, lungs, left auricle, left ventricle, city, suburbs. Every molecule not delivered and used, every molecule not sacrificed to wear and tear, not sequestered, not lost to bleeding, returns to the heart.

If a fluid as thick as blood—to bring this properly to the laboratory—is in a cadaver forced into one of the two coronaries, it does not appear in the territory of the other; if thin as water, it does; which could argue that one side of the heart has no immediate blood connections with the other but that connections might be established if given the chance. And this could argue that there is less likelihood of attack if a block has developed gradually, allowed for opening or channeling. Surgeons have joined vessel to vessel. Surgeons have encouraged vessels to grow into the heart from outside the heart. Have shut off neighboring vessels so as to divert their blood to the heart. Many ideas have blossomed in surgeons' and researchers' imaginations, are blossoming, will blossom.

That gradual opening might also suggest compensation for the gradual closing as we get older, the chance to get us older still, a little.

A distinguished counselor on the heart used to boom out in public lectures that if stricken in his next sentence, please no colleague rush up with an oxygen tent. Enlivened his lecture that way. It also gave him a transition to the subject of control of coronaries. He always had thought that lack of oxygen was the key to opening, so naturally if one of his coronaries got plugged, he would want every other one open, would want no "wretched, rash, intruding foole" to kill him with oxygen. That old counselor loved life, and having lived long enough, knew too he could change his mind. "And had he ever said that lack of oxygen opened coronaries?" He never had. Nor was he contradicting that in other places in the body, not lack of oxygen, but presence of carbon dioxide seemed to do the job. That booming counselor, like the other counselor Polonius, is today "most still, most secret, and most grave."

Everybody knows angina; it tormented his uncle, then killed him. A sudden severe pain, feeling of suffocation, called also breast pang, occurs where coronaries are narrowed and stiffened, the attack itself excited by some sudden blood demand during physical effort, or some strong emotion which tends vessels to spasm. Angina worried the horse-and-buggy doctor, who treated it with nitroglycerin under the tongue. Worries our cardiologists, who, among treatments, have sometimes implanted an electrode in the skin over the chest and put an electric switch in the patient's pocket. He could control his bouts, somewhat.

Experiments come and experiments go. Observations come and go. Years back it was reported that we human beings were selected for this terminal notoriety well in advance, at the time we were pieced together as embryos, soon after boy meets girl, when the chromosomes line up. Thirty-four percent (that was the figure) of hearts had right and left coronaries balanced, 48 percent right dominant, 18 percent left. The last were the relatively doomed. However, in our day everybody goes on seeing his doctor, bribes his doctor to let him have one more round out at the golf links, two more rounds with the planet around the sun.

VI

BLOOD
Five Quarts

WORK:
Red Wine

What is the work of blood?

To be literal, it is to provide the 1,000,000,000,000,000 cells of the body, if it is a human being's, with a pantry, and with a place for piddling. They reach their hands into the pantry, satisfy their needs, drop their wastes and the products of their work. 1,000,000,000,000,000 cells—streams of blood running through—yet our body can strike us as so small when we put on hat and overcoat, walk out over the vast heath alone.

It seems malapropos to call the blood an organ, but it is an organ, and is also called an organ, the organ of the blood. Its molecules move in all directions. They move quickly. They mix. They equalize.

Possibly that is it—the blood's work—to equalize. First to equalize its own content, as it does moment to moment, strew its molecules around, scatter them inside the body, scatter a comparatively small number beyond the body. Equalize the body's water. Equalize the body's acid-alkali. Equalize its heat. Generally maintain the body's constancy, Claude Bernard's constancy.

Odd to consider a man dipping water from a cool spring, drinking, the water entering him and his blood, some of it promptly departing again in the sweat of his brow, joining a cloud on a hot afternoon and as rain contributing to the level of the Pacific. Nonsense! Of course. But in its crazy way, true. What lately was part of the outside world, and soon will again be part of the outside world, has had temporary residence in the blood.

CONTENT:
Wine Taster

Forty-one million five hundred thousand leeches were imported into France in 1833. Each bloodsucking leech, when its great moment came, drew its sample, disgusting wine taster. That sample was not for laboratory analysis, not for study of the blood's content, the heyday of leeches having passed before the heyday of laboratories. As many as fifty leeches might suck from one human body at one time, though usually it was only five or ten. Physicians were themselves sometimes called leeches. This morning, thousands of physicians have drawn samples, or left written orders for various sizes of samples, various techniques, various body parts. From the neck vein of a five-day-old infant that at the prick of the needle opened its grandmotherly eyes, squalled, closed its eyes, complained, and the syringe filled with blood. From the earlobe of an eight-year-old boy who jerked away his head and a scarlet bead hung off his ear. From the blood-forming bone of an elderly gentleman, a special sample. A still more special sample is drawn from the right side of the heart of a woman lying with propriety in the dark of an X-ray room, a thin "spaghetti" tube having been inserted into her arm vein, pushed until it was in her right ventricle, and from her also a large syringeful. A lean cardiologist is watching.

Each of these techniques has been a first step on the way to investigate an item of the blood's content, normal content, abnormal content.

A sample will be spun at high speed in a test tube, the cells thrown to the bottom, the fluid staying on top. The fluid is the plasma. The cells are red cells, white cells, platelets. Plasma + red cells + white cells + platelets = blood. That seems too ridiculously simple in the light of what the professional hematologist knows of the blood's complexity, but the division is useful, is practical. The innocent-looking platelets, in recent years the subject of wide study, are important in the defense of the body.

But the physician may have wanted a different statement, wanted a blood count, a red cell count. He jabbed the ear of the eight-year-old, put the tip of the pipette into the drop, and the blood went up the pipette. The wine taster who tastes the wines of France also uses a pipette. The physician took from that eight-year-old's red river a

measured amount, diluted it in a pipette, a measured dilution, needed to be careful because any error here would be multiplied in the
later steps, and blood counts have always the possibility of a 10-per-
cent error. A dab of the dilution he placed on a microscopic slide
etched with squares of an exact size, square millimeters. Then over
the dab he laid a thin glass, producing in that way a film of an exact
thickness. With the low power of the microscope he now counted
the reds in several squares, struck an average, multiplied to get the
number of reds in a cubic millimeter, multiplied by the dilution to
get the number in undiluted blood, finally jotted down that eight-
year-old's red count.

Five million four hundred thousand reds is average for an adult
male, half a million fewer for an adult female; 7,500 whites is average; 250,000 platelets. In the living body the numbers of microscopic and electron-microscopic items can be grasped only by persons
who habitually think in numbers. Blood is about 55-percent plasma,
45-percent cells.

If the physician had let that drop just drip from the ear, it would
have clotted; after a time a fluid would have appeared pressed from
the top of the clot, the serum. This is a clear amber. Clot plus serum
would be still another statement of the content of *vin rouge*.

All of the body's construction materials are in the blood, all the
demolition materials, solids 20 percent, water 80 percent. Carbohydrate, fat, protein, salts, water: that would be a gross list. An eminent
American chemist died and left his list: water, oxygen, carbon dioxide,
hydrochloric acid, other acids, alkali in cells and plasma, protein in
cells and plasma. At that time it was thought there were two proteins.
Later four. Later thirty-five. The increase to thirty-five came during
the bloodletting of World War II, a cheap fluid then.

Xanthine is in it, hypoxanthine, creatinine, amino acids, ammo-
nia—ugly names that get friendly to those who deal with them every
day. Metals: iron, copper, sodium, potassium, calcium, magnesium,
cobalt. Products of digestion. Products manufactured by the body.
Foreign items. Lead gnawed off a toy train. Arsenic contributed by
an impatient legatee. All molecules, atoms, ions, all of earth and air
get in, and the content of the stars. At least, it is believed that the big
universe from who-knows-where to who-knows-where is built of the
same materials. Stars in our blood. Poetry, you say. Our body operates

on prose, you say. Whether you are wrong or right, the blood is a compromise between what can be packed in and a fluid that can be pumped. It must contain everything that comes and goes to and from the cells of the body, and yet it must flow. The *vin rouge* is not thick, not thin, just right.

PERNICIOUS ANEMIA:
Poor Vintage

Blood can be sick. Pasteur said of the wines of France that they can be sick. Much about healthy blood has been learned from sick blood. "Look into the blood," the old doctor said to the young one by way of helping him to become a research worker.

A person may be born with sick blood, a sick gene, or the blood may just get sick; for instance, some atrophy of the stomach. The person has blood-production trouble and this can be confined to a single so-called formed element: to wit, the red cell. The red cell begins its development in the bone marrow, but in this sick blood the development does not go the whole way. The red cell does not reach a normal maturity.

In the year 1929 there was a physician who was sick in this way, and every day he took a sample of his blood, studied the red cells under the microscope, saw how the number of the immature ones was increasing, and because of that sign, and because of other changes in his body, in his nervous system, had some masochistic satisfaction in watching the grim specter approach step on step. His disease was pernicious anemia. In 1929 pernicious anemia was no better understood than cancer today.

A smooth sore shiny tongue, a yellow pallor of skin and lips and of the whites of eyes, loose bowels, a cringy stomach, a growing enfeeblement, peculiar sensations in fingers and toes, tinglings, numbness, possibly disturbances in locomotion—all of this in a person middle-aged or older would, today, make the dullest physician suspect pernicious anemia. How that disease, inscrutable then, was followed till its cause had been found is one of the vivid chapters in the history of the blood.

An American research worker was inducing anemia in dogs by bleeding them for long periods of time. Anemia is a reduction either in the number of red cells or in the amount of their coloring matter. The worker always bled the dogs short of killing them. His research had to do with the effect of food on blood production, and one fact noted was that if liver was fed to the weakened dogs, they improved. Two other American doctors took that hint and fed liver to patients suffering with pernicious anemia, and their improvement was dramatic. Several years elapsed. It had long been known that in pernicious anemia the stomach did not properly secrete hydrochloric acid, this state called achlorhydria. With surgical removal of a stomach for cancer there is also achlorhydria, since the stomach secretes the acid, and there is also a severe anemia. Still another American set himself to that. Feeding acid did not help. Feeding a variety of foods did not. Feeding juice from the stomach of a healthy person did not. However, if a healthy person were to eat a mixed meal, or a meal of beef muscle, allow this partly to digest, have that removed by suction and eaten by the sick person, that would help. The blood picture would change. Or if the sufferer were to eat the beef and then swallow the juice of a normal stomach, that would help. Or if the beef were incubated with normal human gastric juice and eaten. Any of these procedures would be appetizing to a dog.

Evidently there was a factor that it was necessary to eat, and a factor that it was necessary for the stomach to manufacture. If those two conditions were met, something was formed that made a something absorbable in the intestine, this then entering the blood, traveling to the marrow of the bones, that then did the important work of maturing the red cells. What was absorbed in the intestine might instead travel to the liver, be stored there, to be let out as needed (an exceedingly small amount), and this explained why eating of liver cured or improved patients.

More years elapsed. More research. The food factor turned out to be a reddish chemical, a ton of liver yielding only twenty milligrams of it, vitamin B_{12}. It had cobalt in it. Cobalt was one of the metals in those original oceans back at the beginning when the vast line of the living was beginning its odyssey. More interpretations. What was plain was that the pernicious anemia patient had been cheated somewhere, was a chemically deprived person. Proper treatment made up for this. Vitamin B_{12} may be the most powerful therapeutic agent known. Within hours the blood picture changed, within days the marrow changed, and the physician could say that his patient would be seen on Thursday at Rotary again.

A happy history, the whole of it, because of the step-by-step march of the discoveries, a march that has kept on and has led and is leading in still other directions. A warming history to an American because so many of his countrymen were part of it. Within a few years secrets hidden through centuries slunk from the caves, and our knowledge of the organ of the blood had a renaissance.

RED CELL:
Color in the Wine

Five million four hundred thousand red cells are in each cubic millimeter of blood, on the average.

A red cell is born in the marrow. To study that birthplace the physician thrusts a needle into the breastbone, draws out a specimen, calls it taking a sternal puncture. Breastbone = sternum. He examines that specimen of marrow under his microscope. This amounts to an examination of the manufactory, investigation of the winery. Throughout the active part of its life a red cell has no nucleus, but throughout its maturing it does. Each year we find that it has been responsible for some additional kind of important living labor. Its shape is a disk. A rough life this disk leads, bumped by the crowds, squeezed with thousands into one end of a narrow capillary, then squeezed through, then squeezed out the other end, a subway rider's life. Nothing that did not have an elastic framework could stand it. The disk is concave top and bottom, which has engineering value, gives the red cell (erythrocyte) the largest possible surface for the passing in and out of molecules, which is its vocation. Also, the concave top and bottom give the disk the capacity to swell and shrink, which it may do several times a minute, so it has one size in the arteries, another in the veins, this the result of the continual back and forth of its physics and chemistry. Gases come in, go out, water comes in, goes out, salts in and out; the wonder is that the poor sac survives as long as it does, though finally, like all cells of the body, it breaks down. 35,000,000,000,000 is an estimate of a man's red cell population; 10,000,000 are born every second, 10,000,000 die every second. When a red cell has made the round-and-round enough times, it begins to leak and continues to leak until nothing is left but its thin wall, which is called a ghost.

Several methods measure the life span. About twenty days. Born in the marrow with a nucleus, keeping its nucleus throughout about four days, during which time there are three or four cell divisions, it meanwhile creating color from its nucleus, then getting rid of its nucleus. Now it is pushed off into the crowd. Travels with the crowd. The hurly-burly of its life never stops. Batted by the mechanics of the circulation, by the chemistry of the body, by the ceaseless adjusting to keep Bernard's internal environment constant. Then, pushing one last time, it goes straight toward a scavenger waiting for the ailing. That scavenger, a cell, is an important character in the drama of the blood, and in the drama of body defense.

SPLEEN:
Wine Cellar

Decrepit reds are unsentimentally disposed of. Excellent funeral arrangements exist for this. The spleen is a chief functionary.

A small organ, the spleen weighs 150 grams, 5 ounces. In an adult human being it is five inches long, lies left of the upper end of the stomach, left of the pancreas, above the left kidney, in a bend of the large bowel that appears to be bending around it. It alters during work. It may be abnormally enlarged by disease. In one case eight pounds of spleen were drawn out through his incision by a surgeon, that spleen having required removal because, among other reasons, it had been abnormally destructive of the platelets of the blood.

Besides being mortician and graveyard, the spleen is a manufactory. It manufactures lymphocytes, may pass lymphocytes back and forth between it and the bone marrow. In the fetus it manufactures red cells; therefore it is not surprising that later in an emergency it may take up that work again.

Joseph Barcroft and some others were en route to South America, to Peru, for the purpose of studying the changes in a human body when it climbs a mountain. To pass the time at sea they made routine measurements of the volume of their blood and the depth of its color. Both the volume and the depth of the color rose as the ship steamed into tropical water, into the Panama Canal. Manufacture of new red cells could of course not occur that fast. Somewhere in the body there must have been a reservoir that could release them. This steaming into tropical waters first called Barcroft's attention to the possibility of there being blood-stores, and an organ that he immediately suspected was the spleen. The various blood-stores that he speculated about were skin, liver, lungs, and the spleen.

Barcroft knew that carbon monoxide is a poison, because it suffocates red cells, chemically prevents them from getting oxygen, makes them therefore unable to perform. So he placed rats where they had to inhale carbon monoxide, then at intervals killed the rats, found that the poison did delay getting into their spleens, "a lag of about half an hour." Unpoisoned reds would be in the spleen after there were no unpoisoned ones in the general circulation. Therefore, a rat without a spleen ought to have a smaller chance of survival, because the reds in its bloodstream would be quickly poisoned, and it had no storehouse from which to draw unpoisoned ones. That proved true. The reverse also; if the poison once got into the spleen, it was slow to get out.

Having concluded that the spleen stored reds and released them as needed, Barcroft was pleased to learn of experiments that tested the effects of exercise. During exercise more reds are needed, because more air is needed. Forced swimming is exercise. If rats are stalked, hit on the head during forced swimming, their spleens ought to be smaller than if stalked during sleep. That proved true also. A smaller spleen was found in hemorrhage and in emotion. In emotion the body was spending in all directions, so red cells were needed. A dog's spleen shrank, Barcroft found, if a duster on which a cat had lain was brought near the dog's nose, shrank to half if the cat mewed, to a quarter if the dog chased the cat.

How could Barcroft know the size of a spleen inside the living body? How did he measure it?

He had several ways. One was surgically to loosen the spleen without damaging it, bring it outside, stitch it under the skin. Another way was surgically to fasten metal clips to its edge, the clips blocking X rays, and Barcroft making the measurements on X-ray plates. He estimated that a dog's spleen is able to store roughly one-fifth of all the dog's red cells, 2,000,000,000,000 of 9,000-000,000,000, and also can contract and push them out. A human spleen has less smooth muscle than a dog's, probably does less contracting, less storing, but it does contract. Barcroft stated that surgeons at operation find the spleen smaller, which one would expect because surgery includes anesthesia, hemorrhage, emotion, consequently more red cells needed, and the spleen would be letting them out, perhaps pushing them out. A surgeon asked about this admitted that when down in a human abdomen he occasionally gave the spleen a slap, and it got smaller.

WHITE CELL:
Creeper in the Wine

The white cell is different from the red, different look, different life, different work, is a member of a smaller population. Leukemia is a cancer of whites.

Seventy-five hundred is the average count, but the range is wide, four thousand to ten thousand. This means that for every six hundred reds or thereabouts, one white.

Various echelons of experts study the whites, know them in detail, sometimes tedious detail. Experts are experts—one says that, then quickly feels guilty, because one remembers leukemia; no years of even dull effort would seem too much when one thinks of that group of frightful white-cell diseases, some already comparatively curable, and if the cures were truly successful, all cancer might be attacked from that stronghold.

One of the birthplaces of the lymphocytes is the lymph node, and these take their name from that birthplace. Other birthplaces are bone marrow, connective tissue, spleen, thymus. The large lymphocytes transform stepwise into the small lymphocytes, so that there are not merely the two old types, large and small, but an almost continuous series of types.

The whites that are born in the bone marrow grow up there, reach maturity, leave the place of their birth, make the Grand Tour, die. The marrow serves also as a reservoir for whites.

Some whites creep in our blood quite as if they had ideas of their own. Millions and millions of private lives. The number flowing or creeping in us may be different in an hour, different tomorrow, decidedly different if we have a cold in the head. Surely they give us another picture of ourselves from that returned by our mirror this morning in the bathroom. Creepers in the wine. The polymorphonuclear neutrophil will appear to climb a pole, think better of it, continue off onto a blood vessel wall, think better of that, go to something floating in the stream, hang on to it, leave it, keep that up, a sluggish microscopic acrobatics.

The eye bent over the eyepiece at the top of the barrel of a microscope may see crazy sights.

Life span of the whites? May be brief, may be long, for one type some say three weeks, some for the same type say sixty hours. Either way, we need not be hematologists to be curious at this massive birth and death. For another type, one week, two weeks. For the lymphocyte, some forms, the life span was once thought hours, now is thought years.

A member of a type has been fished out, grown in the laboratory, watched from birth to death, its span directly clocked.

The white cells' graveyard? For the final rites the lymphocyte returns to its birthplace, the stuff of its nucleus being preserved for making new lymphocytes; what remains is dumped into the *vin rouge* to stock it with protein. The graveyards of other types are spleen, liver, bone marrow. Scavengers wait in each of those localities. Scavengers are always cleaning the streets of the circulation. Scavengers are often quaint characters. They are cannibals, and the cannibalism can be followed with the microscope. It entertains a man.

VOLUME:
". . . So Much Blood in Him"

Last century two decapitated criminals were bled. Previous to the decapitation they had a sample of blood taken from them, and it must have seemed a tantalizingly tiny sample under the circumstances. It was diluted one hundred times to provide a standard for color comparison, afterward. Then the decapitation. Then each criminal was given the same consideration. All the blood that would was allowed to flow from his body and head, then his blood vessels were washed, his tissues chopped up, and the color soaked out of them. Washings and soakings were diluted until the dilution had the color of the standard.

This, the quantity of it, gallons, was divided by the number of times the standard had been diluted, and that figure told how much blood those miserables had had in their bodies during their lives. The method was not very inaccurate. It proved our blood to be one-thirteenth of us. Five quarts go the round-and-round. Less in women than in men, and varying man to man, woman to woman, but for a one-hundred-and-fifty-pound man, five quarts. The method was spoken of as direct, as it was.

There are indirect methods. A known portion of dye is injected into, say, our judge. Time is allowed for the dye to color his plasma, four minutes or thereabouts, long enough to mix but not long enough to leak into the tissues. A sample of his blood is drawn and spun in a centrifuge; two layers form, plasma on top. The color of that plasma then is matched with the colors in a set of color tubes, and this tells straightaway how many times the injected dye has been diluted. If three hundred times, the volume of the judge's plasma would be three hundred times the volume of the injected. Plasma of blood is to cells as fifty-five to forty-five, approximately, must be determined for each person; once determined, easy arithmetic gives the answer, the volume of blood of our judge. The dye is special. It must not, as said, and does not too much leak into the tissues, stays in the vessels, is taken up by the plasma, not taken up by the red cells, not taken up by the scavengers that loiter in spleen and liver and other places of what is called the reticuloendothelial system, has a color easy to distinguish from most body colors. One would suspect the discovery of such a dye required patience. It did.

In our radioactive day there are radioactive methods, and the methods seem beautifully accurate, now.

We are more comfortable in our minds if science is practical, and anyone might ask if there was practical use in determining blood volume.

A person has a hidden hemorrhage, is bleeding somewhere inside him, a peptic ulcer possibly, when there might be free blood in his abdomen. The hemorrhage caused an immediate fall in blood volume, but the red cell count did not change because, though he had lost the blood, there had not yet been a diluting by water seeping into the vessels from the fluid around. So there was less volume, but the blood itself was still unchanged, the red cell count unchanged. Soon

the water does seep in. The blood volume is reestablished by the diluting, but the diluting has driven down the red cell count. So, had there
been a hemorrhage, and had the other signs and symptoms not been
sufficient for a diagnosis, knowledge of these shifts in blood volume
and red cell count might have saved a life.

Lady Macbeth was so impressed by the volume of the blood of
murdered Duncan that she began to talk of it in her sleep. "Yet who
would have thought the old man to have had so much blood in him?"
It was only 8 percent of Duncan.

VII
LUNG
Rock of Earth in Raiment of Silk

AIR:
Nothingness Around

The Old Testament has always left in us the feeling of the primordial, where thoughts of an ancient function like breathing belong.

At the beginning of the world—how out-of-date that sounds—Jehovah created the heavens and the earth, and the earth was a rock, a small rock among the many small and large rocks that were out there, all the nearer ones lighted by our sun. Rather a cold rock, His earth. Over the shoulders of it He therefore laid a raiment light of weight, a thin chiffon, almost nothing, air. Within that, generation upon generation would live their lives.

Mountaintop or down where the rivers flow into the sea, man and hawk and ass and jackal nibble at the molecules that are the air, a number of kinds. Hundreds of times more are oxygen than carbon dioxide, two thousand times more are nitrogen, and some rarer molecules sprinkled in. Then there are water molecules, many, and the glory that Joshua saw of an evening when he looked "unto the great sea toward the going down of the sun" was owing to water molecules. The proportion of the kinds is the same on mountaintop as down where men slosh in wet sand at the edge of the sea, but the number is greatest at the level of the sea, fewer and fewer as a man jogs up toward that mountain above mountains where Jehovah dwells.

Plainly this nothingness at which man and hawk and ass and jackal nibble is not nothing. It is something.

A maximum of oxygen enters at each breath, a maximum of carbon dioxide leaves, and of nitrogen about the equal enters as leaves. One hundred and fifty times more carbon dioxide, about, goes out than went in, most of it having come originally from the cells where the hell of the body is always smoldering and producing carbon dioxide.

But how could one know accurately the quantity and kind of molecules farthest in? One could not if one insisted on being literal, but an old and honored way of measuring added learning. A long rubber tube was fitted with a mouthpiece. Near the mouthpiece a side tube led to a sampling flash. The man who was allowing himself to be studied in this fashion breathed in, breathed out, then on command breathed out with full force; the last to come out of him, from that farthest in of him, was trapped by the sampling flash, analyzed. The nothingness meanwhile paid no attention. Was the same chiffon.

BELLOWS:
Tides In, Tides Out

Imagine a single breath having started. It is scraggy Abel we are watching. Abel's diaphragm and the other of his breathing muscles have started on the next contraction.

Abel's diaphragm is a thin strong muscle partition between chest and abdomen. The single breath advances. The contraction of the diaphragm advances. The floor of the chest lowers, and in consequence the total space in the chest increases from top to bottom. The rib muscles at the same time contract, and the ribs that were directed downward sweep like hoops to the horizontal, and in consequence the total space in the chest is further increased, but now it is front to back and side to side. So, more space in the chest in all directions, more space for the lungs to expand into, which they do.

After the contracting of the muscles (inspiration), there is a corresponding relaxing (expiration), less space in the chest, less space in the lungs.

What we are witnessing is the quiet breathing of our daily lives. In, out. In, out.

During the taking in, the entrails are pushed down, and to make room for the entrails the belly wall relaxes its tone, its tenseness. It bulges. In, out. In, out. A bellows truly. When a bellows has opened, there is increased space in it, and something is drawn in. The infant opens its bellows twice while the midwife who sits like a watchdog by the crib opens hers once. She opens hers sixteen times a minute, about. Breathing is slow in sleep. Fast in fever. Smooth and silent on most days, but not smooth and silent if there is threat to the source. Whosoever has witnessed a surgeon laboring inside a chest to remove a cancerous lung never forgets it, the surgeon as busy as a sailor on a ship in a storm, never forgets either the sight of that lung when it was hoisted through the grievous wound, three lobes if on the right, two if on the left.

When a lung is inflamed and swollen, it has less space in it. When a failing heart lets blood dam in a lung, the lung has less space. When we lie down and the entrails press up against the diaphragm from below, less space. A sick man, breathless, has more space when helped to sit up, breathes then more easily, his abdomen lifted off his lungs. (This is really no figure of speech.) A nuisance to be breathless even for a short while. A woman in the final third of pregnancy may be breathless, a happy breathlessness if this is her first-born and her lord is sympathetic.

ALVEOLUS:
Entry

The ultimate place where the molecules of the air of earth come and go is the alveolus. This is far into the lung.

Jehovah did not bless every creature with lungs. The one-celled living its life in pond water keeps in good health though it has available only the oxygen dissolved in the water; indeed, it can be larger than one cell, just so its girth does not exceed one millimeter. If more than one millimeter, whether living in air or in water, it cannot get enough oxygen by such physics. Man, of course, cannot. No bird of any sort. Oxygen never would reach the tissues of man's body, and that was the reason Jehovah laid in tubes. These start at mouth and nose, continue through windpipe, which divides, right, left, those divisions divide, so on, narrower and narrower. At the narrowest is that final entry, the alveolus.

A cast was made of the airways of a human lung, and it looked like a tree from which the storms of autumn had blown the leaves. If the eye follows along those narrowest divisions, it comes to what is called a bronchiole. This divides, and its divisions divide, and at that point there are ducts, then a sac, and the sac pouches, each pouch has outpouchings, and those terminal outpouchings are the alveoli. There the gas molecules enter the body and exit from it. Each alveolus is microscopic, but because there are so many, great surface adds up from all the walls of all of them.

Imagine one alveolus. It has thin walls. Outside it are those other thin walls, of the capillaries. Blood is flowing in the capillaries. Through the two sets of thin walls the breathed-in molecules go from alveolus to blood, and the breathed-out leave blood, leave alveolus, leave lung, become part of the nothingness around. Around the jackal, too. We and the jackal are on this bold rock together and ought not to forget it. We have gasped together. Aristotle did not forget it. Thoreau did not.

The total length of the capillaries in the five lobes of the two lungs of a man comes to one thousand miles. The wall surface of the alveoli comes to one thousand square feet. To arrive at that latter figure, first the total number of alveoli had to be counted, as far as this was possible, then multiplied by the wall surface of one alveolus, which is measurable. What it means is that a veil of capillaries one thousand miles in length and a veil of walls one thousand square feet in area are the essence of the lung lobes of a man. In order to achieve the one thousand square feet, back in the womb in the embryo the beginnings of a lung everywhere fold, and at last when all the foldings are added together there is a vast surface, a veil of walls for all the molecules of the gases to pass in and to pass out. Surface to meet the body's needs when it lies down to sleep. When in the fresh of morning a woman rises too late and rushes to the well for water. She breathes hard. She feels the distance to the well greater than it used to be. That vast surface, blood on one side and air on the other, is only just sufficient.

CAPACITY:
"What Ye Have Need Of"

Esther. Joshua. Ruth. Nehemiah. Moses. Each of those grand characters without cease breathed in a pint of air, breathed out a pint, from maturity to sepulcher, took what each had need of from the immense and awful nothingness, gave back to the nothingness. Scraggy Abel too. A pint, approximately. In, out. In, out. The animals, too, each according to its needs. In, out. "The wild asses stand on the bare heights, they pant for air like jackals. . . ." While the Hebraic scholar intones in deep tones the Law, in, out, a pint. If the Lord Jehovah has said that the enemy hordes shall fall upon the mountains of Israel, shall be left to ravenous birds for food, and if the soldier of the Prince of Rosh heard Jehovah and was filled with fear, more in, more out. Fear, like strife, increases ventilation. If at the end of a usual in, Esther keeps drawing in until she has drawn in all she can, two quarts will have been added to the pint. Esther is a goodly proportioned woman. If at the end of the usual out she instead pushes out all she can, it will be something more than another quart, and yet even then some will be left in her lungs, impossible to push out. In a stillborn, air may or may not be found in the lungs of the little corpse; it will be found if that one ever took a breath. Important sometimes to establish whether it did, whether air was ever in this creature's lungs, whether the scribes should write into their ledger that the crazed mother was not in agreement that No. 4,932,769,432 ought to have lived.

PLEURA:
Like a Coat without a Seam

The pleura, the tough membrane covering the lungs, has two layers. Someone might think it is like a coat without a seam, but it is more like a paper sack that has never been opened. Its inner layer is fitted close to the surface of the lung, and where the pipelines for blood and air arrive and leave the lung, the inner layer appears turned back to become the outer, this now fitted close to the inside of the chest and to the top of the diaphragm. No space is between the two. During breathing the two rub over each other, oiled by a lymph.

In the fetus the lungs fill the chest, and the chest moves, but only the slightest movement. Then the infant is born. The chest expands sharply, the lungs expand with the expanding chest, air is sucked in. The two-layered sac accordingly is stretched, and to the end of that human being's days the lungs will always be expanded, and the pleura will always be right in contact with the inside of the chest and the top of the diaphragm.

A rip through the pleura alters this. A knife stab or a crush wound might cause a rip from outside. Blowing night after night against the resistance of a horn might cause a rip from inside. Occasionally a rip occurs in what appeared to be a healthy person, but was not quite, a sudden piercing pain announcing the rip. Previously there would probably have been a blister in that layer against the lung, then some strain; it may have been no more than a deep breath during a physician's examination, and the blister burst. Air now leaks out of the lung and seeps between the two layers, creating a space between them. At the breathing in, some air leaks from the lung into this space, torn flesh prevents it getting back at the breathing out, so it accumulates. If swiftly, the person dies, but it is apt to accumulate slowly, eight hours, ten hours. The character and the place of the sudden pain would have told, to a person who knows about such events, what happened.

A surgeon while operating had sudden pain, fell down. A fellow surgeon knew. (Spontaneous pneumothorax.) With a syringe he drew out the accumulated air. Had that not been drawn out, the lung could no longer have followed the moving chest, would have been stopped by the cushion of air, giving the lung a rest, but rest that did its owner no good.

Tuberculosis formerly often required that a lung be given forced rest. If only one lung is sick, a surgeon can cut a window through the ribs on that side, air can rush in and collapse that lung, and it can stay collapsed for as long as necessary. (Open pneumothorax.) Instead, the surgeon might push a hypodermic needle between the layers of the pleura, inject air, which presses on that lung, collapsing it. (Closed pneumothorax.) If watched with the X ray, that lung still moves, but small movements. Gradually the injected air is absorbed, and more air must be injected. A remarkable thing has been done: some of the earth's raiment has been transferred to the pleural space where air ought not normally to be, but saving this man from death, as air once let him come alive.

HEMOGLOBIN:
Man's Blood and His Green Pastures

Billions of oxygen molecules enter the body, travel in the blood, arrive at the cells. Billions of carbon dioxide molecules leave the cells, travel in the blood, leave the body. That traveling in the red cells depends much on a pigment that has iron in it, hemoglobin. Hemoglobin can be called a pigment that breathes.

Another prime color worn by our earth, since Genesis, is green. "He maketh me to lie down in green pastures." Plants draw from the sun and are fruitful because of green.

The green chlorophyll is a small but complex molecule. Photosynthesis is the process that enables the energy of the sun in the presence of chlorophyll to combine carbon dioxide and water, release oxygen, produce carbohydrate, which are first steps up the evolutionary ladder to human life. With chlorophyll, and sun, and air, plants create themselves, and we eat them. Mars has a mosslike green arriving and departing through the Martian year. Mars has little or no water, the thinnest of atmospheres, carbon dioxide in it, and up over Mars during the Martian day there shines the sun.

Genesis rewritten would read: sunlight of heaven, elements of earth, a green pigment, and there was oxygen, and there was life, and there was a plant. On a later day among the seven days there was an animal, and the animal ate the plant. "I have given every green herb for food: and it was so." Still later there was that woman, and the man and that woman sat down to supper and ate plant and animal both, and it was so. The man was an early man, and he made this and he made that, fire and tools, and he wished directly to use the fire of the sun, and he was skillful with tools, yet nothing quite worked to his satisfaction, so he was a probing tentative man.

Hemoglobin is heme plus globin. Heme gives the color, and upon it depends the taking up of oxygen. Heme is ancient on earth, may be as old as half the earth's calculated life span. Before hemoglobin, before chlorophyll, heme was already part of the minutest life.

The hawk has hemoglobin. The jackal has. The ass has. The lugworm has. We have. Heme apparently is the same from species to species, but globin varies, so different animals can have different hemoglobins. Within one species of sheep there were found three separate hemoglobins, and by their hemoglobins alone would the members of that species be assured some degree of individuality.

Abnormal genes produce abnormal hemoglobins, which produce abnormal bloods, which produce abnormal bodies, and we say of these people that they suffer inherited blood diseases.

Hemoglobin is the bulk of the red cell, and when the hemoglobin begins to change, even though it was born entirely normal, we might say that the red cell is getting old and soon will die. If enough alter and get old, a man's breathing alters.

Besides the number of the red cells in the vessels, the state of the hemoglobin in those cells helps account for our complexions, the rose of our lips, the peachblow of our cheeks, the bright colors of the healthy and the pallors of the sick. The size of the load of oxygen that the hemoglobin at a moment is transporting explains the scarlet of arteries, the blue of veins. The high pink of the corpse, brought to the morgue from a garage where the unhappy one let the engine keep running, is owing to hemoglobin combined, not loosely with the oxygen of fresh air, but tightly with molecules of the exhaust.

What a rich rugged drama, in act one to have created the earth, in act two to have made it come alive, in subsequent acts to have bestowed the dowry of frailty upon that life, incorporated the life and the frailty of a human being in a woman, over her countenance powdered this fickleness of color. John Donne wrote unwittingly of hemoglobin:

Her pure, and eloquent blood
Spoke in her cheeks, and so distinctly wrought,
That one might almost say, her body thought.

Our little old earth is both strange and sweet but a bit childish to be so preoccupied with two colors. "They shall be full of sap and green," saith the 92nd Psalm. "The life of the flesh is in the blood," saith Jehovah. An old doctor two thousand years later said: "It's all in the blood. Look into the blood." John Donne wrote wittingly: "I observe the physician with the same diligence as he the disease."

REGULATION:
Order Is Beauty

Order is everywhere in the act of breathing. Order is in all that Jehovah does.

In the quiet of evening, in the harassments of day, deep down, order. This ordered rhythm of breathing is affected by all that the body does, by changes in the blood flow through the lung, by changes in the composition of the blood. The rhythm is tailored and retailored, fitted to this one body's constantly changing needs.

As with the beating heart, so with the respiring lungs, all appears softness with precision, and man and jackal breathe without knowing. Everywhere is softness with precision. If we watch the mother hawk standing at the edge of her nest and looking down into the open mouths that also appear to gasp and pant, softness with precision.

Breathing is affected from beyond the body, is affected from within the body, to the end that the body shall always have the exact molecules in the exact quantity. Attempt to change that? Better not attempt. Better bow your head and admit you are driven by ancient law. Or try to deny the law. Hold your breath. Keep oxygen out, keep carbon dioxide in, use up the one and pile up the other, and though you are firm in your resolve, a next breath will crash through. Breathing is affected by all. Allow a young girl to walk dreamily into the village, a robber to be affrighted in the forest, a horned owl to hold a struggling rodent in its grip, the breathing of all four—girl, robber, owl, rodent—will keep pace with what each is doing.

The question, often asked of the heart, is here asked again. Whence this rhythm, whence these heavings, whence these waves that on occasion swell to billows? Is the regulation all within the parts that deal with breathing? Is it the steady push of a gas, or an acidity, or a lack of a gas that produces an intermittence? Is it an even flow of energy into the breathing muscles, regularly interrupted? Is there a piling up of tension to a point of breaking, then a break, then again a piling up? Is the nature of all body rhythms essentially the same? All earth rhythms? Is a colony of cells inside the controlling system and known to be in command of breathing—is that itself rhythmic? Is it?

MIND SIGNS:
High and Low

Joseph Barcroft took a hand in these problems also.

He investigated upon human beings, upon himself, always had a healthy recklessness, once remained twenty minutes in an atmosphere that had two hundred times more carbon dioxide than ordinary air.

Another investigator was with him. Both agreed that they would not care to repeat the adventure. Their intellects were badly confused; Barcroft's wife, not knowing what was wrong with him, kept him in bed. From similar adventures, years of them, a principle emerged: if ever the body gets too much or too little of anything, the mind shows the signs first.

Too much or too little of the gases of the air can be easily arranged. Too much carbon dioxide can be directly breathed from a tank, and too little follows from the taking of a succession of big breaths. The big breaths sweep the normal carbon dioxide from the lungs, hence from the blood, hence a more alkaline blood, hence lower calcium, and lower calcium may cause convulsions, and trouble in the brain. At the same time the vessels of the brain are narrowed, less blood flows, less oxygen goes to the cells, hence blunted mind, faulty memory, dimmed eyes, dimmed senses. The taking of the big breaths may become a habit, give a person satisfaction (the psychiatrists come in here); soon the breaths seem not big enough to him, so he takes bigger, appears an addict, suddenly faints.

Instead of too much or too little carbon dioxide, there might be too much or too little oxygen. Whoever has gone up a high mountain remembers how exalted he felt; meanwhile the valleys below were too beautiful to believe; then when he returned again to the edge of the sea and read the poetry he wrote up there, he saw how giddy he must have been. Mountain climbers and all the newspaper readers who climb with them have heard how in high places the mind slows, the power to think weakens—the oxygen cylinders right where they could be pulled over and used for resuscitation, and not pulled over. Should the person decide to stay in the mountains, not at that killing a height but high, he would be taking bigger breaths day and night, and because of that the form of his body would be altered. His chest would get barrel-shaped. The hemoglobin in his blood would increase. Chest and blood would be rebuilding on a new and grotesque scale so as to nourish the brain that underpinned the mind. That blood would be thicker. It would need more red cells. More would be born and mature. Such thick blood would be harder to move. This would give trouble to the heart. If the pressure of the atmosphere were to drop farther, say to one-fifth of what it is at sea level, there would be pain, and the mind signs might be severe to the point of unconsciousness. The color of the skin would be redder and darker.

Physicians have had experience with too much oxygen. Body signs appear. The airways of the lungs may be irritated, plugged with mucus, the man may suffer what seems a chronic cold in the head. The lungs may be damaged. The brain may be, and in consequence the posture and the locomotion may show signs. Also there may be twitching of lips. Spasms. The most distressing as the most alluring signs, however, are those of the mind. Inattention. Blurring. Blackout. Physicians speak of oxygen as a poison, and it can be for animals. Barcroft wrote of the death of two larks placed in a chamber where there was excess oxygen. First, both became excited (human beings become excited too), and this was followed in both larks by convulsions, a quick death of three hours in one, a slow death of several days in the other. The story as told by Barcroft is scientifically interesting, humanly distressing. Larks belong in the pure air, but it should not be too pure, and not too much oxygen. We are creatures of earth and air but made to function best where the composition of the gases and their tension in lungs and blood are not far from what they are at the level of the sea, whence we came.

CHEYNE-STOKES:
Trouble

The words that follow were written a century and a half ago, anno Domini 1818.

> For several days his breathing was irregular, it would entirely cease for a quarter of a minute, then it would become perceptible, though very slow, then by degree it would become heaving and quick, and then it would gradually cease again. This revolution in the state of his breathing occupied about a minute, during which there were about thirty acts of respiration.

The words were John Cheyne's, one of the goodly company of Dublin physicians whose glory would shine through the generations. John Cheyne does impress us. Besides being the first to leave a careful description of this breathing—periodic breathing—his timing of it comes close to our timing. Hippocrates, physician on the island of Cos, said of it that it was "like a person recollecting himself." But no one before John Cheyne carefully described it. Indeed, uncounted physicians, uncounted members of families from tribe to tribe, could not have missed hearing it echo through some house. The departing was making his concluding suckings from the nothingness. He took a breath, took another bigger, another bigger yet, stopped, no-breath.

That is the one type. In a different type there are small breaths in place of the no-breaths. The house may be so still. During the big breaths the departing may have waked and babbled a bit and been quarrelsome, then during the no-breaths was already not of this earth. When the cycles are long, it is apt to mean a failing left heart, when short, brain damage. Physicians know the facts.

If the physician is standing by a bed set apart from other beds in a place for the sick, and if the physician says of a man's breathing that it is periodic, he means that in every cluster of breaths there is waxing and waning. One breath is larger than the breath before and so for a number of breaths in succession, then each smaller than the one before. It is a large rhythm rising and falling and carrying along the smaller rhythms of the single breaths. Or a cluster of unusual breaths is followed by nobreaths, this followed again by the unusual breaths. In the breathing chronicled by John Cheyne, called Cheyne-Stokes, there are nearly always the no-breaths and usually the waxing and waning. On a mountaintop a similar breathing may be normal in sleep. A dog may need no mountaintop but breathe after this manner in front of an open fire in a house low down where the rivers run into the sea. The newborn may. A man in good health can bring it upon himself of his own will. One manner of explaining lays all to the chemistry. Another explaining lays more of it to the ancient controls low in the brain, those cells known to have charge of breathing, the effect of level upon level, the effect of hierarchy.

During sickness the sensitivity of these cells may be depressed. The probability is that their sensitivity is altering in health too, but sickness alters them seriously, their state helping to sink the man into that final trouble that affrights the watcher because a common pattern of the living creature has become an uncommon one.

As for the solemnity of the sound echoing through the house, that arises from man having heard this breathing just before the awesomeness of death for as long as any man can remember. A man may not hear the snorings that crash through rooms a whole night long, may not hear the noises of a living city, but the breathing of the dying he hears. The physician witnesses this solemnity a hundred or a thousand times. He knows it by ear, by sight, by understanding, anticipates when it will begin. Then one evening he is thinking: now it will begin for me. The physician's hour.

VIII

DIGESTION
It All Works like a Hungry Dog

TUBE:
Beyond the Lips Is More than Teeth

In a zoo at midnight a good-sized live pig was pushed inside the door of the bachelor apartment of the python. The pig was nervous. After the conventional preliminaries that began with squeezing the holy breath of life out of the pig, the python dislocated his jaw, part of the ritual for any of his better meals, and in went the pig, slowly. Whereupon at his upper end the python bulged. Next morning he bulged less and farther down. The pig was melting. Life was making life out of life as the pig advanced through the great snake's digestive tube.

Man also makes life out of life. His digestive tube begins with teeth in the vestibule; he flattens his lips over them, puts a socially acceptable facade over the ravenous.

His tube has many diameters, wide, narrow, valves at strategic points, turns and twists named and unnamed, a purse-string double muscle at its ulterior end. It is the scenic way at his private amusement park. The stations are the following. Lips—the doors to the outside. Mouth—the vestibule. Pharynx—the passage through the throat. Esophagus—the gullet. Stomach. Duodenum, jejunum, ileum—small bowel. Cecum—the word means blind pouch. Ascending, transverse, descending colon, sigmoid, rectum—large bowel. Sigmoid means S-shaped. Rectum means straight.

Thence out into the sunshine or the warm rain.

Flesh-eating animals have a comparatively short tube, four times the body's length; vegetable-eating a comparatively long one, twenty times. Man's tube is on the short side, establishing him as the brother to the lion, the tiger, the other carnivores.

FOOD:
The Daily Bread

About a hundred years ago it could for the first time be said with confidence that there are three fundamental foods: carbohydrate, fat, protein. The human body needs the three. Lacking them it sickens and may die. Later, water and minerals were added as necessaries. Later, vitamins. A sturdy pot roast (one part of a complacently grazing living cow) plus a seasoned salad (from the pregnant soil) plus a dish of velvety raspberries (that yesterday shimmered like dewdrops on a shrub) meet all requirements of body and brain. Yesterday shimmering and living. Today? Well . . . Tomorrow? Living again, just changed their address.

Digestion is the processing of food from the state in which it is devoured to the state in which it is delivered out of the digestive tube into the blood.

Milk is our earliest food. Some shiver to remember it; some righteously drink it all their lives. Milk contains the three, carbohydrate, fat, protein; also minerals and vitamins. Cow's milk has too much protein, not for the cow, but for us, and in cow proportions, and since human infants do not grow as fast as the calves that are hurried off to market, and since protein is the stuff of growth, there is the practice of diluting cow's milk for our babies, never completely satisfactory.

Eggs are high in protein, fairly high in fat, very high in calcium, a good supply of vitamins—everything ready for those fledglings whose necks we shortly have wrung. Fish is high in protein, calcium, vitamins: salmon and mackerel are high in fat, whitefish and sole low in fat, proportions that we do not think of for their value to the fish because we do not think of fish as swimming in the sea but as hooked out of the sea and eaten. Meats differ in the amount and kind of fat and protein. Cereals have a coating of an indigestible substance called cellulose and a career that began in a happy living wind-blown plant. Flour has protein. When soaked with water it makes gluten. Add yeast, living, and this acts on the carbohydrate, forms carbon dioxide, which blows up the dough. The pan is pushed into the oven. Heat cracks open the starch granules of the flour, soluble starches and near sugars run out, and the surfaces exposed to the greatest heat are caramelized and browned, and that is your big sunny loaf of bread.

Vegetables and fruits, both living, have almost no fat. Potatoes, living, are high in carbohydrate; bananas higher; prunes higher; raisins higher; desserts high, often high also in fat, sometimes high in protein, as cocoa, sometimes high in calcium, as chocolate. Alcohol may accompany the meal. If beer, it is comparatively high in carbohydrate and digests slowly.

All of the delight of all of the above reaches not much farther than the mouth, reaches the point where we think, isn't that nice, and the next instant think that something agreeable has gotten away from us.

DEGLUTITION:
A Swallow in Three Acts

A parrot's beak can crack a Brazil nut. A dog's grinders are claimed to build up a pressure of three hundred pounds. A debutante's molars have perhaps a third of a dog's power. The debutante's jaw drops. It lifts. It protrudes. It retracts. She swings it from side to side. In short, she chews. Her lips return the mouthful to where her pearly teeth can get at it a second time, a third time. At last, she has the mouthful mounted on the top of her tongue and it has a name now, bolus. An unsuspecting bolus sits on her tongue, waits.

Then the trap is sprung.

Swallowing starts.

What happens is without division, but a physiologist of the nineteenth century described swallowing as occurring in three stages, so we all do.

The first stage carries the food past the anterior pillars—flesh pillars one on each side of the entrance to the narrows that lead from mouth to throat. The second stage carries it through the throat, pharynx; the third, down the gullet. A hunk of beef, a jigger of whiskey, a safety pin—all go the same way.

The tongue deserves a word. During chewing it operates like a masher, also executes the sudden thrust that drives the bolus backward. The tongue is muscle, largely. Muscle is operating all along the way. The teeth are clenched, by muscle. The cheeks are pressed against the teeth, by muscle. The lips are sealed, by muscle. The bolus is rammed against the debutante's pink palate, by muscle.

Not far down is a treacherous crossing of roads. Air and food must both be gotten into the body, air through the two entrances, nose and mouth, food through one, mouth. Air then continues over Route Windpipe, food over Route Gullet. Engineering problems were bound to arise. Food must not slip back into the mouth, not into the windpipe, not into the nose. Accordingly, the mouth is closed by tongue and pillars; the nose by lifting the soft palate, bringing together the pillars, fitting the uvula in between, and contracting the top of the throat; the windpipe by jamming the voice box against the base of the tongue, pulling everything forward, and thus opening the gullet behind so that food can get in. The V-gap between the vocal cords is sealed, by muscle. Breathing has stopped except for the slight start of an inspiration which creates suction in the now air-tight gullet and—zip, bang, slush—down plummets the mouthful. For fluids swallowing is quick, seconds. For breakfast gruel not so quick, still seconds. For solids slow, sequential muscles of throat and gullet and sequentially pushing the solids along. An infant swallows as perfectly at birth as a man after seventy or eighty years of patient rehearsing.

ENZYME:
Busy as Fleas

Through ages there were evolved molecules to speed up digestion and other action. They were whips. Fleas. Protein fleas. No scientist would call them fleas. Call them giant molecules that activate. Many proteins are such fleas.

The scientist calls them enzymes. They make a molecule accomplish at low temperature, such as our body's temperature, what without enzymes it could accomplish only at high temperature.

One asks then, anticipating the answer, what is it that truly breaks up the deliciously dressed goose? What changes food molecules to debutante molecules? Spaghetti to the dog Paddy? What decomposes, recomposes, with such nightmarish regularity? What makes our digestion chemically move? What makes our body's life chemically move?

The chemist says enzymes. He says much about them. He says that they may start an action in one direction, and when it has gone far enough, reverse the direction. He suggests that the myriad of such changed directions might provide one definition of life. They provide the acceleration in the drive that maintains the constancy of the internal environment. They keep each special cell at its special task. They give the all-around push to that "sum of the forces that resist death" which would be too slow and therefore not powerful enough without them.

Platinum placed in a test tube of chemicals, as any high school student knows, speeds action. The platinum adds nothing to what goes on there, just changes the rate. A substance that does this is a catalyst. Some have thought that a catalyst merely attached itself to the surface of what it acted on, but others, that it always took part in the action, disengaging itself when the action was done, and others, that this differed from catalyst to catalyst. The last is probable if for no other reason than that complexity is the way of the world.

An enzyme is such a catalyst. An enzyme changes the rate of actions without itself being used up. It is an organic catalyst, or, in this region we are talking of, a living catalyst.

As for the enzymes of the digestive tube, they begin in the mouth and follow one after the other, each with its fixed labor, right down to the tube's nether end. A special enzyme stands behind each step in an action. This step could go on without this enzyme, but not fast enough, not joining with a multitude of other steps, so as together to create this that breathes, that runs, that sits, that listens to a heart beat, takes a blood count, watches a peculiar type of breathing, a peculiar moment of digestion, then goes home from the laboratory, scrubs its face, and rides all dressed up to a party in the evening.

Enzymes of the human body work best under the conditions of the human body, the dog's under the conditions of the dog body. Drop the temperature inside Paddy, the Irish terrier, and her dinner will not digest. Contrariwise, raise the temperature enough and her enzymes will be destroyed, she with them. Take an enzyme from an acid locale where it was evolved to work, say, the stomach, place it in an alkaline locale, upper bowel, and it will not work. So on. The sweep of the enzymes, the march of them side by side in the cells of our body, each keeps its place each does its job.

Emil Fischer long ago fancied any enzyme and what it acts on as key and lock. A particular key for each particular lock. What the enzyme acts on is called substrate, and the substrate is believed to move around the enzyme, find a position on it, an active center, one or two or three of these, attach itself, speed reaction there; the substrate is broken down; products are produced, the accumulation of the products presently delaying or even stopping reaction; the enzyme is always released for further work. Three Frenchmen found how the active centers of an enzyme might interact with one another, produce something that performs a different kind of work, and they indicated as parallel to this the rearrangement of the hemoglobin molecule when it is oxyhemoglobin and when it is reduced hemoglobin. Their discovery brought them a Nobel Prize.

In our mouths the substrate would be food, and our saliva has an enzyme that acts on one of the three basal foods. The products produced by that, along with the food which has been chewed but is still undigested, are carried onward in the tube, where other enzymes continue the digestive work. Meanwhile that enzyme back there in the mouth goes on being methodically freed, more of the basal food methodically attached, products methodically produced.

A final idea, or only another idea, is an enzyme inhibitor, a molecule that blocks the action of another molecule. Consider an enzyme inhibiting a step in digestion. Digestion is blocked. Wrong chemicals collect. May damage the creature. May kill him. In that case the inhibitor would have been a poison. When scientists in World War II were scheming chemical warfare, they found inhibitors that could kill in that way. Whereupon scientists of the enemy, whichever was enemy, searched for and found inhibitors of the inhibitors. An enzyme might be poisoned also by ordinary drugstore poison. Cyanide in the gas chamber kills the doomed because it poisons enzymes, and the cells of that body cannot use the oxygen brought to them. If a man is poisoned in the gas chamber or at his wife's dinner table, he thinks it personal, which is unreasonable.

Let a mathematician go on with this account. He should have no trouble now making clear how enzymes can build one man unlike another, in his anatomy, in his histology, in his thought. A smart idea, enzyme, occurred to Nature on a cool morning when for once she had slept well.

FISTULA:
Portraits in a Corridor

This that now follows ought not to be read with the book by the side of the luncheon plate, because—well, it ought not.

An unpleasant word, *fistula*, comes from Latin, means pipe. Here it means any unnatural connection between the inside of the digestive tube and the outside of the body. In Pavlov's laboratory in Leningrad one might see a succession of fistulas, the following all in the same dog, according to one of Pavlov's co-workers. A salivary fistula—the duct of a salivary gland brought out through a slit in the dog's cheek. Next, the gullet cut across—two openings, each brought out and stitched into the skin of the neck, and if that dog was fed ground beef, a moment later the beef soaked with saliva flopped from the upper opening into a dish, a farcical situation, called sham-feeding, a caper played on the dog, the beef never reaching the stomach unless it was refed through a funnel into the lower opening. Next, a small-stomach fistula—a tenth of the stomach separated from the rest and that tenth given a surgical exit through the abdominal wall. Next, the other nine-tenths given an exit. Next, a small-bowel fistula—a loop of bowel lifted up and sewed into the dog's belly, and an exit to the outside cut through the belly wall and the loop. A pancreatic fistula—the duct of that great digestive gland brought to the outside. All these fistulas were in a single dog, a healthy dog, a contented dog, if Pavlov's co-worker's memory was accurate.

In a human being there sometimes has been need of a surgical fistula, as where lye was swallowed and scarred shut the gullet, or boiling clam chowder, and therefore food for the rest of that person's life needing to be fed via a fistula made surgically into the stomach. Rarely, but sometimes, there has been an accidental fistula.

The idea to use such for digestion study, like all ideas, was young once. It was the dewy morning of June 6, 1822, at Michilimackinac, Michigan Territory, or Fort Mackinac, a combination trading station and army post occupied by United States troops after the War of 1812. The population was French, American, Indian. A youth that morning shot himself in the left side, a mistake, the muzzle of his musket only a yard from him, the wound not killing him though "literally blowing off integuments and muscles of the size of a man's hand, fracturing and carrying away the anterior half of the sixth rib, fracturing the fifth, lacerating the lower portion of the left lobe of the lung and the diaphragm, and perforating the stomach." His name was Alexis St. Martin, a name that would live long in the romantic history of digestion. Alexis was employed by the American Fur Company as a voyageur, transporter of furs and men across the north and northwest. About twenty-five minutes after the shot, a surgeon, William Beaumont, arrived, dressed the wound, pronounced the case appalling, the stomach "pouring out the food that he had taken for his breakfast." The following day there was fever, difficult breathing. Beaumont gave Alexis a cathartic, bled him, believed he would die. Five days later the wound was sloughing. Eleven days later the sloughing was extensive, the wound appeared healthy, the fever subsided. Four weeks later the appetite was good, Alexis's digestion satisfactory, bowels moving, but the opening into the stomach did not heal.

Here was an inquiring surgeon's chance. Here one man could look straight into another man, and Beaumont looked. He looked every day of the week, whenever Alexis was willing, which was not always. "I had opportunities for the examination of the interior of the stomach, and its sections, which had never before been so fully offered to anyone." He looked for what actually was there. "I had no particular hypothesis to support." Watched water pour in at the top, remarks that this was "a circumstance, perhaps, never before witnessed, in a living subject." One catches Beaumont's enthusiasm. With a glass rod he touched the inside of the stomach, recorded. He said that the lining had "a soft, or velvetlike appearance." He recorded that changes in the lining accompanied changes in Alexis's mood, that with "fear, anger, or whatever depresses or disturbs the nervous system—the villous coat becomes sometimes red and dry, at other times, pale and moist." He drew out samples of gastric juice, sent them for analysis to the best laboratory of that day, was as scientifically enlightened about all of this as anyone then could be, began it in 1825, continued till 1833, published his findings in a book entitled *Experiments and Observations on the Gastric Juice and the Physiology of Digestion*. A reader can open that book at any page and be rewarded and amused. Beaumont made his patient his complete responsibility. The patient became a declared pauper. Beaumont housed him and gave him his daily bread and had to keep an eye on him because the pauper regularly got bored at the proceedings and absconded. The covenant bound Alexis to "serve, abide and continue with the said William Beaumont, wherever he shall go or travel or reside in a part of the world." (Sounds like the marriage covenant.) Alexis did not respect covenants and did not enjoy the travels of an army surgeon. Letters are extant, the surgeon trying to imbue the patient with the scientific spirit, and the patient drunk. Why not? With a hole in your stomach?

The coolness to use a seriously wounded man for digestive study, the experiments, the observations, the book, are the most unique and may be the most far-reaching achievement of any American who has studied the human body. Those interviews with the stomach are dignified and innocent. They are an investigation into body and head of the highest animal, a prolonged series of clinical procedures, reports on the relation of gastric juice to food, to fasting, to the hour of the clock, to the state of mind of Alexis, which means, to the state of mind of a man who owns a stomach and properly forgets that everything in his daily life affects it. But the science of medicine has not, a whole large division of it having concentrated on the effect of mind on body, to the point of illness.

STOMACH:
More than a Storage Bin

That great baby, man's favorite pouch, which he always speaks of slightingly as he does of the beloved, lies between the lower end of the gullet and the upper end of the small bowel. Food comes in at the top, cardia, called so because it is near the heart, goes out at the bottom, pylorus, Greek for gatekeeper. A mobile organ, it rubs gently against its neighbors down there in the dark, lined on the outside by a silken lining, on the inside by another silken lining but pleated and that has the open mouths of thirty-five million glands. Some stomachs are small. Some seem so because of their worried state. Some seem dropped because they dip down into the pelvis—this is perfectly normal. If only one could say that a steerhorn stomach went with a steer, but it may go with a dove or a man. The stomach usually is J-shaped, and no matter what shape, it may change with its minute-to-minute mood, or life, which can be hard. The first layer of lunch is spread around the inside wall, automatically relaxed to be receptive, a second layer inside the first, a third inside that; meanwhile an animated luncheon conversation is going on all around the table, all around the stomachs. Starch digestion, which began in the mouth with that starch-digesting enzyme of the saliva, continues below in the center of the vat. But why do we say vat? What is the hidden self-consciousness that tempts us always to speak jeeringly of our great reservoir? Reservoir it is. The digesting it does could be done well enough farther along; stomachs have been cut out and nothing too bad happened. Protein molecules start their breakdown in the stomach, which has the enzyme pepsin to attack the large protein molecules, split them to small.

Large to small is the pattern for all digestion. Pepsin requires an acid world, and the hydrochloric acid coming from the glands of the stomach is strong. William Beaumont knew about the hydrochloric acid, though how it is produced he did not know, and we today still are not sure. The stomach has the enzyme rennin, which curdles milk; cannot in the strong acid of the adult stomach but can in the alkali of the infant, where curdling is important. Some fat is broken down—fat slows digestion. While the attacking molecules and the food molecules are engaging each other in chemical combat, muscles of the stomach wall are acting on the softening lunch. A nagging. A maceration. A pushing forward. A steady pressing on the late filet mignon or goose like the pressing on a toothpaste tube. Nothing that once gets into the digestive tube is allowed to loiter. Periodically a wave travels in the direction of the pylorus, which is an easygoing gatekeeper and permits food, chyme now, to pass through and onward into the small bowel with almost no argument. Yet how the pylorus is opened and closed we again are not sure. Water is in and out almost as fast as it is drunk. Beaumont watched water. Food when in chunks delays stomach digesting, and emptying. High alkali in the small bowel delays. Sight and sound and smell of food hastens. The arrival of food in the stomach with its consequent distension hastens. So does the arrival in the upper bowel.

Fear and gloom may stop all of this and gaiety excite it, the stomach's inside color changing, as Beaumont observed, the acid changing, the quantity of total gastric juice changing. Medicine today on every side stresses the effect of emotion on physiology, and every wife knows that her husband's rages may transmute an excellent dinner into an indigestible one. "If only that man would lie down for half an hour before and after dinner—not get so upset." They who have watched the inside of dogs' stomachs would say that wife was giving her husband sensible advice.

GASTRIC ANALYSIS:
Authorized Prying

To obtain a sample of gastric juice requires a fine finical procedure; some persons say it is nauseating. "Everything came up, nose, mouth, eyes, but I kept it in!" That was a medical student proclaiming his success. He had passed a stomach tube into himself, passed it into his mouth, might have into his nose, where high up it would have turned a corner, gone down his throat, continued conventionally into gullet and stomach. He railed, reeled, regurgitated. A distinguished expert said that people swallow stomach tubes as they live their lives, and if they make a great fuss, it tells the physician something he ought to know. One especially long tube arrives at the stomach and continues, duodenum, jejunum, ileum, continues, continues, is assisted by the physician and by waves of muscle contraction, continues, continues, until it has been known to appear on the West Coast. Logically to complete that picture, the two emergent ends might be tied in a knot and thus a human body strung on a rubber clothesline, which is stated primarily to remind the reader that a tunnel does run through him, connects outside with outside, as the Midtown Tunnel connects New York with New York.

Stomach tubes are used every day of the week to obtain samples for gastric analysis, samples farther on for other analyses. Sampling of human cavities has gained a lively place in the last decades, all the quiet areas washed out, the washing scrutinized with the microscope for anything anyone can think of, helping diagnosis and treatment and sometimes prolonging life. Tubes are pushed into the windpipe to obtain washings from a lung. Into the sinuses to the head. Into the genital organs of the male to obtain prostatic washings. Variously into the female to obtain vaginal, uterine, tubal, ovarian washings. Into the rectum, and up. Washings have long been obtained from the inside of the female breasts. Into any veins. Stomach tubes, to return to those darlings, are used if a person has swallowed what he should not and now would not, to feed patients who cannot otherwise be fed, and to find whether hydrochloric acid is where it ought to be, or is not, and how much. The tubes are always long enough and with some to spare—stated to allay the nightmare of the other end slipping in. If the gastric juice happens not to be flowing freely, it may be encouraged with tea and a cracker, which is good; a swig of alcohol, better; an injection of histamine, best—effective, that is. Commonly several samples are collected, curves plotted; in this way the juice of the single Johnny is defined, and when reports come from ten or a hundred cities and villages, the juice of the typical Johnny is defined.

A toy balloon may be fitted to the end of the tube and the combination swallowed, watched with the X ray, and when the balloon reaches a desired point, it can be inflated via the tube, a writing device hooked in, and communications begin to come back out. A hungry stomach contracts, as we all learn, and records of hunger contractions are printed in all the textbooks. Even two balloons a distance apart have been included in such a scheme, the two inflated, a portion of the tract thus isolated, sealed from the rest, and the content of that portion sucked out. Even three balloons, the first maneuvered into the small bowel, the second between bowel and stomach, the third left in the stomach. Happy day! Whatever has been learned from this is a tribute surely to the investigative calm of scientist and subject.

VOMITING:
Arrangement for Bringing Things Back Up

From time to time there occurs a reversal in the direction of digestive events. A household matter. In the easily embarrassed adolescent years, to have it occur may depress one for a week, but thoughtful old age will sit above and watch, even achieve a disinterested admiration for the mechanics, coupled with a touch of annoyance at a speck that did not miss the nightshirt. It was 4:00 A.M. A fever was burning. The man had been attempting to get rid of a cold in the head by washing down a pill with a swallow of milk, and the return was so fast that he thought petulantly of Wimbledon and Forest Hills. Just had time to get his red beard out of the way.

A human body was undoing something done to it, or a mind something done to it, or even on occasion a sick man was wishing to impress his nurse. Whichever, the digestive system was performing an age-old act.

The preliminaries often are hesitant and full of false promise. A spasmodic breathing usually starts them. A dog starts them by eating grass, and in a dog the human being can observe the mechanics unemotionally, the wise bitch simply letting everything happen. The mouth fills with what is mostly water, some of this running down into the gullet, watering the walls; in a man, his mind now goes definitely along with the mechanics. The gullet relaxes. The stomach relaxes, except its exit, which vigorously contracts and stays contracted. Thus there results a long limp pouch that begins at the mouth and is shut tight below. The diaphragm lies on one side of the pouch, the abdominal wall on the other. At an instant, diaphragm and abdominal wall contract, the pouch is caught in the wedge, squeezed, the contents going the only way they can, up and out. This is the body's technique. As when everything is in the conventional direction, windpipe and nose are automatically sealed off, mouth opened. Two incidental facts. If the abdominal wall should be paralyzed, rare, the act is not possible, one side of the squeezing wedge not able to perform. If the stomach has been surgically removed, the act is possible, bowel instead of stomach caught and squeezed. In an experiment more than one hundred and fifty years ago, a pig's bladder was substituted for a stomach, the act induced, studied; the conclusion one can draw is that investigative curiosity ran rampant in that day, too.

BOWEL:
Twilight of a Morsel

Every morsel of every lunch in every nation of every continent of the seven continents is at this moment journeying to its twilight, and over the same scenic way, mouth to gullet, gullet to stomach, to small bowel, to large, to the Matterhorn or wherever.

The small bowel is temperamental. Often if has been described as a narrow long factory. In that factory one manufacturing process follows another, the processes overlapping, muscles supplying the automation.

A surgeon will incise a human abdomen, and loops of small bow-el bulge through his incision. Whoever sees this sees it forever, that belly full of bowel, everything appearing to want to do something and not knowing what, push some direction and not knowing where. This can be observed at one's leisure if a researcher will incise the belly of a rabbit, fit a window, as is done, and allow the rabbit to recover. Looking in at that window, the eye recognizes movements. After a while it recognizes types of movement. A type that travels as fast as five inches a second—called rush. A type that propels the food methodically forward—peristalsis. A type that gently rocks sections of bowel to and fro—pendular. (Marked in the rabbit.) A type that over a length of bowel causes a sudden series of indentations that remind anyone of a string of sausages, then in a few seconds each sausage divided at its middle to produce a new string—rhythmic seg-mentation. An internist relates that through the delicate abdominal wall of a delicate woman he has been able to peer straight at the motor handling of her food, especially if surgery had reason to thin an already thin wall, as happens.

Inch-long strips of small bowel frequently have been experimented with. A rabbit has been killed by a blow on the head, a yard or so of bowel scissored out, flushed, sliced into strips, a strip suspended in warm salt water with oxygen bubbled in, then the spontaneous move-ments of the strip recorded, and their variations. A man was hanged at San Quentin, and human bowel similarly scissored out. It proved us all rabbits, rather.

While the muscles of the digestive tube are supplying automation, the glands of the digestive tube are supplying juices. Special juices for each food. Special enzymes in each juice. Per day, the total manufac-ture of digestive juices may be two gallons, more than the total blood. A single salivary gland has been known in five minutes to pour an amount of saliva that weighed as much as the gland—saliva is the first digestive juice. Then comes stomach juices. Then small-bowel juices.

A small-bowel gland may look as if a microscopic finger had been pushed from inside into the bowel wall, the resulting pouch lined by one or more special kinds of cell, each manufacturing its enzyme or chemical, this welling up where the finger was pushed. Sometimes the finger has side fingers. Sometimes the cells are lumped and the juice flows from a duct. As to the cell itself, when at rest there are granules in it, these visible under the microscope, and when the cell goes into action, the granules dissolve and the juice oozes through the cell's wall.

Besides the small-bowel juices from the millions of tiny embedded glands, juices are added from two huge outlying glands, liver and pancreas. From liver comes bile, from pancreas, pancreatic juice. Per day, the manufacture of pancreatic juice is between a pint and a quart. When pancreas has been boiled and brought to dinner, some persons eat it as sweetbread. In life it lies behind the stomach. It manufactures enzymes for all three foods—carbohydrate, fat, protein.

Concerning bile, many know it as gall, and know the gall bladder, had an old girl friend who lost hers. Previous to that sad event her bile was stored in it—hung under her liver like an eggplant and held an ounce or so. Bile is let onto the digesting food through a duct guarded by a sphincter, a circular muscle. If the bile is not under pressure enough to push open the sphincter, it is diverted to the gall bladder. Tension in the sphincter varies. Show a dog a bone, and a pressure as low as sixty millimeters is able to push open the sphincter—the bone has made suggestions to the sphincter via the dog's thoughts and feelings. Contrariwise, twelve hours after a meal, when chyme (that partly digested food dropped from the stomach) has advanced beyond the first part of the bowel and no bile is necessary, a pressure of one hundred millimeters may not be able to push open the sphincter, and after a fast, three hundred millimeters. While in the gall bladder, salt and water were drawn from the bile, may concentrate it ten times. Bile helps neutralize acid from the stomach. Bile helps the handling of fat, all aspects—emulsification, digestion, delivery to the blood. Per day, the manufacture of bile is between a quart and a quart and a half.

One has the impression that every part of the digestive machine is forever humorless. Humor, indeed, stops with the last mouthful, when nature becomes grim and each morsel marches obediently to its twilight. At the end of the ileum—lower right abdomen in the region of the appendix—the passageway is guarded by a different sphincter, and there the erstwhile meal goes from small bowel to large. Thus has the gravied goose, after humiliating pummeling by muscles of the digestive tube and insidious disintegrating by juice, been reduced, reduced, reduced. *Sic transit gloria goose.*

X RAY:
A Man Will Photograph Anything

A timetable for these events would be a handy thing for a man to carry around. It would include stations along the road, hour of arrival, hour of departure, illustrated by scenic photographs in black and white, X-ray photographs. The timetable would keep a man informed on what was happening when.

Years after William Beaumont was dead, that other distinguished American physiologist, Walter Cannon, fed a cat a substance opaque to X rays and at intervals took photographs. He was a raw medical student at Harvard. That was the beginning. Soon a like opaque substance was fed to men and women and boys and girls and babes in arms from Tokyo to Moscow and on to Tokyo again. If a tumor was leaning somewhere or an ulcer had etched a crater, this would be seen, but if the tumor or ulcer was simply a persistent idea, as often, this would not be seen, and nothing is such a holiday as a nonfatal diagnosis. Henceforth the diurnal doings of man's digestive tube were as vulgarly exposed as everything else in his private life.

For his brilliant discovery Walter Cannon, though he added other discoveries, never received a Nobel Prize.

The subject drinks what looks like chalky milk flavored with chocolate, if anyone likes chocolate. The X-ray specialist has his timetable and is ready, snaps. Or he just watches. He sees the black and white against the ocean green of the fluoroscope. Gruesome sometimes, funny sometimes, that opaque bolus waiting on the top of the tongue, then diving down and starting its creeping advance toward limbo while the next bolus waits. Occasionally the X-ray specialist hurries the advance by kneading the belly wall with his fingers. He may imitate the digestive process more exactly by coating actual food with the opaque and in the dark watch the meal ride by. The first of it leaves the stomach almost as quickly as it enters, the last not under four hours, fourteen hours approximately to that gate between small bowel and large, thence a climb up the right side (ascending colon), a crossing (transverse colon), a sliding down the left (descending colon), followed shortly by a reappearance in the sunshine or the warm rain. Twenty-four hours for the total journey, or forty-eight, or less or more, even a week, even two weeks in the stingy with a penchant to constipation. All of which knowledge and much associated knowledge has been bequeathed to our civilized society without the nicking of any man's belly or any rabbit's or cat's or dog's, and still Walter Cannon never received a Nobel Prize.

ABSORPTION:
Faithful to the Last

But now, the true good of this all, the human intent, the animal intent, the digestive intent? It is absorption. Up to now the processes could seem to have been for the sake of digestion, but nay, brother, they were for thee. It was the getting of everything into thee.

When the raspberry from the shrub, the sugar from the maple, the lard in the cake (from the nice white pig), prefaced by the sandwich with the thin slice of ham (from the other white pig), had been divided and divided, there arrived that fated moment when every molecule would have the grand privilege of becoming a part albeit humble of a man. Billions of molecules. Think of the dull places they might have gone.

Absorption is the passing from the inside of the tube, through the wall of the tube, into the blood, into the cells of body and brain, and, for all we know, into the mind.

How?

That how is far from fully clear. Generally it is probably the same process, but detailedly it is probably a different process for different foods in different parts of the long tube. In each part it is the chemistry and physics of the food, met by the chemistry and physics of the wall, indeed by the chemistry and physics of the parts of the wall. The wall is cells. There is the entry into the cells, a membrane. There is the substance of the cells. There is the exit from the cells, a membrane. A molecule might, and does, enter the wall by interacting with a molecule in the wall, perchance advances by acting with one molecule after another, climbs out by a similar kinetics on the other side, enzymes paddling everything everywhere along the way. At the selfsame instant at the selfsame spot a variety of food molecules may be using a variety of transports, and it seems bewildering, and from all that is known is bewildering, and bewilderingly fast. Water molecules also.

Salt molecules also. We are as busy as mice even when we are quiet as mice, and so are the mice.

Protein, a large molecule, is broken down to amino acids, small molecules. Carbohydrate, large, to glucose, small. (This action on carbohydrate formerly was thought to take place all in the lumen of the tube, but is known now to take place also in the wall.) Fat, large, is broken down to fatty acids and glycerol, small. It is the small that are absorbed. Fat might slip through by simply dividing into fine fat, might chemically disguise as soap and leak and sneak through, might chemically produce a molecule that dissolves in water and this slip through with the water might use that trickery of stepping-stone to stepping-stone, molecule to molecule, slip in as fat, slip out as fat, even be seen afterward as fat in the lymph stream, the lymph spaces, the blood of the great vein of the liver.

Everything in the digestive tube that is valuable to a gentleman or a lady or a dog must get into his or her hungry cells—lung, eyebrow, thumb.

A tithe of absorption can take place at the entering terminus, the mouth, as nitroglycerin for an aching heart is absorbed when simply laid under the tongue. Some absorption takes place at the other terminus. Much takes place in the large bowel, much in the stomach, especially water and alcohol, but most in the small bowel. If a small bowel is cut open (in that rabbit) and its inside wall lighted and magnified, juttings become visible, thousands, villi, and during absorption these are in continual motion, a slapping in all directions, a panorama concentrated on business. Some scientists have thought the slapping to be a way of bringing a maximum of bowel to the broken-down food molecules, and unquestionably the villi do present a surface far greater than if the bowel were smooth. Other scientists have thought the continual motion was to pump food-laden blood and lymph out of the villi to make room for more food-laden blood and lymph. The villus appears most nearly a cone overrun with blood vessels, and one realizes that lumen and vessels are closer than one thought. Food and blood are closer: that is, world and creature, spaghetti and liver.

A man does not eat his dog—that occurs to one. Why? What makes a man fastidious at that crossroad? Grenfell, the explorer, was able to bring himself to eat his dog—and was praised for it, for his judiciousness in saving his own life. Perhaps the rest of us would do the same this noon, except that we can see there is enough roast beef and chicken. One is disgusted with man. One is sad for him, too, always pulled between the Sermon on the Mount and something that can be tracked to how his particular conscience defines self-preservation. Anyhow, man does not eat his dog.

DEFECATION:
Reflective Interlude

Dust to dust after the right dust has been saved—that would be a condensation of this total story.

Over the funereal final scene the bile casts its color. Red blood cells break down, their hemoglobin remainder is green (biliverdin), which goes to the liver and adds a red (bilirubin), those two hues tinting the dust that changes in color from day to day.

The chyme that hours ago left the stomach was transformed into what arrived at the end of the small bowel, continued to be transformed, passed to the large bowel, continued to be transformed, passed onward, hindward.

What passed into the large bowel was apt to be swept along in a wave several times in the twenty-four hours, the sweep beginning often in the middle of the transverse colon, thence into the descending, into the sigmoid, and beyond—out of the digestive tube and out of the pelvis and out of the body into the world. When sufficient had collected in sigmoid and rectum, and the pressure there had reached forty or fifty millimeters of mercury, a man might find himself reflective and sensible of a fullness somewhere. This might be followed by a wish. A stupid man might look as if he had a thought. However, the sweep did not have to go sedately from thence to thence for there to be the wish. A single peristaltic wave might shoot whatever it was into the beyond, the wish come right along, or the wave might have begun farther back, where small and large bowel join, encouraged by something still farther back—breakfast! It is a common way for the body to salute the day.

The ulterior end of the digestive tube was locked by that purse-string double muscle, the inner muscle responding to the laws of pressure, the outer to man's command. Assistance was given by that same abdominal force that assisted at childbirth, as Galen back in Rome already said, the force being in the same direction, down. The pressure can rise to two hundred and eighty millimeters, twice the arterial blood pressure. Small wonder, with such pressures inside our bodies, that blood vessels may burst inside our skulls, where it is not a good thing. The bathroom remains the most dangerous room in the house.

Defecation postures are similar and different and familiar over the scale of animals. A dog, having circled in the well-known manner, thrice perhaps, assumes temporarily the posture of the kangaroo. A horse practically pretends it is not taking a posture, but its tail gives it away. Man sits. Man is the thinker.

IX

METABOLISM
Procter & Gamble

THE CHEMISTRY:
Demolition and Construction

So there are the products of digestion. They are absorbed through the gut wall. Chemistry is implicated, physics implicated. The molecules are delivered to the blood, to Bernard's fluid around the cells, go through the walls of the cells, into the cells, to their different destinations in the cells. In each cell—it can be plant cell as well as animal cell—the molecules take part in demolition, the tearing down, and in construction, the building up. All this comes under the head of metabolism. That word is from the Greek, means change. A seed grows to a sapling, a zygote to a baby, a baby to a man, the man falls in love with a woman, stubbornly stays a man, as his riding horse his riding horse: the innumerable shifts of energy that make those possibilities possible are metabolism. The infinitude of change forever taking place in us, and in those who inhabit air and earth and sea with us, is metabolism.

CALORIMETRY:
Is Hot

Heat given off is a way to measure the total metabolism.

Heat flows from every hotter part to every cooler part, then flows to the usually cooler air outside of the body, where it is measured.

In the classical experiment, a space was walled in. A man was no more cramped in that space than he would have been in a roomette in a Pullman. He had a bed. An electric fan. A telephone. A window. There might be a bicycle hoisted on a metal frame, the wheels not touching the floor, so that the man could pedal and his body still stay in the same spot, exercise by the clock, so-and-so many revolutions against so-and-so much friction in so-and-so much time. How much heat was given off? The amount was a measure of the metabolism of exercise. Or the man might lie on the bed, not move, and the amount was a measure of the metabolism of rest.

Biochemists called the roomette a respiration calorimeter.

That calorimeter had no heat leaks. There was no loss or gain except as due to the chemistry of the man pedaling, or lying. Double walls, cold air forced between them to keep a steady temperature, and water flowing through coiled pipes in the walls, the temperature of the water read as it entered and left the pipes, and from these data the heat gain for any period was easily calculated.

BMR:
Needs Air

Oxygen consumed is another way of measuring total metabolism.

The common BMR, basal metabolic rate, measures the amount of oxygen breathed in over a period of time.

When a physician takes the BMR, his subject needs to be basal. That means, his muscles slumped almost as in sleep, his heart pumping at a necessary minimum, his lungs filling and emptying at a minimum. This basal level could be forced lower by drugs. Whoever is taking the BMR keeps in mind its intention, to find how much oxygen this quiet body consumes. He takes it in the morning. Before breakfast. A minimum of emotion. A peaceful morning following a peaceful night. The best place is a hospital bed, though a half hour of bed rest is adequate.

The physician makes his patient comfortable, he says, pinches his nose shut with a clip, stuffs his mouth with an airtight mouthpiece, through which he receives oxygen. If he has fallen asleep, the nurse wakes him because the BMR of sleep is 10 percent too low.

The subject lies. He breathes regularly. In the classical instrument, the oxygen comes from a bell that falls and rises with each breath; a pen is attached to the bell and its movements are recorded on paper wrapped around a rotating drum. That produces a tracing. Since each breath removes some oxygen from the bell, it drops. Meanwhile the carbon dioxide and the water manufactured by the body and breathed out are gotten rid of inside the instrument. After six to eight minutes of such breathing the clip is taken from the nose, the mouthpiece from the mouth, and by the slant of a line drawn with a ruler through the falls and rises on the tracing it is simple to calculate oxygen consumed. That figure is then corrected for sex, age, sea level, and temperature, and the new figure is referred to printed tables where oxygen consumed is expressed as heat—heat produced per square meter of body surface.

Why body surface?

That used to be answered glibly, more surface exposed, more oxygen consumed. It is true that oxygen consumed, when related to surface, does turn out to be much the same for large animal or small, elephant or mouse, the smaller always having a relatively larger surface. A mouse, bulk for bulk, ounce for ounce, has greater surface than an elephant and burns relatively more intensely, as a mouse should. Shape enters in. A tall thin person has more surface than a squat person, and if the two weigh the same, the tall thin, pound for pound, burns more, like the mouse. Any animal (mouse, elephant, snake, guinea pig, man) has an irregular surface difficult to measure, and there are other difficulties. Notwithstanding, the day-to-day, hour-to-hour, life-to-death drives that play on the BMR, particularly the glandular drives, thyroid, pituitary, have become better understood because of what has been learned from the method.

The physician expresses the BMR as a percentage, so-and-so much above or below the average for people of the same sex, height, weight. For the infant just out into the world the BMR is low, in a few weeks is twice what it was, in two years may be as high as it ever will get, drops again in the old. The old burn less, therefore eat less, and when any one of them eats more, we think we know an additional fact about him, his mind no longer in control. Usually the BMR is higher for male than female. During the years of the possibility of conception a woman's stays steady, drops at the change of life. A single cigarette may raise the rate 9 percent. Coffee raises it. Tension. If a patient is afraid of her physician, worried by her illness, by the fancied dire meanings of the test, by the bill her husband will get, and if during the test a riveting hammer is rebuilding Doctors Building, that BMR is worthless. Intellectual activity on the whole has not much effect. Emotion has. A high rate is +70, a low—40. An Englishman, a Frenchman, an American, a Russian—each may behave as exaggeratedly as each thinks the other does, but their oxygen consumed and heat produced do not behave exaggeratedly. The living machine pays not much attention to nationality, race, status.

CARBOHYDRATE:
Coal in the Bucket

Carbohydrate is eaten at every meal, often between meals, chocolates, pretzels with beer, a snack at midnight when we are more asleep than awake—but we are wide awake on the scales next week. Carbohydrate is one of the three fundamental foods: carbohydrate, fat, protein.

During digestion carbohydrate is broken down in mouth, stomach, bowel. Glucose results from practically all such breakdown. In us the glucose is delivered through the lining of the bowel into the blood, thence hither and thither through the body, finally into the substance of some hungry cell.

Glucose supplies two-thirds of our body's energy, is our readiest fuel, coal in the bucket.

Our economical body does not allow glucose molecules to float around with nothing to do, so converts them to glycogen, which can be stored, principally in liver and muscle. When the body needs glucose, the right number of molecules are released into the blood, carried, picked up by the cells.

A well-fed man has been found to have three to four ounces of glycogen in his liver, one storehouse; half a pound in his muscle, the other storehouse: fuel enough for the body for a day. Whenever in this chemistry oxygen is not immediately required, it always is eventually, as for the burning of any tallow.

Everywhere in us sugar is burning, a special burning—that of the living.

FAT:
Coal in the Bin

Fat or the fatlike—the lipids—we also eat at every meal. Fat and what metabolism makes of fat reaches into all the systems of the body. Fat is part of the permanent structure. But it also is fuel, also is storage.

During digestion fat is broken down to fatty acids and glycerol. Fat = fatty acids + glycerol.

Those fatty acids are burned eventually to carbon dioxide and water, supply heat and energy. That useful word *burned* must of course not be read in the everyday sense. The process is not that direct, the molecules are not sped up to the speed of a bonfire.

Of the fat we eat, some remains undigested, travels the mileage of the digestive tract and is dumped out of the body in the feces. It might of course just be carried in the blood. The destination of a single fat molecule is a single cell. This cell has been calling for fat. The fat arrives, enters.

By the time the fat molecule has been entirely used up, it has gone in and out of the bowel, in and out of the blood, in and out of the liver, in and out of the blood again, into a cell, and enzymes everywhere spurred the journey. Glucose was coal in the bucket; fat is coal in the bin. This fat is not merely dropped into the bin but chemically fused in, huge amounts often. It is a savings account that the body can draw on much longer than on glucose. Fat can keep yielding energy after the other foods have quit, wherefore starvation can be mercilessly slow, the victim dying when the last fat has been burned, the end of the tallow truly.

Much of the stored fat is in the bowel neighborhood, a great apron hanging down in front of the abdomen where it contributes to insulation. Joints are protected from jarring by cushions of fat. Kidneys rest in beds of fat. Half our fat, however, is under our skin, prevents heat loss from our snug inside world to the cold outside.

PROTEIN:
Stone

Protein is more than half of the solid substance. For long years it was the most prominent in research, then lost first place, now has it again.

Protein, like the other two foodstuffs, is eaten at breakfast, dinner, supper. During digestion it is broken to amino acids. Amino acid + amino acid + amino acid = protein. It may be an enormous chain. Only in recent years has something been known about the shape of the protein molecule. Such a molecule might be fifty or more times bigger than some other big molecule and still not fall within the seeing power of the ordinary microscope. Electron microscopy and X-ray photography and the computer have been necessary for the establishing of the architecture. Keystone is the word used for protein, building stone the word for amino acid.

We eat it, digestion breaks it up, the amino acids get through the gut wall, circulate in the blood, which is a river flowing with amino acids besides everything else, the amino acids picked up where an old cell needs repairing or a new one needs building. A sick man has been kept alive with an infusion of amino acids introduced into one of his veins, which should not be surprising. Protein demolition occurs in stomach and bowel. Protein building occurs in the cells everywhere.

Amino acids are in the blood that enters the liver, but not in the blood that leaves the liver. They are stored there, or converted to blood proteins, or torn to harmless by-products, especially urea. Urea leaves the body in the urine.

Besides the maintenance of the tissues of the body, and their growth, protein can supply us with glucose. No wonder it has been called the perfect food—perfect for a dog, not quite for a man, a dog's metabolism encouraging the table manners of a dog, filling up on meat, whereas a man to get enough energy from protein would have to fill up with an amount that would choke his engine.

Enzymes are proteins. Enzymes whip along the building of proteins. So, protein drives protein.

The actual manufacture in the cells takes place in the ribosomes, small bodies visible with modern microscopes. The manufacture strikes one as a jerky Chaplin film, atom jostling atom, but all of it over in a minute, literally. One amino acid is squeezed in here, another squeezed out there, the giant molecule throwing itself together in a way that would be cockeyed except that everything is, or nothing is.

To risk an extravagant figure of speech, the protein molecule is the stone of a Gothic cathedral. Our body is an infinitude of cathedrals. Impossible to overstate our body, possible of course to state it wrong. Each protein molecule is three-dimensional, has height, width, depth, and this grand dignity the vulgar digestive juices methodically demolish.

If one extricates oneself from terms, and if one remembers always that the chemically complex is the summing of the chemically simple, then it does help somewhat to fathom how green grass can become rust-colored cow.

VITAMIN:
Mortar

In the eighteenth century, while Americans and Frenchmen were writing political documents, Englishmen and Japanese were experimenting with nutrition, on seagoing ships. Captain Cook made a three-year voyage, included fruits and vegetables in the diet of his sailors, lost four men, phenomenally low when a ship was a pesthouse. Lord Anson in 1740 had three-quarters of his crew down with scurvy, then the dietary acumen of the Scotsman James Lind and the forcefulness of Sir Gilbert Blane got a ruling through the British Admiralty that all sailors must take lemon juice. Scurvy withdrew like an apparition come daylight. In the nineteenth century, Dr. Takaki, a young Japanese, persuaded his government to send out two ships, identical lengths of voyage, identical crews, the sailors of the first ship to eat their customary polished rice, those of the second to eat meat and vegetables. One hundred cases of beriberi (pain, irritability, paralysis, swellings, failing heart) flared on the first, sixteen on the second, and each of the sixteen had stuck to his polished rice. Later it was proven that beriberi could be cured after it was started if the diet was supplemented with an extract of the polishings.

Experimentation soon ran worldwide.

These molecules do not get to be part of the body. The body cannot manufacture them. They must be consumed. They do not supply energy. They are necessary in the chemistry.

There are twenty, as far as is known.

A woman visited the clinic with heavily pigmented skin, ulcers in the webs between her fingers, numb spots on the underside of her thighs, bedsores, other agonies. What she had been living on exclusively was beer, calories enough but lacking vitamins. If one sets about it in earnest, one can achieve vitamin deficiency right in the middle of our sophisticated cities. In unsophisticated Australia, in the interior where the aborigines slice off an end of coconut and drink the juice, there is no deficiency, the juice being loaded with vitamins. The absolute amount of a vitamin that a human being or an animal requires has no doubt a level, but the level often is not known, and there is reason to believe that the level differs from animal to animal, person to person.

When an embryo at a period of its development has via its mother been deprived of a vitamin, there might be malformation. Chemistry of the unborn has a timetable increasingly understood.

The following list of vitamins might as accurately be read off a package of dog biscuits, since God's dogs require vitamins too and in right amounts if they are to be watchdogs and not corroded tramps that scratch at back doors.

Vitamin A—necessary for the health of tissues in and around eyelids and eyeballs, for a healthy retina, healthy lens, to keep off night blindness, necessary for the nervous system, for normal epithelium, tissue that lines the surfaces of the body and its cavities and forms glands, and to protect pregnant mother rats from having offspring with small brains, harelips, cleft palates. Man ought to admit his debt to the rat, at least to the white rat. Liver and the oil of liver contain vitamin A. The vegetable kingdom has the chemical forerunners, the precursors, the vegetables fed by the sun, getting their quanta of energy, and the animal eats the vegetable, man eats the animal, saves his eyesight.

Vitamin B—a complex that began as one but that proved to be many, up to ten, among them thiamine, or B_1, necessary in the metabolism of numerous plants and bacteria and all animals.

Riboflavin, or B_2—keeps up a man's weight, keeps the angles of his mouth from cracking, keeps off raw scaly bleeding lips, an inflamed tongue—is found in yeast, milk, peas, grain, particularly the germinating.

Vitamins, listed this way, could begin to loom as large as anything in the world.

Nicotinic acid, or niacin, or anti-black-tongue factor, staves off a disease suffered by millions, with weakness, depression, diarrhea, scaly skin and thick tongue, no appetite, no clearly identifiable ill health either, known around the world as pellagra. In our country in the southern states during the depression there were every year thousands of pellagra deaths, not recognized as nutrition deaths, the mystery about them as great as the vitamin mysteries of the eighteenth century. Later they were no mystery at all.

B_{12}—cobalt in its molecule, necessary to human life, contains the intrinsic factor that prevents pernicious anemia. C—prevents the loathsome scurvy. E—prevents sterility. K—prevents capillary hemorrhage.

Bacteria have been knitted everywhere into the story of higher creatures. B_{12} was synthesized in the bodies of bacteria. Some bacteria have gone on killing us, some have fed us. In an experiment an animal has been kept alive without bacteria, though normally all lead infected lives. A human infant may be born sufficiently sterile, but then in a few hours the game begins, and the bacteria begin providing vitamins, the parasitic guests begin repaying the host who is supplying them with a warm wet roomy place to live and breed. In cows, bacteria are killed in the upper bowel, where absorption is active, disintegrate there, the ruminant easily absorbing all the vitamin needed. In us, the hordes are in the lower bowel, where absorption is less active, but our machine usually absorbs enough if it is working efficiently. Nicotinic acid, biotin, riboflavin, inositol, choline, thiamine, and vitamin K come to us in significant amounts because bacteria multiply in our bowel, die, and our body benefits from an extract of corpses. The administration of an antibiotic may halt an infection, but it halts also the bacterial luxuriance of our bowel—fewer corpses, less vitamin.

LIVER:
Kitchen

Procter & Gamble, who wash the world with Ivory soap, have the inmost part of their chemical combine in northwest Cincinnati. Busy around the clock. A colony of steamy buildings, looming conduits with the dimensions of Roman aqueducts, railroad tracks running through and beyond. An iron fence encapsulates the whole. The air smells. The air tastes. Interlocking chemical processes proceed with clocked speed, and many clocks and processes at the same time. Here there is no need for fancy. The facts are fancy. Everything begins more or less with tallow and ends with molecules bottled or boxed or wrapped or barreled, supplying a thousand factories, a million kitchens, a million bathrooms. A huge squat heterogeneous chemical aggregate. One thinks: laboratory. One thinks: stew kettle. One thinks: metropolis. One thinks: digestion. One thinks: metabolism. One thinks: liver.

Largest organ of our body is our liver. Complex in one way in the fetus. Complex somewhat differently in an old man. A gland that secretes into the blood and into an outflow duct. A gland that has greater bulk than all other glands of the body added together, weighs forty to sixty ounces in the adult, which is its dry postmortem weight, heavier when wet with blood, only three to four times more blood in the entire body than in the liver.

In the X-ray room its shadow heaves with the diaphragm at each breath. It extends across the top of the abdomen, the spleen on its left, the ribs encircling it, provides a roof for stomach, large bowel, small bowel. Blood vessels enter, leave. The physician reaches his fingers under his patient's ribs to find the liver's edge, decides whether this one has kept its shape and position, or whether it is two finger-breadths below where it ought to be. Does it have its natural texture? Is it too hard? Too soft? Tender? Would this patient be better off in liver respects if he gave up his fine wines?

A massive chemistry of innumerable reactions steams in that stew kettle. Gross analysis shows the liver to contain fats and the fatlike, and amino acids, blood proteins, metals, many chemical species, plus liver enzymes known and suspected, each assigned to its duty, a fact that the researcher trusts long before he knows the nature of the duty. Sometimes an enzyme deliberately tantalizes him. He goes to the slaughterhouse to pick up a liver, wants it as near normal as he can get it, therefore stands ready with his pail of ice while the animal is being slaughtered, lets the liver drop from animal to ice, hurries to his laboratory, and before he arrives has lost the enzyme he meant to study. It has degraded into something else.

Samples of alive human liver are routinely obtained by pushing a needle either from underneath through the belly wall or from above through the chest wall, a knife concealed in the needle. After the sampling, the patient is kept quiet for twenty-four to forty-eight hours, a bloody bleeding liver being a nuisance or worse. The reason for the sampling is to examine for cancer or some other liver possibility.

Magnificent is the liver's power to rebuild. Some organs have no power, but the liver, even when masses as large as 80 percent are gouged out by accident or in an experiment, rebuilds to its old contours. Fabulous numbers of liver cells regenerate within a day, within an hour.

In the living liver, cell leans against cell, the functioning cells of the organ, microscopic geometric form leans against microscopic geometric form. When a block of rat liver is frozen and finely sliced, and the slice is viewed with the light microscope, the cells are polygons, usually with a single nucleus, a wallpaper sameness that heightens the wonder of the unsameness of the chemistry. When viewed with the electron microscope the cells prove loaded with mitochondria, those microscopic powerhouses, as many as a thousand in a single cell, packed with enzymes, different enzymes, working in assembly-line trimness, each step concluded in a fraction of a second.

One hepatologist has summed the obligations of a liver to a man. He claimed hundreds, which would justify that figure of speech "metabolic metropolis." It chemically crashes, chemically creates, chemically demolishes, chemically constructs, chemically breaks down, chemically builds up, is busy with all three basal foods—carbohydrate, fat, protein.

Digestion broke protein to amino acids, the liver breaks the amino acids, and finally there is left waste nitrogen in a form that can be removed from the body in the urine.

The liver takes part in both blood production and blood destruction. So, it takes part in the metabolism of iron. Liver and spleen and bone marrow hold about one-fourth of the iron of the body. The liver manufactures blood proteins, stores them, slips them into the blood at the rate they are needed.

When that great organ fails, the doctor says with cold logic, liver failure, and liver failure may result from any of many diseases, the molecules that a healthy liver rids us of now piled up in the blood. A person who loses his liver outright dies in a day.

When the liver cells have manufactured bile, it is collected in bile capillaries. Those capillaries lead into larger conduits, those into still larger, and the largest run off between the great lobes of the liver, finally into the first part of the small bowel. Into it the bile flows intermittently. It helps there with the absorption of fat, a fact long recognized and still not entirely understood. Its free flow, at the proper time, must not be obstructed, as it could be by tumor or inflammation or cirrhosis. If obstructed it dams back into the blood, the whites of a man's eyes get yellow, his skin yellow, his urine yellow, also his bed sheets, this recognized by physicians from before Hippocrates. The man has jaundice. A severe jaundice makes people turn in the street and look at their old acquaintance. Physicians speak of a creeping jaundice, a bad prognosis.

Why has Nature put these labors all under one roof? Why has she given the liver so much to do? Why so many small factories in this huge chemical combine? Capitalist instinct? Merger instinct? Economy of space? Not likely.

An infant's belly may jut because of its liver, the mother may worry. The adult liver is most often up under the ribs where its immensity does not show. An infant physiologically and medically is in some respects a different animal from the adult, hence pediatrics. Sickness may make a liver large at any age. Sickness may shrink a liver, atrophic liver. Sickness may fill it with fat, fatty liver. Normally, every one of us has liver to spare.

PASTEUR:
Le Maître de la Maison

Now might be the time to hang another portrait in the corridor.

A citizen of the world. Much of what we know of the detail within detail of the living, we owe to that inflammable irascible grizzly Frenchman who, like Beethoven, had a pailful of water of genius spilled over his zygote.

It must have been exciting to Louis Pasteur when he discovered that two crystals, tartaric acid, both the same chemical, dealt differently with a beam of light. He had studied crystals. Probably he did not foresee that the study of crystals, crystallography, would become a powerful aid in the chemistry of the future. Certainly he did not foresee that the way crystals let X rays (because he did not know X rays) pass through them might have something to do with our understanding of the very foundation of life. As for the tartaric acid, it came from the crust that forms in wine vats, and wines were and are important to

France. The one crystal caused the beam of light to be rotated to the left, the other crystal caused it to be rotated to the right.

There was the further fact that a green mold fermented the one crystal, not the other. Louis Pasteur was recognizing that a difference in a crystal's pattern could determine whether fermentation would or would not occur. It was Pasteur's first intense scientific penetration.

To him it seemed that chemistry of fermentation was chemistry of life. Yeast cells had been studied long before Pasteur, and their part in bread and wine was known, both bread and wine holding prime places in the human being's daily routine. But in Pasteur's mind it was their chemistry that got the emphasis. We today use the word *enzyme* much more frequently than *ferment*. It was found that the juice of such cells, pounded up and broken, induces the chemical change as well as do the cells themselves. It was leading the way to the molecular depths. We know that Pasteur's yeast cells were packed with enzymes.

Louis Pasteur when a schoolboy was merely average as to grades. That happens so often where there is a mind with a deviant talent. Like other Frenchmen, he was born with a pencil in his hand, was a competent draftsman. By the time he was fifteen he had produced drawings of persons that let us see how he could look straight into the faces of those he loved and not look sentimentally.

He decided, the schoolboy, cautiously as was his nature and in spite of his excitability, that though his grades were indeed average and would let him slip into the École Normale, he had better wait. The grades were lowest in chemistry, again that irony. He caught up. He took a prize. What he was, first, second, last, was a passionate experimental worker, inspired usually, not by some great idea waiting to be cracked open, but by some reality, some trouble, some worry of Frenchmen, his countrymen, as when their industry ran into disaster because of the failure of their wines, those products of fermentation.

His knowledge of how to work matured as did his knowledge of how much work a human being can do. And he had an eye that could see microscopic shapes.

Human eyes had been seeing such microscopic shapes for a long time, but not seeing them with that penetration. From this time forward such shapes would be seen more and more. Human acumen arriving at the shapes of molecules and the patterns of atoms would become commonplace.

Pasteur had noted juttings on the one tartaric acid crystal and not on the other—noted this with the microscope. He had studied the yeast cells—with the microscope. He was being led to suspect, then to accuse, then step by step to convict, as troublemakers for animal body and human body, those tiny objects sometimes found somewhere in them, and that were like those tiny objects that moved in pond water—this also via the microscope. So the chemist Pasteur was increasingly the microscopist Pasteur. Those two increasingly worked hand in hand. This scientist spoke of beer and wine as getting sick. Healthy wine had only yeast cells, sick wine had other cells, and that appraisal, sick wine, sick beer, healthy wine, healthy beer, was a physician's language and a physician's manner of thinking.

The silks of France got sick. The distress of the silk workers swelled to a national cry and then an international cry, the sickness having crossed borders and spread over nations, and now the chemist and the microscopist and the physician and the patriot joined hands.

Everyone has heard the story. It is part of everyone's freshman course in the history of science. Pasteur saw that there were spots on the sick silkworms, spots on their eggs, and he could separate the unhealthy that had the spots from the healthy that did not. What added first to the bewilderment, then to the allure, of his five years of study was that the silkworm disease was not one disease, that the state of the external environment in which the worm lived could increase or decrease the worm's susceptibility to disease. And this was Pasteur learning principles of pathology.

Bacteria had become the whip that lashed his genius. The microscopist had become the bacteriologist, and before this bacteriologist was laid to rest, he would have sown and reaped an immense portion of this scientific field that he had done so much to create. Bacteria would be hunted all over the world. Bacterial diseases would be run down one after another, cures found. Within three-quarters of a century the accomplishments would be staggering, but as staggering almost would be the fact that Pasteur's part of the field was already beginning to recede, and that other field, that the chemist in him had done so much to set into motion, was catching up and running past.

The picturesque details of this man's journey, much of it in Paris, everyone knows, accurately or not. Pasteur's hatred of the enemies of his country. His affection for his family. His solicitude for his friends.

His modesty. His fire. His first apoplectic stroke, that might have killed him, that destroyed much of the right side of his brain, paralyzed the left side of his body, deprived him of the use of his left hand, made him dependent upon assistants, but his thinking unslowed. His shattering discoveries and his universal acclaim occurred after the stroke, in the final twenty-seven years of his life. He lost a child to typhoid, a bacterial disease that would be conquered. Here is a translation of what he wrote at the time: "I give myself up to those feelings of eternity which come naturally at the bedside of a cherished child drawing its last breath. At those supreme moments there is something in the depths of our souls which tells us that the world may be more than a mere combination of phenomena proper to a mechanic equilibrium brought out of the chaos of the elements simply through the gradual actions of the forces of matter." This in his French is as quiet as the Gettysburg Address. It had a different humanitarian theme from Lincoln's but was clear like that, full of feeling, simple. The two writings were set down at about the same time.

X
KIDNEY
Shrewd Housewife

URINE:
Drain Water

We drip, drip, drip, all day and all night, three to four times more by day than by night, drip from the kidneys into the silent pool of the bladder, an underground dripping into an underground lake. One thinks of small boys, male and female dogs, summer camps, sleeping cars in Italy, hospital end-rooms. In the seventeenth century, in tragedy they spoke of urine, in farce of piss. Every time we urinate we are present at the evidence of the destruction of red cells.

All vertebrates, all the loftier beasts, form urine, and these are the same who have the complicated nervous systems. The maintenance of the constancy—especially the constancy of the human brain—requires this sensitive elimination.

The kidneys drain us. This is what one thinks of first, drainage. But there is that other fact: they retain us. They keep us as near as possible to what we were yesterday. Our correct quantity, our correct quality is saved. Correct glucose. Correct nitrogen. Correct water. Correct acid. Correct alkali. The creature *sui generis* remains behind while his urine drips off into the rivers. For generations, years, days, hours, minutes, we are scrupulously kept what we are by the scrupulous elimination of what we are not.

Since the amounts and the concentrations vary during the twenty-four hours, the physician asks for a twenty-four-hour sample. He wants an average of the day.

In the adult the twenty-four-hour sample contains two ounces of solids, a flexible figure as every physician knows, dissolved in a quart and a half of water, also a flexible figure, as every beer drinker knows. A beer drinker is busier than other men, and if he is a big drinker he may be beerlogged for hours, never loses his beer as rapidly as he imbibes it. From the smell of him one thinks his beer must be coming through his skin. On a hot day he sweats more, and since there is only so much fluid in him, his kidneys are bypassed, so he urinates less. But there is a minimum.

Our physician presents us with an empty bottle as we rise to leave his office. "If you'll just return that to me in the morning." How crisp and clean he looks. Clean kidney, both as anatomical fact and chemical fact. It scrubs us morning and night. Shrewd housewife.

NEPHRON:
Little Helper

A kidney is built of nephrons. Each is microscopic. Each is a complete urine-making machine. Each labors to wash us.

A nephron is a complex microscopic tube about an inch and a quarter long. It has a tuft of capillaries thrust into its shut upper end. That shut end is cupped. The capillaries appear fitted into it by the hand of an artisan, are in loops, fifty loops approximately, with many interconnecting capillaries, some broad, others narrow.

Glomerulus is the name of the combination of tuft of capillaries and cupped end of tube. The glomerulus starts the urine making. Tubule is the name of the tube that leads off from the glomerulus. It continues the urine making.

The tubule has a twisted portion that straightens, makes a hairpin loop, twists again, becomes a collecting tube, the walls throughout thinner in some places, thicker in others.

Wrapping the outside of the walls are more capillaries, all carrying blood, and passing back and forth between them are the products and the by-products of the work of the nephrons, and the raw materials upon which that work is done. It is one of the body's busy traffic centers.

We are told that if all the tubules of both kidneys were spliced, they would make a tubule fifty miles long. The importance of this is the huge surface presented to the fluids.

At its nether end each tubule is open, and there the forming urine forever moves—toward the sea.

From the collecting tube the urine goes via a delta of channels into a gross tube, ureter, and through this the final urine flows into the bladder. A ureter can be seen with the naked eye. Each of us is blessed with two. They descend from the kidneys into the pelvis, approach the bladder, and enter it; via yet another tube, the urethra, the urine sprinkles the earth. One thinks again what one thought helplessly while trying to contemplate the circulation of the blood—so much of us is tubes.

MICROPUNCTURE:
A Sneaker Sneaks In

How does the single nephron work? There have been thousands of experiments, direct and indirect.

First—the direct.

A single nephron was studied while it was manufacturing. The technique was ingenious. It made use of the dissecting microscope. One looked through the eyepiece at the same time that one was manipulating the microscopic field. The tip of a hypodermic needle, attached to the barrel of the microscope, an exceedingly fine tip, was plunged into some part of a nephron, a sample of the urine-water drawn out, this analyzed. One knew therefore what the nephron at that point was doing. An inspector's eye was inside the factory.

Frog and salamander (mud puppy) were the chosen. A frog kidney is not kidney-shaped but ribbon-shaped, silky, frail. First, carefully, it was lifted out to the side of the frog, and no blood vessels were broken. Next, carefully, frog and kidney were brought onto the stage of the microscope. Next, a desired spot of that ribbon was gotten into focus. A number of glomeruli, some active, some inactive, would probably be visible in any field. Because red cells are squeezed out of shape as they enter a capillary and recover their shape as they exit, the experimenter's eye is helped to recognize where the capillaries are.

A quartz tube is beveled to a fine tip to provide that hypodermic needle, this fitted to a glass tube, the two together fitted to the dissecting microscope and the needle directed, plunged in. The spot can be precisely selected. Guided by the controls attached to the microscope, that tip could enter any part of the nephron. If glomerulus, that could be blocked off from the rest of the nephron by pressure applied by

a microscopic glass rod. If tubule, that could be blocked for any desired stretch between two glass rods, the stretch cleaned, dried with a
frail blast of air, the quartz needle pushed in, a vacuum created, the
urine-water sucked out. To analyze such samples, chemical methods
of parallel delicacy were required. Without the direct experiments,
hundreds of them, on frog and salamander, the present-day edifice
of kidney knowledge, so intricate, so reasoned, would have lacked the
foundation it needed.

CLEARANCE:
Checking on the Housewife

A man's kidneys are not studied by a microscopic hypodermic needle
poked into one of his nephrons. He wouldn't like it. For man no direct
method is practical. Fortunately there is an indirect method, famous. It
was the result of mathematical thinking. It is called clearance.

A clearance is the number of cubic centimeters of plasma that can
be cleared of any item in one minute. Plasma is whole blood minus
blood cells. The two kidneys must be imagined as putting their entire
effort on those cubic centimeters, freeing them completely of the item,
clearing them.

Two specimens are necessary: (1) of the urine; (2) of the blood;
then a chemical analysis of both to find the amount of the item in
both. The urine sample is a ten-minute sample because it is easier
to collect and then to divide by ten. How much of the item—sugar,
sulfate, phosphate, et cetera—is in a one-minute urine? How much in
a one-cubic-centimeter plasma? Divide the amount in the one-minute
urine by that in the one-cubic-centimeter plasma, and this is the clearance. It tells what those kidneys can do to the blood passing through
them. It states kidney work in numbers.

With the micropuncture technique the investigator went into the
animal kidney. With the clearance technique he stays outside. Clearance studies have been performed in clinics and laboratories around
the world, and on many members of the animal kingdom—monkey,
fish, dog, man. With many modifications of the clearance technique,
the student of the kidney in the twentieth century has won urinary
triumph. He has learned how every man produces his rain. He has

learned how his two kidneys with their two million nephrons operate. And he has learned all of this without poking into any man to draw out samples.

RESERVE:
Thrift

We have more than enough kidney for most purposes. Every organ of our body has its extra. This is built in. It is insurance against wear and accident. Masses of an organ can be lost, work capacity reduced, and still sufficient left for life and pleasures and vicissitude. A margin of safety, a reserve, is part of the inheritance of every healthy organ.

For the fetus in the womb, draining and retaining were part of the job of the mother's kidneys. Hers were serving Michael from the moment he was conceived to the moment he clambered out and his own kidneys took over. Her nephrons were performing for his. Essentially, she was urinating for herself and for him, and nobody thought this untidy.

Throughout the pregnancy Michael's nephrons, when they had developed as far as being nephrons, were idle. They were increasing in number, growing in size, doing no urinary work. The alive cells of the filter portion were changing from cubes and columns, which they were first, to the thin flat sheets necessary for filtering. The glomeruli were crumpled together. The kidney had possibly a tenth of its adult size. In the adult they weigh five ounces on the average. It was only late in pregnancy that Michael's nephrons were now and then for a second or two roused into activity, to settle back again into their before-birth waiting. During about a month after birth a number of nephrons continued to increase, each kidney needing time to get its million, its full development.

The two million would be including this human being's kidney reserve, which he would need as the years went by and his body was regularly losing nephrons.

Michael's fetal bladder did contain some urine, and Michael's fetal nephrons did manufacture that urine, and the ingredients, especially urea, did appear in the waters of the mother's sac. This would not necessarily have gone into the sac via Michael's urinary machinery, but there is evidence that it did, that he was amiably urinating a bit down there into that fluid world around him, as a small boy into a swimming pool. If the prebirth urinating has any future meaning for Michael, it could possibly be to exercise his developing kidneys, have them ready, because what they are going to do for him is as important as anything in his bodily life.

A large fraction of a human kidney now and then has been destroyed in an accident, and soon the remainder was manufacturing as much and as excellent urine as before, the individual nephrons having expanded. An entire kidney has been surgically removed for tuberculosis, tumor, stone, and the person's life after a few shaky days has gone on as if nothing had been removed. More serious and dramatic, one kidney has been removed, then a cyst found in the other, a sizable fraction of that kidney destroyed, and still the life has gone on, the reserve reduced. A physician many times in his career watches the progressive loss of reserve during illness, may know the day that he ought to say good-bye to his patient, and does not say it.

UREMIA:
Adieu House and Garden

In spite of the reserve, in spite of the durability, in spite of the evolutionary foresight put into their construction, the kidneys may fail. This may be rapid, may be slow.

Because through life they have been cleaning the body, removing excesses and waste, retaining the correct amounts of everything, renal failure is serious. Blood arrives at the nephrons as usual, but these have lost their discrimination; the tissues of the body are left to get along with molecules that should have been removed, and lack molecules that should have been saved, and if this keeps on—disaster. A surgeon has extirpated one kidney, then after the operation had to face the embarrassing fact that the other was an inherited freak and useless. Such inherited freaks are rare. A fetus has survived the whole way to birth lacking kidneys, a gene gone wrong, and after birth the mother could do no more kidney work for her infant; it died. Usually failure has been less dramatic. Usually it has been a late stage in a degenerative or inflammatory process, nephritis. High blood pressure may have hammered the blood vessels of the kidneys until the flow through them was so reduced that the remaining nephrons were insufficient.

When kidneys fail, nitrogen is dammed back, rises in the blood, largely in the form of that urea molecule manufactured from protein in the liver and eliminated by the kidneys. It may be twice the usual amount. Three times. Thirty times. Symptoms appear. That sickness is called uremia—urea in the blood. The evidence seems to be that the symptoms are not caused by the urea. At least, urea can be administered in huge doses with only transient upset. It is not a poison. Crystals of urea may come out of the skin like a frost, like flaky snow, making the skin itchy. Items other than urea are dammed back in the issues—acid, potassium, et cetera. We know that potassium can be a poison and injure the heart, kill. Calcium balance is upset. Whatever the poisoning, or group of poisonings, or imbalances, the verdict remains the same; the chief organ for waste removal has had its efficiency reduced too low, and this creature will not see another Easter moon.

In recent years a man-made kidney has been used to remove substances; the sick man's blood run through a "sandwich" of cellophane and plastic, substances leached out of the blood into a chemical bath. Substances can also be added. The man-made kidney is gauche, is not like nature's, not working with that slowly evolved perfection. Remarkable, though, that man should have been able to build any substitute whatsoever. There is today an artificial kidney that cleanses the blood of fifteen patients at a time, making survival less high priced. Hundreds of hospitals have such kidneys.

A loop of bowel has been lifted through the abdominal wall to where it could be flushed with salt solutions, the concentrations of these juggled in the light of kidney knowledge, wastes cast out of someone's blood, the moribund kept alive for weeks, exceedingly rarely, of course, and why keep that one alive? In an experiment, the inside of a dog's abdomen, after the kidneys were removed, has been washed with salt solutions, the concentrations similarly juggled, the dog kept alive ninety days. The abdomens of human beings have been washed with solutions, are washed (dialyzed) for various reasons, the persons kept alive for lengths of time. The examples merely call attention to the fact that it is possible for the body to get rid of wastes by other channels than the kidneys. Even natural diarrhea gets rid of wastes. Everyone who reads the newspapers know that kidneys have been transplanted from one body to another, precarious once, common today. There have been thousands of transplantations of organs in the last decade and a half. Liver, lung, kidney, heart—all have been transplanted, but only the kidney with proven long-term success.

Normally, lungs and skin rid the body of wastes, but the riddance via those passageways is either not sufficient or not special enough, has not the discrimination of the kidneys. Those bean-shaped organs alone are the shrewd housewives.

When a person was subjected to too much or too little oxygen, too much or too little carbon dioxide, it was brain and mind that showed the signs that most concerned patient and physician. So in uremia it is brain and mind. There are twitchings, jerks, weakness, nausea, hyper-irritability, or the reverse, depression, abnormal thinking, stupor, loss of consciousness, coma, and at the end that Cheyne-Stokes breathing that reverberates through the house.

XI

ENDOCRINE
La Comédie Humaine

HORMONE:
Lighting the Show

That word *hormone* was first used at the turn of the century, referred to something that happens in the digestive tube, and seemed unique.

What happens in the digestive tube?

Acid molecules are carried with the chyme, the digesting food, from the stomach to the first part of the bowel. There the world is alkaline. Chemical action occurs. New molecules are formed, are carried by the blood to the pancreas, deliver instructions to that organ to release its digestive juice. Meaning that a messenger-molecule is manufactured in one place, carried to another, and there it rings up a curtain on a special performance. This discovery was epochal. The year was 1902.

Hormone. Internal secretion. Endocrine.

Those are overlapping terms. Each has its emphasis. *Hormone* emphasized that a chemical is manufactured in us, circulates in us, and somewhere excites or suppresses an action. *Internal secretion* emphasizes that a chemical goes from the manufacturing cells straight into the blood, no outflow pipe. *Endocrine* emphasizes that the manufacturing cells are lumped in an endocrine gland. One creature may manufacture an endocrine, another use it, as cortisone from the cortex of the adrenal gland of an animal slaughtered in a slaughterhouse in Chicago accomplishes a spectacular improvement of performance in an arthritic actress in Alaska.

La Comédie Humaine. That was Balzac's half-bitter title for his shelf of novels. Our body with our mind does in some moods seem not a half-bitter comedy but a half-grotesque carnival. *Le Carnaval Humain.* Each creature performs on the sandlot of the universe. Each creature has within him endocrines playing supporting roles, by day, by night, under sun, under moon.

A tornado rips across the sandlot, flings down a tent pole, a dozen tent poles, broken bones, horror, death, disgust, exaltation, disappointment, bewilderment, heroism, cowardice, and then for an instant even the unsentimental and the sentimental mutter, Theater of the Absurd. Most days, no broken bones, tent poles all where they ought to be.

But—take a walk on the sandlot.

Read the words, the signs over each show, or first read the list at the entrance. Man printed the words. Let us pretend that God took a day off and fabricated the sideshows, the endocrine glands. Also, if anyone's scruples do not permit him to use that equivocal word *God* let him feel free to substitute that seemingly less equivocal word, *Nature*. Nature built the sideshows, or evolution did, or chain reactions.

PITUITARY
THYROID
PARATHYROID
PANCREAS
ADRENAL MEDULLA
ADRENAL CORTEX
MALE GONAD
FEMALE GONAD
THYMUS
PINEAL

What names! What a Sign Painter!

HORMONE AND METABOLISM:
Performers Must Eat

Insulin is a long-familiar hormone of metabolism. It is hormone, internal secretion, endocrine. Is manufactured in the endocrine portion of the pancreas, islets there, small islands of cells.

Chemists have torn apart the insulin molecule, and one skillful chemist has put it together again, synthesized it in the laboratory, and this achievement was a cool refreshing spot in this hot half-century.

The insulin that a man injects into himself is more economically gotten from animals, but it *could* be put together, and that quite properly
is a satisfaction to the human spirit.

In the human pancreas there are a million islets, yet added together
they make probably less than two one-hundredths of it, the rest of the
gland manufacturing that juice of digestion, pancreatic juice—exo-
crine, not endocrine. It flows by an outflow pipe along with the bile
into the gut.

Insulin, some believe, does its work at the walls of the cells everywhere, opens pores, lets glucose in, glucose then performing its role
in the hurly-burly of enzymes that every cell of the body is. The
result here is energy for cell work. What finally is left of the glucose
slips off-stage in a disappearance act, water and a gas, the water out
the drainpipes of the kidneys, most of it, the gas out the chimney of
the lungs.

It is those human beings who lack insulin, some machine part
out of gear and stricken with diabetes, who have helped our
understanding of this endocrine, and have helped our general
understanding.

The first extraction of insulin from the islets was in 1921 by Fred-
erick Banting. He was not a chemist. He was a medical student. Banting had been thinking of a career in pediatric surgery, was reading
surgical journals, and in one read what suggested to him that if the
pancreatic duct, the outflow pipe for that digestive juice, were tied,
and the flow of the juice blocked, there might result changes in the
cells of the gland like those found in the patient about whom he was
reading, who had had the pipe blocked by a stone. Suppose, however,
that the islet cells showed no changes; after all, they did not have an
outflow pipe; nothing was blocked. Suppose the endocrine was still
in those islet cells? Suppose it was extractable? Banting persuaded
another student to join him. It was summer. The professor of physiology had gone off on a vacation, grudgingly had given permission
for the experiments, had no faith in them, but allowed ten dogs and
eight weeks.

So, a summer's work for a medical student assisted by a medical
student, the pancreatic duct tied off, a nervous period of waiting, a
rising excitement, the cells that manufactured the digestive juice de-
generating, those of the islets not, an extract drawn from them, and

when this was injected into a dog rendered diabetic by cutting out its pancreas, that dog was helped.

Diabetics all over the world were helped. Many of the tangles of hormones and many of the tangles of metabolism were brought nearer to where they might more freely unravel through an indefinite future.

Of every fifty persons between ages sixty and seventy, one has diabetes. Two percent of the population has. Of the 2 percent, 10 percent are children. If an identical twin has it, his brother is almost sure to. Of every thousand persons between ages twenty and forty, one has it. He may die. We say that we do not care whether we live or die, but usually prefer to put it off.

Because the diabetic is not able properly to process glucose, it rises in his blood and spills over into his urine, hence sugar in his urine, the energy tied up in that sugar lost to that man's cells. He feels weak. His brain lacks its principal fuel. His senses dull. Is not hearing as well as he used to. Misadventure follows misadventure. When the glucose spilled over into the urine, water spilled with it, so water was lost. Thirsty diabetic, he drinks and drinks, and for that the elegant word is *polydipsia*, and for all the urination, *polyuria*, and for his hunger, *polyphagia*.

A syringeful of insulin might have cut across that line. Did so that July day in 1921, when a dog, rendered diabetic by surgery, was cured, briefly, by injections of an extract of the islets of another dog. Since then an injected hormone has been restoring human bodies to their normal 1-2-3, prolonging their survival.

A freshman medical student was injected with insulin. This was a human experiment. It was a sizable dose. The results were prompt. He trembled. He flushed. He turned pale. He sweated. He could not conceal he was frightened. He was aware of the wrong inside him. He was hungry, as one would expect—eerie in the X-ray room to see a stomach contract with hunger contractions because a hormone has been injected. He yawned—eerie to see a hormone drop a jaw. In this student—the experiment may have been carried too far—there was sudden great weakness. Abdominal pain. Mental clouding. Dizziness. If aid had not been forthcoming, which of course it was, there would have been delirium, confusion, spasm. A gumdrop might have been enough to deliver that student safely to his classmates. A solution of glucose into one of his veins would have been quicker. The overall informal conclusions were clear. Tamper sufficiently with the role of 777 atoms in a molecule and the routine comedy has an ambulance clanging at the door, turns life to tragedy, gives it a passing interest.

HORMONE AND GROWTH:
Hercules and the Midgets

A body must grow. It must grow sizable enough to elbow its way through the carnival crowds.

In the beginning—in the dark of the womb—there was no skeleton, then after a time a small skeleton, then larger, bones larger, cartilages larger. Vitamin D, parathyroid hormone, busier.

Over the growing bones were growing muscles, and outside of them was the growing skin, and inside of the growing body walls were a few guts and other matters, all growing. This deaf-dumb-blind ambition to grow! For every cell, every molecule, plans would have been sketched on the chromosomes, instructions delivered, enzymes accumulated, proteins accumulated, and after more time there would have been a bouncy vice-president of the United States, and before him there would have been Moses.

The endocrines concerned ooze into the blood from pancreas, thyroid, adrenals, male and female gonads. And the brain has its control on the endocrines, the parts of the brain generally but not detailedly known, and the how not detailedly known. Finally, there is a special endocrine, growth hormone. It supplies overall growth drive and subsidiary growth drives. It seems everywhere on the sandlot. Experimentally it is injected into a white rat at time zero, then the increase of amino acids, which are the building stones of proteins, watched for in the blood, and the body watched, with no end of variants on the experiment. Growth hormone may, as was thought for insulin, control the cells' entrances and exits, lets things in, keeps things out, is another stage doorkeeper. But it is also superintendent, day laborer, curtain raiser, lamp damper—any functionary that will help expand this gala enterprise. After the ends of the long bones have become sealed to their shafts and the growth of the body has become stationary, growth hormone still goes on being secreted into the blood, this continuing into old age.

The manufactory for the growth hormone is the anterior part of that three-ring show the pituitary, that marvelously intricate endocrine gland. The pituitary is fitted into a saddle of bone at the bottom of the skull under the middle of the brain, which saddle has another saddle inside it, of tough membrane, so that the pituitary may seem the most protected organ of the body. It must be important to *le carnaval*.

Besides growth hormone, the anterior pituitary keeps at least five other hormones in production. Some of these have the job of keeping still others on their jobs. The play is interesting and complicated. One easily understands why the anterior pituitary has been called master gland, but it would take from the interest, and the facts, if one failed at all points to remember that all the endocrines relate to one another. The anterior pituitary gives the dominant endocrine orders, is the manager-producer-director's office. The glands that receive the orders are called target glands; a special pituitary arrow is aimed at a special target, hits the target, in fact can't miss, and the target obediently sends out its own brand of hormone. This is to the end that every body scene and every body act shall receive its right cue, right place, right proportion, achieve for the total performance a harmony quite miraculous in this cast playing its repertory on the time-space of hot and cold stars.

At the top of this, towering over everything of the body, is of course the head, the brain, and over that the mind, wherever and however that is.

Long before Adam and Eve, millennia and millennia, there was the growth hormone, but not till millennia and millennia after Adam and Eve was it extracted from the pituitary of an ox. Two scientists did it. Americans. They injected their ox extract into a rat, and the rat forthwith grew to a great rat, 50-percent increase of weight, not water, not fat, but the virtuous materials out of which rat is made. That an ox could do that for a rat! An ordinary dachshund became a great dachshund—a dirty trick to play on a dignified dachshund. Textbooks print that dachshund's photograph, a puffing four-legged square-piano construction. This is the sideshow that behind separate curtains exhibits giants and dwarfs and acromegalics, the oversized, the undersized, the missized, the growth extravaganzas.

Grossly, what the growth hormone is exciting is the steps to protein production, increase of polypeptides, increase of amino acids, increase of that ordinary element nitrogen. But ridiculous at the point we have reached to call anything ordinary.

There were rats before Rome and Greece and Egypt, and whichever was the first mother rat, she manufactured molecules of rat growth hormone, and these alongside the molecules of the other hormones, and the molecules of growth factors, et cetera, were pushed through the placenta from mother to fetus, wherefore the fetus grew, in time had grown a pituitary of its own, this producing growth hormone and the other hormones, hence on and on and round and round and rats and rats and mice and men. Remove growth hormone, alter nothing else, the rat stopped growing. Return growth hormone, the rat started again.

Those two scientists, when they got intimate with ox and rat, extracted a growth hormone that was crude. Today, the extracts are purified, but still not all the questions are answered, and never will be if the answers must be final answers, science forever only moving from answer to answer.

If in the course of these late considerations someone suddenly felt an ecstatic enthusiasm for the growth hormone, let him not forget that health, and the other hormones, insulin, thyroid, parathyroid, adrenal cortex, gonads, and the growth factors, and the state of the world, and happiness and unhappiness, all contribute, also what she ate for supper, because, though perish the thought, the growth of the nose of the beloved does depend partly on spinach.

STRESS:
An Act Is a Strain

Thomas Addison, British physician, described a disease, afterward called Addison's disease, where the patient had an irritable digestive tract, anemia, a peculiar patchy darkening of the skin, weak heart, enfeeblement so profound that he might not be able to lift his arm. The first description of the disease came the same year, 1855, that Claude Bernard described the secretion of sugar from the liver into the blood, the first known internal secretion.

"I am dirty but I am washed," said a woman in the clinic, stricken with Addison's disease. Even the gums around her teeth would make anyone understand why she said dirty. It was more a bronzing. The effort to breathe was almost too much.

To enjoy the carnival one must at least be able freely to breathe, freely to lift an arm, freely to meet the stress of an ordinary day, and, if one is lucky, freely to throw energy away.

ADRENAL MEDULLA
ADRENAL CORTEX

They are parts of a two-in-one gland, the adrenal. It has a center and an outside, a medulla and a cortex, the two parts for some thrift of nature fitted together, though their work not any less separate than that of the other endocrines, the combination of the two sitting each like a dunce cap on top of its kidney. In the human being the adrenal is about one-thirtieth the size of the kidney. In some animals the two parts are in different places.

The following would be a loose way to describe their separate work.

For sudden stress—hormones of the medulla. For prolonged stress—hormones of the cortex.

Professionals speak of stressors. Stressors we will have to meet until the planet goes up in flame, if that old story turns out true. Specifically, stressors would be pain, hemorrhage, heat, poison, threat, combat, wounds, fracture of a bone, infection, the taking of an anesthetic, also the prick of a hypodermic needle, the climbing of a high mountain, any other activity that threatens the supply of oxygen. But moment-to-moment life requires moment-to-moment adjustment to stress. A baby was born at the carnival when no one was expecting it, police siren screaming, another stressor for everybody, for the mother too, the father too. Birth always has been. An obstetrician states that late in pregnancy the mother's adrenal glands get squashier and larger, are readying themselves for times ahead, for the supplying of greater quantities of hormone.

Throughout history the carnival has been readying itself. It has been tightening the tightrope, anticipating that the acrobat might break his precious neck, and now and then an acrobat does. When that kind of stress for our body and our head utterly ceases, we cease. Each of us knows this, anticipates it, has written his will well ahead of time, though he rewrites parts of it, a phrase every day of the week, never allows himself to forget. We believe in our death no more than in our birth, but at the back of our heads we do not forget, always have half in mind the dark figure that stalks round and round just outside the carnival fence. The dark figure helps maintain the right tension. Life must be lived with the right tension. And it must be lived now. There is an end to everything. We are on borrowed time. Children do not know that, therefore though children are gay and pretty, their gaiety is thoughtless and has no interest in it. Their bodies go up and down, but their minds are not matching. Every adult knows. Some adults have the intelligence to enjoy the cost of maintaining the right tension. Despite that, they are apt to avoid looking directly at the dark figure. If they accidentally do, they look away.

The best-known hormones of the medulla are adrenaline and no-radrenaline. The first stimulates carbohydrate metabolism, stimulates the pituitary, stimulates the nervous system, stimulates whatever goes into action to meet the sudden. The second, noradrenaline, is released far and wide among blood vessels, narrows them, hence controls their size, hence controls blood pressure and the quantity of flowing blood. Size of blood vessels, blood pressure, quantity of blood must be precise, so the body muscles may contract properly to meet the stress, and the heart muscle pump properly. Precise fueling is necessary for the sudden.

As for the sudden, it need not be too sudden. An evening stroll may be enough to alter the adrenaline level of the blood, even the antici-pation of a stroll. A police dog after surgical removal of his medullas sat for a motion picture, maintained his composure, and would have continued to if nothing too severely stressful happened.

Had it been the cortexes of those adrenals instead of the medullas that had been removed, the dog would have died.

Thomas Addison tracked Addison's disease as far as the adrenal glands. He said the cause was tuberculosis, but allowed that it might be malignancy or atrophy. Our clinicians and pathologists have gone farther, have tracked the cause to the cortexes of the adrenals.

A galaxy of hormone diseases has in our time been tracked to those cortexes. The diseases are mostly rare. One disease is almost the re-verse of Addison's, an excess of cortex hormones pouring out of a secreting cortex tumor. Bizarre sex signs appear. A child may have a look of an adult, and the desire. A woman has hair grow on her chin and other unwomanly places, her voice drops to baritone; she applies at the carnival office for the job of the bearded lady, her body changed almost to a male's, her temperament remaining female, though that may have become male too, partly because she has sanctioned with her mind what nature has visited upon her body. A man may experience an abnormal virility, or, by the paradox of hormones, be transformed into a hermaphrodite, in which case he is neither sex wholly, or, turning the dial half-around, both sexes somewhat. Everything is carnival. The trick is to keep seeing the sandlot as sandlot.

For years that cortex of the adrenal was the Miss Hormone Gland of the century. During World War II research was whipped up, it was said, by reports that the Germans were injecting cortex hormones into their aviators with excellent results. Twenty to forty compounds were extracted from the cortex. Some were steps in the gland's manufacturing process, some were artifacts produced by the chemist while he was cooking and baking.

The discovery that the adrenal cortex was one of the glands fired by the pituitary preceded the discovery that the pituitary was fired by a region of the brain. Later, one of the hormones was recognized as fired, not via the pituitary, but by the brain directly. How exasperating! Everything is always insisting on being somewhat different. Nothing is final. Nothing is settled. No peace for the acrobat. "More light and light it grows . . . more dark and dark our woes." But more and more light it does grow. Only one has the illusion that though it does grow more light, there is more carnival to light. The acreage appears bigger. It is still the same old carnival, of course.

HORMONE AND SEX:
The Glow and the Fact

The femaleness of the female, the maleness of the male, the mind side, the body side, before puberty, at puberty, after puberty, the twenty-eight days, the fertilizing of the ovum, the planting of it in the flesh of the woman, the ensuing ten lunar months—sex hormones play on every part of that. They guide the chemistry. They guide the physiology. They guide the behavior.

A few hormones wind that clock, they say. The thing is utterly impossible, but let us hear what they say. If anyone wishes to buy a ticket for this sideshow, please go that way, next corner, left. Be warned, however, that once one buys that ticket, one may forget mother, father, wife, seven children, the commandments, burning of hell, reward of heaven.

Ensconced in the floor of the skull, the pituitary is fabricating, storing, releasing its target-directed arrows. Three are aimed at the female sex glands, which when hit send out other hormones, whereupon the female body and head act in never simple ways, the female finding herself haggled over, cooed at, jealously watched, angrily watched, happily watched, or suavely turning the tables on him, uneasy when she watches, uneasy when she doesn't.

Three pituitary hormones bear up under unbearable names: follicle-stimulating hormones, FSH; interstitial-cell-stimulating hormone, ICSH (male), or luteinizing, LH (female); prolactin. From such reasonable straightforward capital letters and unreasonably bulky words, the story unfolds.

Three hormones, three condensations of witchery, three shapes in space have accosted other shapes in space, and in consequence more and more shapes accosted more and more shapes, accostings, accostings, accostings, performances, performances, performances, and then, sometimes unexpectedly, sometimes expectedly (that's planned parenthood), at the crossroad a new thing, a wiggling wonder, subsequently often a wiggling bore, made its appearance on the sandlot.

The first of the three, FSH, directed its arrow, perhaps not as straight as a bird directs its flight between wind-thrown branches and leaves, but straight enough at its target, and on arrival excited the growth and maturing of the follicles of the ovary, especially that follicle in which the egg of the month was housed, nudged it, pushed it out, but did that only when assisted by the other pituitary hormones, the follicles having begun now to secrete a female sex hormone, estrogen. This sex hormone traveled through the female body, met, as the sophisticated say, its proper receptors, affecting the sweep of female characteristics, dominating the first half of every menstrual cycle, and, if there is a pregnancy, there ensues the mammoth theater of the endocrines of pregnancy.

The second of the three, LH, assisted by FSH, directed its arrows so as to cause that house vacated by the egg of the month to secrete, to mature, to become tenanted by a reddish-yellowish microscopic sideshow that for the span of its survival produced its own female sex hormones, progesterone and estrogens. The former worked upon the lining of the uterus, prepared it for the egg, which if fertile causes all female hormones to continue in production, the lining to grow gorgeous, till that midnight-to-morning when the curtain rose on the trillionth matinee. If not fertile, the juggling of hormones continued, the lining sloughed, the lady bled, and that explained to her the twenty-eight days.

The third of the three, prolactin, was a variety performer. Among its performances, should there have been a pregnancy, and a delivery, it instructed the breasts to secrete milk to stuff the mouth of that yelling parcel. Meanwhile, in the pigeon under the window, prolactin was exciting the growth of the crop gland, with the production of crop milk, the cells of the lining of the crop having rubbed off and mixed with water, and that mixture pumped into the open mouths of the squabs; a wrestling match between the two, it seems.

Disturbing it was when as college students we first heard that these same pituitary hormones were vouchsafed to the male also. The sandlot seemed bewitched. For the female they produced female witchery, for the male, male witchery, excited the interstitial cells of the testis to secrete male hormone—to help the male keep male, keep swashbuckling.

These and these atoms in these and these arrangements were for females only, but not quite. These and these were for males, but not quite, always some leaning, some cheating both ways, making it comprehensible why a body could sometimes be given so small a shove and fall so far, why a life in the sex regions might so often be so unsettled, why a mind could so often play so wantonly on a body. Everything was so easily tipped to begin with, boy, girl, man, woman.

There was the male gonad (that sideshow) located forever insultingly outside of the male, its hormones sometimes believed to have transmuted septuagenarians into eligible bachelors, flattered young men with lower voices, male bravado, the privilege of shaving, given the rooster its comb and its crow and its bad habits. There was the female gonad (that sideshow) located like some other of the body's most secret secrets in the silliest places, inside in the depths of the female pelvis, hiding there, its hormones keeping the female cycling in her cycles, swindling the hen of wattles but allowing her to cackle. The master gland tapped by the brain (as that had been tapped by the mind) therefore tapped the male gonad and the female gonad, and these tapped the male and the female, who accordingly felt very male and very female, felt exalted, felt depressed, lifted high, dropped low, never at peace until everything was at peace.

THYMUS:
Illusion

For years no endocrine role was assigned to the thymus, in fact no role. Then, increasingly, hints that it was part of the body's defense, some special, some glorified lymph gland. So it would not have been an endocrine gland.

It did, however, go up and down with the ups and downs of the adrenal cortex, an endocrine gland. It enlarged with adrenal insufficiency. Those two, thymus and adrenal, had some kinship. With the thyroid also, the thymus enlarging in troubles of the thyroid, in goiter, exophthalmic goiter.

See again the list. It may have been a mistake to tack this one on. Possibly nowhere else to tack it. One must be cautious.

It lies, the thymus, in the upper chest. It must be important in the infant because it is relatively large. It lies above the heart at the base of the neck. It continues relatively large up to puberty, gets smaller after puberty, is small and was believed mostly to stop working in the adult, whatever that work was, shrinks, atrophies in old age.

A tumor of the thymus in a high percent of cases is associated with a disease that has been claimed to be startlingly cured by surgical removal of this sideshow, if sideshow.

Lately it again seemed sideshow, important to the sandlot, important still in defense, important as an endocrine in defense, in immunity. For a time it seemed to be sending critical cells to the spleen, to the ordinary lymph glands, other places. It had something to do with the rejection of skin grafts, maybe other grafts, maybe big grafts, transplanted hearts, transplanted kidneys.

Indubitably there is still illusion. Indubitably this thymus, with its peculiar cortex, peculiar medulla, peculiar special cells, would for a time produce illusion. However, it has earned its uncertain place in the Sign Painter's list. The Manager-Producer-Director always proves in the long run to have been shrewd.

INDIVIDUALITY:
Clowns Are Serious Actors

A change of costume. A change in the makeup of the clown. The makeup does not hide him. His message comes through. As clowns before him, he is trying with clowning to express that there is pain but there is glory in being born an individual.

In the next three minutes, while the lights are being damped for the night, end of the day, night falling over carnival, let your body relax, your thoughts wander. Here and there on the sandlot you still see moving figures, and it will be darker than this, entirely dark before you lose the clown's false-face whiteface. No need though for the face. The Harpo Marx actions tell all.

Every fly has six feet, as if flies had been tooled by machine tools. Nevertheless, if the human eye is quick, it can recognize the individuality of a solitary fly living with you in a room, the rest of the building silent, 2:00 A.M. An expert on spiders will be disappointed in you if he thinks you do not believe he knows his spiders individually. Invertebrates have it. Vertebrates. When it is our kind, we call it personality. A cat has it—no one will need to convince anyone who has shared a house with a cat. A sporting man counts on it in his English setter, as the English setter in his man. Both have it with such shadow and substance that the two can sit across the table at dinner night after night, or on a couch where there are two butt-polished indentations, and neither ever gets bored. Dogs ought really to be given a longer life.

A book that was a sensation forty years ago disposed of the individuality without a quiver: from pimples to wrinkles, endocrines were it. Like the recipe for a salad, there was a pinch of adrenaline from the adrenals, thyroxine from the thyroid, three pinches from the gonads, et cetera, and the book was popular and had value, but today reads as if the author were Savonarola and spoke from a fifteenth-century

Florentine cathedral. Nothing in life gives a man the right to be that confident, to be that exclusive. Yet, do we of this hard-headed last quarter of the twentieth century proceed so differently from him who wrote the book? Have we even much changed the way to get to any one of the problems? We bring in his endocrines, hope sometimes to stumble on a new one, or a new idea about an old one, dash them somewhere, inject them, isotope them, do a hundred experiments, surgically remove whole endocrine glands, sew them back at a different place of the anatomy, manufacture synthetic endocrines, dose the nature-made or the man-made into human bodies or animal bodies, test the back-and-forth between any two endocrines, any three, delete one, substitute another, have the skills of the century, but in the deleting and the substituting do we not sometimes overlook some small historic fact, as, say, how far off that crust of planet we have pushed the ghost? Yes, the ghost. Hamlet's father's ghost. And not only a grand ghost like Hamlet's father's but ordinary ghosts like Mary Dinosaur and Mildred Dinosaur. What would the fifteenth century have thought of pushing a ghost off the sandlot? You say we do not care what the fifteenth century thought?

Very well, the twenty-fifth? By then the ghost may have been reinstated, each body have its ghost again. That ghost will live in its body as in a rented house and get out of its house, say, by committing suicide. Mary Dinosaur and Mildred Dinosaur have given notice by suicide to the landlady, then arm in arm in their dressing gowns those antediluvian sisters reappear as ghosts. That would be a sight!

An endocrine without a doubt can do drastic deeds to the individuality. A hospital clinic has the extremes walk through its doors. An undersecreting thyroid leaves a fattish, mucous-looking fellow who nods every time he lays down his cigarette, an oversecreting puts a near maniac (says his wife) thrashing about in his bed. The near maniac has a goiter. Flushes. The bedspreads tremble with his trembling anatomy. Talks through the night. Captivates you. Horrifies you. His eyes pop. He breathes noisily. His body wastes while you watch. Not enough adrenal cortex hormone in a human body leaves a feeble female collapsing ahead of time to her everlasting rest. Too much—a hairy Amazon. Too little of one pituitary hormone sends out onto the sandlot that freak of growth the dwarf; too much, that other freak the giant, or, if the too much occurs late, the acromegalic. "That fellow," says the head nurse, "has a secreting tumor of his pituitary up in his skull, but all he can talk about is his hat is tight!" Too much parathyroid creates a humpback, abnormal curvatures of his spinal column, his body shortened, an ugly walk; too little—twitchings, tremor, convulsion, spasm, possibly spasm of the larynx and a tragedian's frightened exit. A testicular aggressor and a castrate apologist are accounted for, or the accounting sought, among the gonads. Get one wrong number in the balance sheet of the endocrines and it can multiply into a mistake so large that a policeman in a cruiser spots in that transient something that he is not responsible for and resists pulling up to the curb.

Most of us dangle between too much and too little of everything, feel a righteous pride in being what we cannot help being, normal. How dismally in hormone balance, how at-home, how integrated is the person we absentmindedly call our friend, how tedious and dull, yet in his twin brother there is running a distinct and golden vein of fairy-tale fantasy. Through Paris strides Madame Defarge, looking at nothing, seeing everything, bold and cold in fury, a type in well-nigh perfect endocrine balance for her type, and, apologetically keeping up with her, her muscular husband, the wine seller, balanced for his type, watching her, watching her. But neither of them and none of the others of the Paris of that day, and not Mary or Mildred Dinosaur of their day, were all endocrines, any fool realizes.

It is likely that endocrines do with the individuality what a portrait painter does at early sittings, lays in the broad strokes, leaves it to later sittings, mechanisms of a different system of the body, to pencil in the singularity, the signs of the life experience, the small hurts, small satisfactions.

René Descartes in the seventeenth century thought the pineal, perched in the middle of the brain, was the seat of the soul. The soul sat up there while the man sat below on his own seat in his pew in a French church. Latin *pinea* means pinecone, and the pineal is shaped like that, and reddish. Sometimes it has been claimed to have growth effects and reproductive effects, even effects that accumulated through successive generations, the off-spring smaller and smaller, maturing earlier and earlier, as they say do some children of India, or some of some ghetto. One present-day cult has assured us that there comes an instant when the pineal bursts, and then one sees the past, sees back, back, back, and if one saw back far enough and forward far enough, that would be the eternal, would it not? Three hundred years since Descartes lived and died, and those molecules there in the middle of our brain, plus the grains of sand that go with them, have not yet let out their secret, provided they have one, provided the pineal is not a sideshow with a show in front and no show behind.

However, before it is completely dark, before night settles over the carnival, it is a temptation to recall certain experiments.

A female mink may display a proper unwillingness toward one male and be quite free with another. Inject atropine, which temporarily blinds her, and that female who has been hospitable to one male takes on all. Here eyes are out of focus, and mink is mink. The discrimination of any female, we know, does go down with mink. It does go down with mink. Which could be brought to ions and atoms and molecules and endocrines, but why? Why so serious, child? This is only the carnival! Come, eat your cake.

XII
DEFENSE
The Investment

NATURE:
Statistically Correct Number of Defectives

Nature often seems a person. Seems animate. Seems against us, for us. Seems not as our science sees her but as ancient myth saw her. Seems testy. Cool. Female. The earth is hers. The creatures are hers, the man-bodies, dog-bodies, cat-bodies. None can escape.

Nature is the unmitigated existentialist. As it is, so it is. She is a mathematician. She is a concoction of gonads and brimstone, a breeder and a destroyer.

She is a defender. She has made her investment and she plans to protect it.

She is shrewd in her defense. Never is bothered when she has turned out a poor job or when a good job has crashed into a poor job. True, she does not want too many defectives. She calculates it. If the defenses are right, she knows that replication will take care of the rest, she will get her quantity. She wants quantity. Pores over the percentages, worm to man, and to hell with that withered nine-year-old with an IQ of 160, a genius but to whose running nose someone else must bring the handkerchief. The withered one is one, adds up to one; let him sicken, let him die. Wastes no emotion on the other either, the football fullback who has broken his back, or that feebleminded middle-aged contraption that cannot ever be said to have belonged to the human family, not from the womb. Like Vesalius, she wants bodies. Not corpses to dissect, of course. She wants living bodies, more and more living bodies; then if ever things get crowded, she will cut down overpopulation quantitatively. A good war. Efficient birth control. Legalized abortion. Faster automobiles. Formerly, she used the Black Death, and in parts of the world still uses famine.

She can afford to be fickle. This evening she bristles, tomorrow morning will purr, in the afternoon masquerade as a typhoon, and just before sunset will take off her powder-blue dawn skirt, change to the gored rose that she will sweep around the horizon, undermine the human mind with poetry. However, beneath all change, or just frankly outside, it is scales, talons, fangs, or a seductive smile.

Through it all she does needlework. A female can think while she does needlework.

Her best thoughts come in the chill of morning, when the haggard sleepless look out of their windows and see the edges of the land cut threatening shapes out of the sky, she lolling in a half-sleep following an exactly sufficient whole sleep. Is pleased with yesterday's accomplishments. No conscience, men say, implying they have.

She is too old for the kittenish mood she is in, flicks open that Pandora box that contains the anxieties, lets more of them out among human beings, to keep them human.

Abruptly she crosses space, as if it were a room, to a man-made microscope, peers into the eyepiece, sees down there a defending white cell in a drop of blood. Laughs lightly.

Men say *blind* Nature. Men make bearable in talk the insult that affairs on this planet are not of more concern to Nature. She behaves always as if there were planets enough. She starts again at her needlework. Without that stitching there would be no surviving living creature. Leaves the fewest possible unpatched places, mends all the holes, until she gets bored, the bitch. The hole was too deep and in an out-of-the-way place, so she let drop the garment. A dead liberal. Dead priest. Dead prizefighter. Badly torn garments she drops. They roll with the planet.

SKIN:
Outer Barricade

A sleeping alligator would wake if one hit it with a hammer. A sensitive creature when approached properly, zoo experts say, the female charming when she watches over her young, suddenly fierce in their defense, in a minute so sleepy, never resists the temptation to go back for five minutes more. By a jungle river an alligator sleeps and sleeps, can risk it because besides the built-in defenses that all of us have, it has that outer barricade. One lies prone in a bathtub, a small one (eight inches long), the boy of the house having given it asylum, looks lazily up through soft bright eyes at the nearsighted aunt who has come to take her bath—this ridiculous screaming hysterical world. Those soft eyes look through slits in armor plate.

Man has his outer barricade too, his skin, gentle, soft, yet doing rough work. Its outermost cells are flat and, by textbook, dead, a dead layer ready to rub off and blow away. Beneath that, the cells are less flat, less soft, more what one thinks living, more lacking in that special chemical, keratin. Horns, quills, fingernails, toenails, claws have the greatest amount of keratin. The enamel of teeth has a substance close to keratin, and teeth from when they first appeared under the moon have been a first line of defense. Thirty-two. Glistening teeth. Who has thirty-two?

Skin is a defense against assaults of many kinds, cold winter, hot summer. The way it tans defends, the dark pigment rising from the deeper layers of the skin to block further absorption of the ultraviolet of the sunlight, defending thus the depths from further exposure to that assailant. How permanently damaging that exposure may be has come to be known only in late years.

Skin is a defense against the inside of us, too, imbalancing molecules being let out, the skin classed with kidneys and lungs among the excretory organs.

A defense, too, against that other assailant that is everywhere, water. Keratin does not dissolve in water—one reason why a water-attacking world cannot get at us, cannot erode us. A tanner's skin, says the grave-digger in *Hamlet*, "will keep out water a great while." And water in the opinion of that grave-digger is "a sore decayer of your whoreson dead body." The grave-digger knew.

For all of these reasons and other reasons, we are not astonished when we are told that the total weight of our skin is three times the weight of our liver. Our hide is three times our liver! Kidneys, lungs appear diminutive. This hide furthermore is greased. A thin sheet of fat lies on it, fabricated by sebaceous glands, and each time a hair moves, it squeezes out a bit of fat, and somewhere a hair is always moving. A polite fat—does not get rancid. An aggressive fat—its fatty acids and soaps can kill bacteria. (The skin normally holds bacteria.)

A friendly fat—moisturizes us, keeps us pliable, causes one human being not to be averse to touching another, helps therefore the male-female situation.

The living machine was perfected by natural selection, tooled to precision by natural selection, provided inside with an array of defending molecules by natural selection, provided outside with this barricade by natural selection, this skin; then, finally, a mind laid on top, or wherever, and all of this shrewdness meant to protect the dab of sometimes complicated nobility within.

HEMORRHAGE:
Breakthrough

A breach in the outer barricade occurs. It may be a jagged breach. "They never make a clean slash you can stitch!" That was a snarling intern, weary, in Receiving Ward, and he spoke without looking up from repairing a man's neck. Three brought him in, and no one of them knew how the thing got started, except they all were drunk. Severed the jugular vein on the left side. Instead, the knife might has slashed one of the four feed-lines pounding into the brain, and that might have been fatal, but the fellow was lucky, Irish. It was only a regular Saturday-night card party. A third of his blood spilled out. If the third had not been put back promptly, that also might have killed him. If more than a third, his chance of being erased from the city directory would have been excellent. Hemorrhages usually are less than a third, owing initially to the mind, the defenses in it, vigilance, imagination.

By the book, defense, against hemorrhage is general and local.

General. If not much blood escapes, the blood pressure does not drop; if much, it does; less push is applied to the circuits by the pump, less blood left in them; on both accounts, less escapes from the wound, so the drop becomes a defense. If a drastic drop because of a drastic escape, survival depends on whether that which still travels the circuits can supply an indispensable organ like the brain, and the animal go on using all the intelligence within its range, go on maneuvering.

Local. The arteries of the slashed area constrict, narrow the sluices, may shut them tight, the hydrants turned down or off. A chemical starts this, is released from the platelets, those particles that look carefree and useless as they bobbed along with the earnest red and white cells. The platelets furthermore clump, glue themselves to the walls of the small vessels, jam the open ends. The inner lining of those vessels at the same time curls up, also jams the cut-open ends. The whole idea is so neat—in that bloody wound that is so messy. Normally, blood vessel walls and platelets repel each other electrically, thus reduce resistance to flow, free the stream. With injury, the electrical state reverses, walls and platelets attract, cling to each other, the flow hindered, slowed. Then, as all the world knows, the blood clots. A cork is stuck in the wound. The breach in the barricade is plugged.

CLOT:
First Aid

Our blood must keep flowing yet must be able to gel on occasion. The gelling must be limited to the wound, to the jaw where the tooth was ripped out, the puncture streak where the IV needle went.

The gel, the clot, must be firm, a firm fabric. It must not be brushed off on the outside if the breach has been through the skin, but also must not be swept along on the inside, not creep along, not get beyond the scene where it is needed. If it does, if it creeps, say, until it blocks some critical vessel, even extends from a man's leg into the cavity of his heart, as happens, still another will have helped that problem of overpopulation.

Clot construction is worth a few moments of anybody's time. Clot construction takes place inside blood vessels too, because they get injured and need the defense of clot.

The manufacture of the clot is so methodic, so cool to its importance, so grim when it is grim.

A drop of blood suffices for an informal observation. During the first minute after the blood has dropped from, say, a cracked-open finger, one sees in the blood, under the microscope, red cells, white cells, platelets. Then, miraculously, fine fibers appear. They are called fibrils. The material for making them is in the blood all along, waits there. With the ultramicroscope aided by the X ray, the molecules of the fibrils have the shape of fibrils. Soon they will have built a latticework, which shrinks, and in the shrinking the blood cells are trapped, and that is the clot. On top a clear serum rises, its color between hemoglobin and bile, amber.

That eye-observable clot construction could have been the same inside a wound, and the wound in the end and at that point would have been plugged with the cork, the leak in the dike mended. This creature's body defenses would once more have shown themselves adequate.

COAGULATION:
The House That Jack Built

The fibrils mentioned are made of fibrin. Laboratories have fibrin in bottles. No fibrin is in the man's blood or he would clot to death, but a forerunner is, fibrinogen. The fibrinogen is dissolved in his blood. It is molecules. Fibrinogen molecules are one of the blood's longest and narrowest and heaviest molecules. The liver manufactures them. Rarely is a liver so sick that it cannot manufacture them.

Nudged by an enzyme, the fibrinogen is converted to fibrin.

Fibrinogen → fibrin.

That fibrinogen molecule, that soluble molecule, has the form of three nodules joined by a thread of insoluble fibrin. What the enzyme has done has been to shave off the nodules, and that left the fibrin. Molecules of fibrin are linked end to end and side to side, produce a tough framework, a matrix. This jams the cut vessels.

The picture here and so far is a visual picture, which is still dependably what we have, but the chemical is underneath, a most, one might say, diabolic chemistry, an evasive chemistry, and the chemists have rudely but also refinedly taken charge. No one can say where it all will lead.

Fibrinogen → fibrin → clot.

For this critical, long-known fibrinogen-fibrin step there is expectedly an enzyme, and this is called thrombin.

Thrombin → fibrinogen → fibrin → clot.

No thrombin is in a man's blood, either, or once more he would have clotted to death, but once more there is a forerunner, prothrombin.

Prothrombin → thrombin → fibrinogen → fibrin → clot.

This is the scheme at its simplest, its grossest, the classical scheme, and the classical language at its simplest and grossest. There are everywhere pro factors. There are everywhere anti factors. Those who study the intricacies, who learn the language, are called coagulationists.

Sometimes no coagulationist scheme for the clotting process, if one includes and plays with the chemistry, seems to remain intact for longer than an afternoon. Yet, on the contrary, if looked at over the last decade or even the last quarter-century, to the nonspecialist the changing schemes often appear not too changed, and when one thinks of the inner magnitude of the body, the changes might never seem enough. Indeed, to follow the changes takes more knowledge than the nonspecialist can ever muster, probably. Fibrinogen and fibrin in the last quarter-century have increasingly been studied directly, and all on the road from prothrombin to thrombin studied directly, for the chemical concerned, not merely indirectly for its blood stream effect. A factor, some protein, some lipid appeared needed, was searched for, was found, another needed, found.

All in all, it is not surprising that there should have been these discoveries and rediscoveries, these interpretations and reinterpretations. One needs relentlessly to remind oneself how fraught with danger as well as happy possibilities the coagulation process is—also how almost countless are the items in our blood. Everything must be so versatile—nowhere in the immense complexity of the animal is it more evident than here. Everything that occurs could so quickly be final—the coagulated man lie there dead. Coagulated somewhere—his heart, his brain, his leg, merely his poor leg.

Laboratories have prothrombin, too, in bottles. The liver manufactures it with the help of vitamin K, and vitamin K exists in edible leafy plants, and exists in our bowel, where bacteria disport and procreate and fall ill and decompose, and in the debris is the vitamin. As of now, four distinct clotting factors require vitamin K. When prothrombin is not available, it might be not for lack of the vitamin in food or bowel, but because something went wrong with the liver, or the bile is not flowing out of it properly, or not reaching its destination properly.

Vitamin K → prothrombin → thrombin → fibrinogen → fibrin → clot.

When a person goes to the blood bank and gives his pint for his friend, his blood is run into a solution. The solution is a citrate, which removes calcium ions, and calcium ions are required for clotting; therefore the citrate is keeping fluid all that bottled blood of the blood banks. (The blood of the blood banks nowadays need not all be fluid. That sometimes inconvenient situation has been met, too.)

Calcium → vitamin K → prothrombin → thrombin → fibrinogen → fibrin → clot.

Normally our blood has plenty of everything required for this "house that Jack built," everything to clot us solid, yet our blood does not normally clot in our vessels, for which thank God.

A person mechanically inclined and chemically naive might have deduced that his blood is not clotting because it is flowing so swiftly. The great Virchow seventy-five years ago thought that it was the slowing of the flow that was the reason for clotting.

Obviously there are other reasons than the slowing. The coagulationists have pointed out many. And, then, there is the big reason back at the beginning, the wound. Jack built the house in the first place because of the wound. In the wound are chemical details and physical details. Among the chemical are enzymes, lumpable under the monstrous name thromboplastin, and thromboplastin has classically a still more monstrous forerunner, prothromboplastin. There is a tissue thromboplastin, but also a platelet thromboplastin, and they operate in harmony.

Among the physical details there is the electrical charge on the molecules. Some have denied that that has importance. Then there is the physical detail that though in and around uninjured vessels there are no rough surfaces, around the injured there are, and rough surfaces start clotting: thromboplastin is formed, acts on prothrombin, thrombin is formed, more thrombin, more, more, more, faster, faster, faster, and the clot builds.

That speed is astonishing, and it may be frightful. In the barbarous days one saw an unpleasant lecture-room demonstration. A Ph.D. would previously have macerated a lung, and now in front of the class would inject a few drops of this concoction into the vein of a healthy, apprehensive rabbit. Hurriedly inside the rabbit there would have been jiggled together the pieces of the coagulation puzzle, and a sight it was (gently, brother, gently, pray) in seconds to see that rabbit die, its blood rivers a tree of clot. From the portal vein, a perfect cast would fall out or could be shelled out.

Prothromboplastin → thromboplastin → physical situations → calcium → vitamin K → prothrombin → thrombin → fibrinogen → fibrin → clot.

And this still is not all. In another year this year's description of some stretch of that road is bound to be outmoded, to need updating. The road of evolution is always needing updating. If to accelerate clotting is an advantage to the creature, the coagulationists have found accelerators; if to inhibit, inhibitors. Impure extracts of liver yield heparin, which inhibits. Drugstores have multidose vials or ampules of heparin. The amount of heparin normally in our blood is small. If a physician infuses heparin into a man because of indications of clotting, say in his coronaries, the infusion may save his life. Too much heparin may kill him, or paralyze him, start bleeding in some dangerous place, mash nerves.

Clotting theory, clotting language, clotting discoveries may appear, as suggested, disproportionately shifty to anyone not caught up in the web of specialist interest, would appear more shifty if he had had time to glance at last month's journals. The chemist has us in his intellectual grip here. His mind is always imaginably somewhere in among those multitudinous molecules of the blood. The mind is always the overriding defense.

BLEEDER:
Flaw in the Material

"Bill looked so healthy and did his work and you wouldn't have thought there was anything wrong with him, but there was." In earlier years the neighbor would have lowered her voice and added, "Bad blood." Everybody knew there were bleeders in that family. The boys. It was said to go back in the family, not all of the boys, but always boys, and the men too if they lived that long, which they didn't usually. The women, though, carried the bad blood. Bleeders were part of the mind of the town. But science has reached even into that small town, affects its thought without its knowing.

Bill's trouble might not have been what the town thought, a bleeder. It might have been that his small blood vessels, when cut or ripped, did not contract as they should. Their inside linings might not curl up as they should. His platelets might not supply his bloodstream with their coagulation contribution as they should. Also they might not clump and help plug the small vessels. Or there might not have been enough of whatever is the living stickiness that sticks living tissues together.

When men rode the sea in sailing ships and were not getting the vitamin C that is in lemon or orange or lime, there was scurvy, and with scurvy there was bleeding, into the skin, pinpoints or blotches, petechiae or eechymoses, blood having leaked through the vessel walls, and that would result in another kind of bleeder. An abnormal liver results in another. One kind crops up among the newborn, their vitamin K not sufficient. Four separate clotting factors depend on vitamin K.

But the spectacular bleeder does owe the dangerous tendency to a hereditary lack. The sufferer has classic hemophilia.

Hemophilia is caused by the lack of a gene that is a defense against bleeding, the lack of an antihemophilic factor, AHF. The blood lacks this defense. Bill is suffering from the defect. The clotting time in his case was much too long. He was born with the lack of the antihemophilic factor.

The lack occurred originally in a male, spontaneously. It was a mutation. That bleeder bled. His sons did not. His daughters did not. But the daughters passed the taint to their sons, and half of them bled.

Hemophilia plainly attaches to sex, a characteristic passed to the affected male via the unaffected female.

The treatment has become increasingly successful. Instead of the repeated transfusions formerly necessary, there can be a regular injection of a precipitate of a concentrate of the AHF.

Hemophilia got its turn-of-the-century popularity because of the way it ran through the royal houses of Europe. "Disease of kings." Alfonso, last of the kings of Spain, was a bleeder, the lack of the gene having been a mutation that occurred among the genes of the family of Victoria of England, and she scattered hemophiliacs around her. The last empress of Russia came of that substantial family, and her son was the bleeder said to have been saved during his bleeding episodes by Rasputin, the evil priest, and the method of treatment seems to have been hypnotism. Subsequently Rasputin was poisoned, then bludgeoned, to make sure he was dead. One geneticist-litterateur has suggested that the Russian revolution with all its sequelae is owing to a hereditary chink in the armor, the lack of a gene.

In a children's hospital, there was one child who had an uncontrollable temper and blood that took overlong to clot, or would not clot, those two viciousnesses egging each other on. Only a Greek dramatist or possibly Eugene O'Neill would have had the language to convey the malice in that fate.

HEALING:
Kept in Repair

In all hospitals, for all reasons, in all cities, in all towns, they bleed, and she darns, Nature darns. For better or worse she darns and they live on. An annoyed intern asks: Why? With a curved needle he has put twenty-seven stitches into this one.

Imagine a smaller wound inflicted in the course of a bad-tempered shaving. Over the razor cut there was now a bright gemlike bead, attractive if one looked at it straight. Along the edges of the wound were redness, swelling, heat, and there was pain, slight pain, a sting when the chin was rinsed with the hot water. Redness. Swelling. Heat. Pain. Those have been the four age-old signs of inflammation, each trifling here because the wound was trifling.

The blade had cut through capillaries, which was bad, but it has stimulated uncut capillaries, and that was some good anyway. Those capillaries had dilated. More blood was brought in. The local defense was aimed at installing fresh supply lines as quickly as possible. And the razor blade had stimulated blood vessel cells. These began propagating. New vessels formed. New joined old. Channeling began. Blood flowed.

Some plasma, some white cells, some fibrin had been left in the cut produced by the blade, and this, besides stimulating the blood vessel cells, had stimulated the connectivetissue cells. Their nuclei went to work. Branches of cells pushed into that debris in the fissure. The fissure was crossed and recrossed by the fibers. Dame Nature was at her darning.

At the top of the wound the flat cells of the skin had meanwhile laid down more flat cells. They had done this neatly. Flat cells from all sides met across that cut. The scar had strong color. When later the quantity of blood in it had diminished, it would be pale.

How long for these events? The original hemorrhage had lasted minutes. The first bridge across the gap was built in hours. The healing was complete in days, or less. Soon it would be impossible to find where that cut was.

PHAGOCYTOSIS:
Home Guards

It might not have been a razor cut but a lawn mower gash. That would have altered the strategy. Myriads of bacteria would have been sown along in. Some were in the soil, some in the layers of the skin, some in the air.

The bacteria, by joining in their attack on us, were only taking sides with their side. So who invaded whom? They wished to live.

Those fellow creatures, those small ones, wish to live, but I wish to also, so why shouldn't I take action? I do. My body does. It has that in its blood that defends my life. Sixty minutes later, so-and-so many million bacteria had been multiplied by such-and-such a factor. That's many. Unseen cohorts, trained to Marine deportment, had come to the defense of the host, us. Enemy cohorts were escalating. They multiplied, multiplied, multiplied, microscopic bacteria. Combat was on. Sickness. Sweat. Dryness. Fever. Chill. Commonly there would be recovery. To the host it was an annoyance, to the physician, a fee, to the scientist watching from the sidelines, a battle worth including in his folios of battles.

It would, then, have been cells that defended, and this meant molecules in great abundance, many kinds, carefully arranged, carefully wrapped in a sac, in a membrane. That was your defending cell. Numerous varieties of defending cells. Numerous services in the cellular defense arm. All citizens were subject to call. The home guards waited till the invader reached their suburb, then bestirred themselves. Others had forever been on the move, insinuatingly. Services overlapped. A phagocyte was any cell of the cellular arm that ate, phagocytized the foreign.

The bacteria, when they broke through the outer barricade, through the gash, asserted squatter's right, clinched their position by a population explosion, as nations, as races. Local healing was not so quick, not so manageable. Leukocytes packed around the inside walls of neighboring capillaries and veins. A stickiness developed. In some vessels flow stopped. Leukocytes came like eels from all the oceans to the Bermudas and on arrival behaved in accredited fashion, thrust part of themselves through whatever interfering barricade, then more of themselves, then all. Soon leukocytes were everywhere outside of the vessel walls, were engaging the bacteria, were destroying some, or many, or every one, ingesting them. A battling leukocyte assisted itself wherever it could, maneuvered a bacterium against a surface, some mat of fibrin, jammed the bacterium against that, digested it there, used the power of its acids or its enzymes. Millions of bacteria, of course. Millions of leukocytes, of course. Ceaselessly leukocytes were digesting bacteria. Ceaselessly the debris was being disposed of. If the cellular defense arm had not been too insufficient, the host not too resistant, the bacteria not too virulent, they would be killed off to the last aggressor, the slaughter delightfully complete.

IMMUNITY I:
Chemical Arm

One woman's defense was stuck into her with a syringe. That was when the police brought her to Receiving Ward. It was 5:00 P.M. Hours passed after that. Then it was 11:00 P.M. She had waited all that time, the interns overworked, and when they had a moment they talked of their hospital status, or their expectations, or, best, talked to a nurse. Interns in a metropolitan receiving ward have an exciting service, but not easy. At last one of them got around to repairing the woman's mangled fingers, so just as well that the molecules of tetanus antitoxin had been stuck into her (forty thousand units into a vein, forty thousand into a muscle) at 5:00 P.M. She had that one bracket of insurance, from the chemical defense arm.

Instead of cells like bacteria attacking us, viruses may, and instead of cells like phagocytes defending us, the chemical arm may, immune sera may, antitoxins may. How these operate is by no means an obvious chemistry: (1) the capacity to recognize that something destructive came into us, and (2) the capacity to neutralize it, to render it harmless. Recognize and destroy. It is one of the potentials of our body more alluring than almost any other.

An opsonin belongs to the chemical defense arm. An opsonin is not stuck into the body with a syringe. The body manufactures it. Normal serum has it. To state the situation melodramatically: a submicroscopic ooze is squirted over an invading bacterium, to ready it to be devoured by a phagocyte when the phagocyte arrives. "I prepare to eat." That is what the word *opsonin* means. Unless the bacterium was so prepared, the phagocyte would not touch it.

So, there are complicated atom structures circulating in us, some murdering for us, some merely getting the victim ready for the murder.

Often our chemical defense arm might seem to have little to do, like the members of the fire department at the corner, occupies its time with leisurely housework, cleaning up the bacterial and viral world in which all of us are forever immersed as in a dirty ocean or a fetid atmosphere. But let there be a five-alarm fire and every defender can be counted on to toil through day and night, and, also, the fire department immediately and apparently automatically expands its membership.

Among the ways of tracking these molecules, the chemist incorporates fluorescent isotopes, tracers that give off a fluorescent light, that dots the scene, and the scientist can view it analytically.

Every defending molecule would be different from every other, because the enemy was different, and that would mean many. Also, in the defending molecules as in the infecting molecules, there can at any time be mutations, alterations in the patterns of atoms, sometimes favoring infection, sometimes favoring defense. Whatever working out of the mechanisms, when danger threatens there are underground chemical units waiting to protect nature's investment. True too for the flea-bitten Saint Bernard. He too has had a long defensive history, a long evolution of immune devices, to keep him one step ahead of the fleas.

Immunologists long have defined for us an antibody as a defending something against a foreign something. The foreign was the antigen. Antibody neutralized it. Anti-egg. Anti-poison-of-snake. Anti-oil-of-castor-bean. Anti-toxin. Anti-bacterium. Anti-virus. Anti-protein.

Anti-the-blood-of-another-species.   Anti-another-blood-of-the-same-species. Whatever excited the production of the antibody was antigen.

Antigen → antibody.

New terms have followed new conclusions. New terms drop off the pens of experts, fill the air, then one day nobody knows where the terms went. Some terms and ideas stay.

IMMUNITY II:
Old Military Plan

Natural Inherited Immunity. An immunity that the creature was born with, a resistance to something that might invade it. The baby just placed for the first time in the crib had this inherited immunity. It had been anticipated in the chemistry of that one cell from which this body came. The defending molecules were already in the fluids and the tissues of that baby's developing self, not the formal antibodies yet, but the molecules that could become defending molecules. Sometimes the defenders were bacteria-killing. Sometimes they were enzyme-destroying. Sometimes they were antiviral proteins. Different creatures had different natural inherited immunities. My dog was not stricken with typhoid, and my human associates were not stricken with some of his diseases; therefore he and we dared come near each other. That may not be friendship, not the whole of it, but at least it is an essential for it.

Natural Acquired Immunity. A person had the whooping cough, and the antibodies produced in him stayed in him, or they somehow kept repeating themselves in him, and he did not get the whooping cough again. Yellow-fever antibody recognized and remembered yellow-fever antigen when that reappeared. Measles antibody might remember for a lifetime.

Artificial Acquired Active Immunity. Antibodies were manufactured in a person who had been vaccinated with killed or attenuated bacteria or virus. The bacteria would have been not virulent enough to infect but virulent enough to excite the manufacture of antibody.

A person might have been vaccinated with something closely related to the trouble-causing antigen. Cowpox virus, for instance, manufactured a cowpox antibody that was effective against smallpox. The person would have suffered a mild attack at the time of the vaccination, antibody manufactured, left in him, he immune for years. This immunity was artificial because he had been vaccinated, and the immunity was active because his body had taken an active part in the production of the immunity.

Artificial Acquired Passive Immunity. Diphtheria toxin was injected into a horse, the horse thus immunized. It had diphtheria antitoxin molecules in its blood. If the serum of that blood were injected into a man, it would immunize him passively. Diphtheria antitoxin seems ancient in this rapidly changing field.

Some of the substances of our own tissues could inside us act as antigens. The body could attack itself. When it did, the result would be an autoimmune reaction. Doctors have thought serious diseases like rheumatic fever and multiple sclerosis might be autoimmune diseases.

The allergies that are suffered by human beings have been considered antigen-antibody reactions. Instead of being immunized, the persons were sensitized. They were hypersensitive. A person might have been born with the sensitization, then on being exposed to a drug, a food, a pollen, react abruptly and it might be violently, and it looked mysterious. The signs might be local, great red hives, or widespread, as after a toxin-antitoxin horseserum injection in someone with an allergic state like asthma. He might die.

Before dying, the wretch might have had convulsions, might have half-suffocated (spasm of the smooth muscle of the bronchial tubes). A similar dying had been visited upon animals experimentally, called anaphylactic shock, inducible by two time-separated injections of a small quantity of egg white.

Paul Ehrlich, chemist, microbiologist, immunologist, lived in a time of light microscopes and zoology, whereas we live in a time of electron microscopes and chromatography and radioautography and high-energy physics and nuclear chemistry. Ehrlich imagined a toxin as a molecule with two hands. One hand had a shape that let it grip some part of a healthy cell, and while it gripped, the other hand could work the cell, demolish it. Many cells, of course. The cells had their revenge, chemical revenge, produced extra hands. Those were the antitoxins. They were distributed through the fluids and the tissues, and when the same toxin next assailed a living body, the hands took the toxin into custody. This could be retold in other terms. Paul Ehrlich's theory is here stated at its barest. It caught the scientists' imagination of the time, caught his, still catches ours, may by A.D. 2100 be all wrong, or all right.

LYMPH NODE:
Inner Barricade

The skin was the outer barricade. There was also an inner, complex, multiple. Important in it were the lymph nodes. These were regularly placed defense posts that provided mechanical defense, cellular defense, chemical defense, and, in some of the cells, chemical memory; the cells remembered an old attacker and were ready. There was evidence that these nodes lived long, or might, remembered the attacker for fifteen years.

The nodes and several related defense structures (spleen, liver, bone marrow) were the fixed fraction of the inner defense. The thymus has recently been discovered to be part of that system too.

In peacetimes a node worked at peacetime jobs, such as manufacturing lymphocytes; the manufacturing sped up if the body was invaded. From the lymphocytes, by a number of cell divisions spread over five days or so, there were manufactured the plasma cells, and these were shown by clinics and laboratories all over the world to be essential defense because they were the final manufactories of the antibodies.

Then the thymus was recognized as a character in the tale.

That gland of the endocrine system sent forerunner cells to the several defense structures of the body, such as the spleen, and followed the forerunner cells with a hormone that matured the cells.

In wartimes—war of foreign invasion—the mechanical aspect of the lymph node had use, could intercept invaders. Having intercepted the foreigners, it could digest them, kill them with its chemicals. Standing guard at the node's passageways were phagocytes, and we already know what they could do.

Seen from the side of a bacterium, the node was a snare, a hazard, a booby trap. In one experiment, ninety-nine of every hundred bacteria that entered a node were destroyed. In another, a node was flushed with bacteria for an hour and not one bacterium escaped into the channels leading out of the node. The node defended against toxin molecules also. And virus molecules were not forgotten, antibody gluing virus molecules together, the mass of them bigger, so that they could not slip into a cell, multiply there, do their destruction.

Those lymph vessels drain our scalp, our face, our abdomen, our genitals, our extremities. Cancer might reverse that, might employ those passageways for its own guerrilla warfare, which was why a surgeon when he removed a breast removed all lymph tissue anywhere in the vicinity, up toward the collarbone, out into the armpits, deep toward the lungs, surgery that might take five and more grueling hours. The surgeon did similar drastic surgery in other stricken parts—stomach, uterus.

The military plan of the bacteria meanwhile was to join with as many as possible other bacteria, infect the node, tear it to shreds, move on to the next node.

Should the invaders not have been mechanically halted by the nodes, not beaten down by the total cellular arm of the body, not by the chemical arm, they were free to swarm. Such a creature was suffering a bacteremia, bacteria free in the blood, or a toxemia, toxins free. Death might be hard to stave off—not as hard as it used to be.

A, B, AB, O:
Bolstering the Wartime Economy

A night visitor groping along gloomily lit corridors in City Hospital plumps suddenly on big white letters lighted from behind that startle him, but it is a gay sign for a gay place. *Incubator Room.* No one can resist stopping to look through the window. A skinny pink baby lifts a skinny pink leg, a pale sleepy baby coughs, a tomato-red-faced baby lets out one horrific yawn. The three are at the beginning of their long-short journey. The tomato-red looks as if it already had high blood pressure. Opposite the Incubator Room is the Blood Bank, and a nurse is coming out the door, has picked up two pints of blood for a transfusion.

It long had been impossible to replace lost blood, a fact that gnawed at the human mind until a way was found to do it. All over the world thereafter in hospitals, drop, drop, drop, blood entered veins, so-and-so many drops per minute, the transfusion equipment checked hour after hour. In spite of popularity, it continues to feel odd, tapping a first body and running the tappings into a second. It is so intimate. "And how do you know whose blood that was?" That also has gnawed some human minds, and the query was right enough. For centuries men had tried transfusion. Sir Christopher Wren, who built Saint Paul's, was one of the early ones. It was tried again and again, man into man, other animals into man, always abandoned for the same excellent reason, death. The bloods somehow were incompatible.

In the twentieth century came the discovery that accounted for past failures. Bloods were different. One blood could not be safely mixed with some others. Bloods fall (imperfectly but perfectly enough) into four large groups. That made transfusions simpler but not entirely simple. Madame Nature always holds something back.

Not only was the blood of one group generally safe for transfusion into persons of that group, but some crossings were safe. A quick test let a physician know in advance whether the person who was giving the blood, the donor, was right for the person receiving it, the recipient. That much had been clarified before World War I, but the triumphs for transfusion came during World War II, altered war, because now after shooting holes into human bodies with unprecedented efficiency, we could replace the lost blood.

The manifest cause of the transfusion deaths had been that the donor's red cells clumped in the small vessels of the recipient, blocked the stream, and organs or tissues did not get their deliveries; and if the person did miss dying outright, he nevertheless was in trouble. Later those clumped cells would dissolve, and that was more trouble.

In the clumping and dissolving, the body could be regarded as defending itself against the foreign. Madame Nature, so far as we can understand her, wants the body to maintain its blood nationality. Best to keep all foreigners out. Madame Nature has made it a danger not to keep them out.

The four original blood groups were A, B, AB, and O. A thrifty man has his group typed on a card in his wallet so that he is sure to have everything ready for his auto wreck. Whoever enjoys matching pennies or working crossword puzzles might have an interest in the matching groups. Blood cells and blood serum must be kept separate in one's mind. Blood cells have on them, or do not have on them, A antigen or B antigen. Blood serum has in it, or does not have in it, A antibody or B antibody. B blood cells have B antigen, but B blood serum does not have B antibody, because if it did, it would destroy its own cells. This probably immediately makes the term *incompatible* clearer, and the nurse will not look ridiculous when she puts a drop or two of serum on a glass slide and adds a small quantity of the blood cells of the person scheduled for transfusion in order to see whether the cells do or do not agglutinate.

The traffic in blood has grown as vulgar as the traffic in money. Blood banks are everywhere. Blood bankers receive and pay across the counter. Blood bankers have a society, the American Association of Blood Banks, with national meetings. International Transplantation Conferences take place every year. There is more accumulated data. There is more effective application. Every twenty-four hours every large hospital transfuses blood in every ward, and human lives are saved. This defense may fail but it often succeeds, in hemorrhage, shock, burns, clotting abnormality, anemias, before and after surgery.

It ought from the start to have been anticipated, and it has already even in this brief accounting become evident, that blood is more complex immunohematologically than the four groups. Besides the A, B, AB, and O, there are more than ten well-known groups, and as a result of the linking of this with that, it was said for a time that the figure was three hundred thousand recognizable blood individualizations.

Rh:
Odd Bit of Investment

Rh, when first discovered, could have been thought essentially a blood group. It had somehow insinuated itself into the known groups. What was quickly clear was that it supplied the explanation for a previously mysterious cause of death of fetuses, of infants, cause of stillbirths, of miscarriages, of brain damage if there was a birth.

The infant came forth jaundiced, or got jaundiced a few hours or days after birth, had anemia, had a large liver, large spleen, then died or did not die but was very sick.

What had happened in the sex situation was that an Rh-positive father (born Rh-positive, genetically Rh-positive) had at a previous pregnancy conceived an Rh-positive fetus in his Rh-negative wife. No Rh in her. While that fetus was in her womb, some of its red cells with their antigen crossed over into her. Her blood produced antibody against them. This stayed in her. At the later pregnancy the antibody passed in the other direction, into the new fetus. Poor foreigner in the womb, his red cells were destroyed and he got anemic. His bone marrow labored to make up for the destruction, whipped up its red-cell manufacture, hustled the cells into the circulation before they were completed, and under the microscope one saw the immature reds. The fetus might not survive. More likely he would. In a children's hospital, if he were lucky enough to be near one, or any hospital, the physicians would start at once on admission to draw his blood, to replace it with blood compatible with his mother's, exsanguinate him, literally, transfuse into him an entirely new blood volume, red cells that that incompatible serum would not destroy, Rh-negative cells, in exchange for his own Rh-positive cells, like putting a new transmission into an unluckily built car. By the time that infant had used up those transfused cells, his personal blood-manufacturing machinery would have swung into full action, the destructive antibodies would be negated, his life saved.

The mother meanwhile had had a danger added for herself. She had acquired a serum sensitized with that new antibody, and she dared not be transfused with Rh-positive cells, because her serum (as at the second injection with any antibody) would rapidly produce more antibody, destroy the transfused, and she would die, or might. Death is wherever life is.

Some psychobiologists choose to regard the fetus, any fetus, as foreign. It is a three-pound, resented foreigner. Eventually he is rejected. That is to say, born. It gives to all of us the tone of Greek tragedy. We are all resented and rejected. The interpretation is bizarre if nothing else. It makes us think thoughts larger than one catastrophic birth.

PHYSICIAN:
Nature Accepts Help

There are millions or billions—many anyway—of years between the first fern and the first man. All surviving and nonsurviving animals fell somewhere between those dates. No one animal ever paid a professional call on a sick animal, so far as we know. An animal helping an animal get well—that would be a milestone on any planet around any star. During the mating season a mate does groom a mate, and that sometimes looks like physical therapy; the pigeons were at it this morning on the Medical College roof, a rough-and-tumble in each other's feathers. But a human being would think he was drunk if ever he truly thought he saw pigeon A fly over to pigeon B to bring it medical help. Psychological help? It could be, but one always ends thinking it was a human interpretation and hard to be sure. Big-brained baboons are claimed to bring medical help, but one reads the accounts, precisely what did happen, and one ends thinking the same, hard to be sure.

Man brings that help day and night, not only to a mate, to anyone who will pay, or won't.

Human disease is old. Fossil bones prove it. All the kinds of body defense are probably old. The physician is old. Hippocrates, the Greek, was merely the first "Father," Imhotep, the Egyptian, merely the first name. The evidence for Imhotep seems to be that he was an architect and an astronomer and a high priest and became a physician only twenty-three hundred years after he was dead.

A dog licks its wound. The physician washes the wound, removes the dead and lacerated tissue, lessens thus at the outset the chance of infection, and should infection in spite of that occur, he supplements the body's inherited and acquired immunities from his vials. He thrusts his vials, the liquid artificial immunity that is in them, each labeled. And he thrusts his syringe. Has reason to.

He knows more than other men about both the defenses built into the living and those he introduces, which fortify a bloodstream with friendly molecules to deal with enemy molecules. Knows about antigens, antibodies, phagocytes, macrophages, blood clotting, transfusion, wound healing, mind healing. Considers how he can make mind healing help wound healing. Knows the capacity for repair and growth in individual organs, how far he can count on that power of the liver to reconstruct after injury. How soon after birth he can count on the infant's kidneys to take up their work, take over from the mother's. Knows the protection that is in the essential chemicals in the right foods, how much or how little and what kind a fat man ought to eat, a skinny woman. Knows even the partial pressures of the gases in the air that we breathe via the lungs, at least thinks a moment about it when his cardiac patient is planning a trip to the Andes.

This now requires a different acumen: he knows at what point he should *not* rely on his textbooks of physiology, pathology, medicine, surgery, his journals, but should rely on himself, on his sharpened senses, his premonitions and intuitions, these added to by all manner of humanly useful detail gathered from his experiences. He is alert to such a fact as the day and hour a life is turning irreversibly toward death, sees, that is, when Nature is beginning to desert her investment; or when Nature has weighed the situation and intends to protect her investment. In the latter case he plunges, invests all he has, and if Nature is in a mood to accept, the patient improves. In short, the physician is supporting his hundred or his thousand pieces of immunological information, supporting with his intelligence, his discretion, his manual skill, and his past mistakes.

To the sick man he gives enforced rest, gets him to agree to lie down two hours in the afternoon, or, the contrary, keeps him on his feet, suggests he walk once up and down the hospital hall, tomorrow twice, the day after tomorrow up and down the street. Says peremptorily that the patient must stay one more day in the hospital. He humors him out of his consequent bad humor. For another patient he reams a prostate, or gets his friend, the specialist, to do it. Or he extracts a splinter. Another friend replaces a heart valve. He himself modifies a milk formula. Quiets a bellyache. Quiets a thought. Admittedly, he may damage where he expected to cure. Admittedly, he may accomplish nothing. Undeniably, he has made life last longer for many, and made it more bearable while it lasts.

If he is honest and reads the journals, he probably acknowledges why and how he succeeds, what, especially in late years, he owes to physics, chemistry, genetics. He recognizes what the laboratories of the world have accomplished, and he looks toward what they will. All the sciences have joined to make medicine effective. Nevertheless, the physician does not cringe before the more narrowly focused intellects around him. He does not know what they know, does not too much worry that he does not, that he has a less exact, but sometimes can claim a wider, spread of learning. He has gone on routinely performing the routine acts. He has established the diagnosis. Written the prescription. Pronounced the dead, dead. Signed the certificate. Consoled the family. May have gone to the funeral.

The physician is the healing process become self-conscious. He has put mind into the healing process. He does not say this, and it may even not occur to him, but he has put our late-coming human mind into forces as old as evolution. He has allied mind with those blind forces. We all have done that, but some of him has done that somewhat more, and a recognition of this has made, sometimes, a philosopher of him. It is a fine good fortune to be a doctor. It also has sometimes done more mundane things for him, made a traveler of him, made him want to go to find what the rest of the planet thinks of life and of medicine, what the latest medical gains may be, on the spot where they arose. He has traveled to a meeting in Tokyo, in Patagonia, met those other colleagues who are enjoying a few days at government expense. Paracelsus in the sixteenth century went to all the corners of Europe and inquired even, as he said, of the barbers and midwives.

Meanwhile the physician will have continued to divide that overall learning into classes of learning, will have strengthened medicine, he believes, by narrowing his portion of it, one hopes without narrowing himself. His technical and theoretical findings he will still have reported in articles and books. Mostly he will have been cultivating his own garden, but sometimes too he will, for an evening or a decade, have been trying to bring together all the gardens, create a single science of what remains a scattered, though now and then an integrated, learning. In his unwillingness to compromise the integration, when he is unwilling, as also in his unwillingness to reduce his day to the calendar of a businessman, when he is unwilling, lies medicine's allure.

For these reasons, when next you pass an Academy of Medicine on a meeting night and see some not unordinary-looking, not ungentle gentlemen in huddles talking, one leaning against a wall, one blocking the front door, one draped half around the mailbox, all rich but not in the high brackets, no one of them poorly dressed, all scrubbed, particularly the bald ones, their cars out in the street illegally parked, if that sight annoys you, suppress your annoyance a moment, remember that they date from Hippocrates. They may not look it, but they date from Hippocrates, and from Imhotep, and Aristotle, and Galen, and Vesalius, and Harvey, and Pasteur, from all who have succored the sick or by their experimentation have supported the physician in his intent to prevent past misery repeating itself in the future, his intent to make human life less pitiable, even make it quite happy for a twilight. Much of the past may be embodied in each of those men, a historic past. He does not often think of this, as most of us rarely think of our historic past. Consider, though, what advantages this one has. Consider what it could do to his mind to live every day near the wonder of the naked human body, that unparalleled machine, but every day also to live near misery, to have the difficult privilege of being every day near pain. On that same day he may have the happy privilege of taking from a human being the worry about his body. Have the happy privilege of assisting at that miraculous moment when a newborn literally pops from a womb.

There is the outstanding one—say he is a surgeon. On some frightened midnight we realize that this man is astonishing, that surgery is astonishing, that the economy of this one's talk has been imposed upon him by the abundance of human beings who with their bodies and their minds are leaning on him. He works as the great work, with patience, with intensity, with what swells to be wildness for one half hour. As he has gotten older, he has grown genially rotund. Last night he made one think suddenly of Johannes Brahms.

XIII

NERVE
Carrying the Torch

SIGNAL:
Marathon

Third row, aisle seat, the stadium in Athens, the Olympics. In a moment we shall learn who won the marathon. The crowd is hushed. Now it stands up.

Third row, aisle seat, that spectator there wishes he had a humbler nervous system, like that of the cockroach that stepped out of the shadow and was stepped on. Wishes he had a less large ovoid skull for his mad racing to race in. Wishes his small body had not been cursed with his huge head.

Purpose of a nervous system?

A multitude of tracks over which there is run a multitudinous marathon. Ten to fifteen thousand million tracks with branch tracks are in the human brain alone. Messages are hurrying everywhere. Each helps to join the north of the body to the south, the east to the west, to blend the total into the affairs of planet, galaxies, universe, the affairs of the inanimate, the animate.

NEURON:
Torchbearer

For years it was thought, by some, that the nerves of the body made a network, all of them connected, nerve filament to nerve filament, and over the network the signals raced. The system was built as one, so could do what it often appears to do, operate as one. The chief defender of the idea of network was Golgi, Italian. The chief opponent was Cajal, Spaniard. Cajal thought the system built in units.

Golgi and Cajal each intensely believed, a volatile Italian, a volatile Spaniard.

So, to Cajal it was units. A unit was a neuron, and a neuron was a nerve cell with its fibers. The signal raced over the neuron, reached its terminals, and at the end of each there was a cleft, a gap. The electron microscope in our day can see that gap. Cajal's eye could not. Across it the signal leaps—at least it gets across—in the direction of die next neuron. If enough signals get across enough gaps and add up their voltages, the torch is lighted and handed on to the next neuron.

The unit, the neuron, can be grown. A speck of sterile brain can be planted in a culture medium such as bacteriologists use to grow bacteria. The speck might also have been put on a sterile glass coverslip with some of the culture medium, another coverslip placed over the top, thus the growing directly observed under the microscope. That medium has been stocked with right neuron food, kept at right neuron temperature, and those neurons look then like the nature-grown. Each has its branches. Each has its insulation. If the original speck was taken from the front of the spinal cord, the fibers grow out straight, as they would have in the body, and if from the back of the spinal cord, they loop. One sees there in the culture medium how those units will make their inevitable connections. They were born to go one way. They were born to generate their particular electrical action.

In the culture medium other cells grow. Formerly these were believed to serve merely as scaffolding. Later this did not seem true. Like heart cells, these other cells beat. Thus there is a slow beat in the food and drink. Beat, beat, beat. Life is different from death.

Golgi's claim of a network lost. Cajal's claim of a system built in units won. With a vigor that seemed not to weaken as he got older, in more than two hundred articles and up to the last days of his life, Cajal hammered the proof of the neuron doctrine into the thick skull of the compact majority.

A single unit?

Billions are in our brain. Always a temptation to picture a drawing reproducible in an engraving. Each has a cell body, and, leading toward that, rootlets, and, leading away, the large branch with those terminal twigs. Rootlets → cell body → branch with twigs. The junction between any twig and the next unit is often called a valve. Because of that valve—synapse—the signal is forced to go in one direction. The overall direction is: rootlets, cell body, branch with twigs. The rootlets are named *dendrites*, the cell body *cell body*, the branch *axon*. An axon may be short, may be long, may reach from a man's spinal cord to the tip of his middle finger. Each neuron operates on its own power, pays the cost for the signal's advance over its lap, and at the end of that lap helps to hand the torch on to the next neuron.

GALVANI:
Thundercloud and Athletic Leg

Twenty-two centuries after the Athens of Pericles, the year 1737, there was angled out of a womb Aloisio Galvani. (Aloisio or Luigi.) He was Italian, had the quality of the Renaissance Italian.

Some envious persons, so many years later, ladies very likely, hinted it was Aloisio's wife, Lucia, who deserved the credit for the great discovery, and undoubtedly she and Aldini, Aloisio's windy nephew, were in the room. Three—and a frog.

One of the three held a scalpel. The frog had been skinned. The large nerve that leads to the muscles of thigh and leg had been exposed. Close by was an electrostatic machine generating electricity, and each time the machine sparked, thigh and leg twitched. From Galvani's careful account it might be inferred that the metal of the scalpel picked up the electricity through the air, conducted it to the shank, which twitched. Galvani would write a whole book on the subject. His first article reporting his experience with frog legs appeared in the *Proceedings* of the Academy of Bologna; he was professor of anatomy at Bologna. After a while the galvanic battery would be named for him.

As for the scalpel, it needed to be grounded, could not be merely a scalpel held in space, and not enough either for the experimenter's hand to touch the handle, which was bone, but must touch the metal that riveted the bone to the metal of the blade. Two dissimilar metals. Alessandro Volta seized on that. He said it was the contact of the dissimilar metals that excited the twitch—physical electricity.

Galvani said it was the frog—animal electricity. Between Volta and Galvani there blew up another of those famous debates of science. Neither debater ever vanquished the other. Galvani sometimes was represented by his nephew, he himself being too gentle and shy, or some other reason, and represented after his death by his nephew. The debate added fire to the combustible talents of Volta and Galvani. Volta laid wetted cardboard between dissimilar metals, and from that came the voltaic pile, the electric battery, a current that flows instead of sparks, the transatlantic cable (likened later to a nerve), and toasters for breakfast, and Broadway by night; the Sphinx and the Great Pyramids near Cairo that had dreamed on the dark desert for three thousand years would be waked by floodlight; the Parthenon in Athens would be lit like a ball park. From Galvani? Galvani would have begun the electrical investigation of the living. There would follow from that the recording of the rhythms of a million beating human hearts; a steady adding up of the number of hearts with abnormal rhythm that was electrically shocked back to normal; the adding of the recordings of the rhythm of who knows how many brains; the stimulating for localization during surgery of tumorous or injured or scarred brains. Electrical studies would grow more complex, more delicate, the electrodes finer, the electrometers more sensitive.

There remains the anecdote of the frog shanks. They were intended for dinner, a tidbit for Lucia, who was ill, were hanging by a copper hook from an iron railing, a copper-iron coupling, and the shanks twitched. "Dissimilar metals," said Volta. "Animal electricity," said Galvani. The arguers still were arguing. The shanks were dead. They behaved alive. During a thunderstorm they twitched crazily, big cloud in the sky was inducing electricity in small shanks on earth, and at the flash of the lightning the electricity discharged and the shanks twitched. They twitched also in fair weather, but not so predictably, said Galvani. Fortunate that Volta and Galvani were Italian, that they did their thinking heated by their passions, because the point they were making was an important point. Electricity, the demoniac thing, and nerve, the life-inspired thread, were promenading together, and no man could be sure whither that promenade would lead.

NERVE IMPULSE:
Swift and Brief

Up in the skull the demoniac has always been ready to haunt the flesh. The number of nerve fibers is enormous, the number of electrical events enormous, difficulties of understanding enormous.

The nerve impulse has been studied for years. Physiologists must all have been pleased when the 1963 Nobel Prize for physiology and medicine went to three students of this small event.

The gross nerve—the transatlantic cable—is a bundle of many nerve fibers. Anatomists work with that cable. Surgeons repair it. Physiologists rip it out of the frog. They also tease out a single fiber. The fiber is wrapped in a membrane. The ordinary microscope cannot see that membrane but the electron microscope can.

When the membrane is at rest, there is an electrical difference between outside and inside. Textbooks have diagrams with plus signs outside, minus signs inside. Now, let that membrane be touched by electricity, or by a hope, or by a disappointment, and instantly plus signs behave as if minus signs had neutralized them, even reversed them, minus signs outside, plus signs inside. For that instant there has been created an electrical hole.

Beneath the stress of our world, beneath the turmoil of our life, there is this methodic production of electrical holes. Promptly, neighboring plus signs pour into the hole, patch it, but leave a hole from where they took the patchings. Pour. Patch. Thus does a hole advance. Thus does the nerve impulse become an advancing electrical leak.

The bulk of the fiber—as against the membrane that wraps it— is maintaining a busy chemistry. It is doing metabolic work. It is keeping the fiber alive, producing the resting state, the difference between outside and inside, so the impulse can upset it, produce the active state.

A species of squid that goes down to the sea has a giant fiber whose bulk has been squeezed out, studied, and the membrane, the sheath-gown, also studied.

The nerve fiber, being living, consumes oxygen, gives off carbon dioxide, produces heat. Heat-produced is another such minute quantity, one ten-millionth of a small calorie for one gram of nerve fiber. The carbon dioxide is another such minute quantity. Thales, the Greek, the measurer, also did not live in vain.

VELOCITY:
Hundred-Meter Dash

How fast is the race of the nerve impulse along the nerve fiber? Over the long tracks from fingertips and toetips or the short tracks through the jungle of spinal cord and brain, what is the velocity?

A century ago scientists thought such speeds never could be measured. Six years after Johannes Müller gave this as his opinion, Hermann Helmholtz made the measurement. Helmholtz freed a nerve in a frog. It was a gross nerve, a bundle of nerve fibers, and he let it stay attached to its muscle. The apparatus was arranged so that marks could be written on the smoked paper of a revolving drum, one mark when the nerve was stimulated, then, after the nerve impulse had run along the fiber and excited the muscle to contract, another mark. Two marks. With everything ready, Helmholtz stimulated the nerve as far as possible from the muscle, five centimeters, got those two marks, then near the muscle, one centimeter, got those two marks. A longer time elapsed between the former two than between the latter two—because the impulse had to travel the longer stretch of nerve. He measured the stretch. After that the calculation needed no Helmholtz.

Thirty meters per second or thereabouts was the velocity for frog nerve, approximately a mile a minute. One hundred meters per second is possible in a large motor nerve of the human body, approximately two hundred miles per hour. The brain is a slow instrument, marvelous but slow, appropriately slow, to keep to the pace of our lives. If our brain were not able to hold back to the contraction time of our muscles, and to the time it takes our joints to get out of each other's way, we should all fall on our faces as we did when we were babies.

A nerve and a track of gunpowder have often been compared. Ignite the powder at one end of the track and the flame burns its way to the other, that burning having a velocity that also can be measured. Each particle of powder ignites the particle next, the flame remaining steady, does not die out as would an echo that in every instant draws its energy from the original sound source. (The nerve impulse does not draw its energy from the original stimulus but from yesterday's breakfast.) If the powder is wetted somewhere along the track, enough to slow the burning but not stop it, once that wetted spot is passed, the burning will continue at its previous speed. It would be much the same for a nerve wetted by a drop of ether.

Velocity has for years been measured on single nerve fibers without a muscle attached. A nerve trunk like that of the frog would have many sizes of fibers. Velocity over thick fibers is higher than over thin. Size and speed bear a strict relation. An almost frightening array of speeds has at every instant been operating inside of you. In a nerve trunk there may be thousands of sizes, thousands of speeds.

Instead of the heavy muscle lever that Helmholtz used, the researcher today has the most weightless lever in the world, a beam of electrons.

VOLLEY:
Not Singles but a Team

In the living body, the nerve impulses do not travel singly. They travel in volleys. It is volleys racing in our brain, our spinal cord, in the far-flung remotenesses of us.

The fact of volleys was first recognized by an Englishman at Cambridge. The crucial experiment included a loudspeaker. Out of the loudspeaker spoke the volleys, click, click, click, for each nerve impulse a click, more and more nerve impulses over more and more nerve fibers.

First, the experimenter converted the electricity of the impulses into flashes of light. Second, photographed the flashes. Third, transformed the photographed flashes to clicks. Fourth, made the clicks permanent on a gramophone record. A similar experiment had been performed also on the muscle in a human arm, on the membranes. First, the experimenter fitted a metal core into a hypodermic needle. Second, plunged the needle into an arm. Third, manipulated the needle until it had a right contact on some muscle surface. Fourth, brought in the loudspeaker. And out of the loudspeaker dutifully spoke the clicks, each click following the click before, soldiers again. If the arm was bent, more and more muscle fibers were included, and more and more impulses traveled over each fiber. The clicks themselves were not different, only faster. A literary physiologist stated this felicitously. "The general effect is that of a machine-gun firing at increasing speed but the noise made by the explosion of each cartridge does not alter."

END PLATE:
Finish Line

A friend of Claude Bernard's sent him some arrows tipped with a gummy substance, curare. The arrows were from South America. They were used by the Indians of the Orinoco River valley to shoot birds through blowguns. Bernard related how in his laboratory in Paris he scratched a rabbit's back, introduced some of the gummy substance into the scratch, disturbed the rabbit so slightly that it did not stop chewing, but shuffled to a corner, settled down, its head drooped, its ears lay back, it turned on its flank, was dead. The death occurred six minutes after the scratch.

Where in the animal's body had that poison acted? In the nervous system, but where?

Bernard's reasoning here often has been cited as an example of how the scientist at his best comes to a conclusion. For twenty years Bernard returned to the problems of curare, using various animals, and students the world over have performed experiments that paralleled his.

A frog places itself at a student's disposal. In both legs, right and left, he exposes the calf muscle and the large nerve that leads to it. Previously with a needle he has mashed the frog's brain against the inside of its cranium (*pithed* is the word), and so thrown the frog into a state like that of a man who has suffered a massive cerebral hemorrhage. Next, the student ties a ligature slightly around the frog's right thigh, stopping all blood flow down the right leg, the nerve not included in the tie, so that nerve still is supplied with blood. At this stage, if he stimulates the nerve on the right side, that right calf muscle contracts, and, of course, if he stimulates the left nerve, that calf muscle contracts. Next, the student injects the curare under the frog's skin. Soon the frog becomes paralyzed everywhere except the right leg, which was tied off from the circulating blood, the poison therefore not going to it, but going everywhere else, including the left leg. Nevertheless, if the left calf muscle, the muscle itself, is directly stimulated with electricity, it contracts, as does also the right calf muscle. Plainly, curare is not a muscle poison.

Next, the student stimulates the nerve to the right leg, and the right calf muscle again contracts, even though the poison has been carried to that nerve because it was not included in the tie. Plainly, curare is not a nerve poison.

When, however, the nerve where there is no tie is stimulated, that muscle does not contract. So curare, which does not poison muscle and does not poison nerve, does poison something between muscle and nerve.

Today, that tissue between muscle and nerve has come to be accepted as a special part of the architecture—the end plate.

Bernard in his time could know only that something had been poisoned in the general neighborhood where nerve meets muscle. He grasped the large facts, was the first to do so, and the pattern of his thinking has been handed down. Bernard's rabbit died of suffocation, because the end plates of its breathing muscles were poisoned and no nerve impulses reached them, so the animal could not breathe. Death was stirless because the rabbit was not able to stir. It had been deprived of the power.

Curare and its substitutes came into wide use and wide thinking. By surgeons, when they wished to diminish disturbing contractions during an operation. By psychiatrists, when in shock treatment they wished to prevent the convulsions that in earlier days fractured bones. By physiologists, when they wished an animal not to struggle. This special tissue, the end plate, has drawn men's attention to the nerve side, to the muscle side, to the baffling detail of structure in both places.

SYNAPSE:
Runner Touches Runner

The finish line need not be an end plate, where nerve meets muscle. It can be a synapse, where nerve meets nerve.

Sherrington, the Englishman, gave the name synapse to the junction, and Cajal, the Spaniard, made it a reality.

There is a gap at the synapse. The electron microscope sees that gap. Messages must get across it. There is delay. There are possibilities of modification of the message.

Many such gaps, such junctions, such synapses, are everywhere in the nervous system. At some points they appear hopelessly matted together. But human ingenuity has kept at them.

In our human brain are an estimated ten to fifteen thousand million neurons. Each has its dendrites, its axon, and the axon has its terminal twigs, hence almost inestimable numbers of synapses. Each synapse has its minute important electrochemical decisions to make. There the traces of the world upon the brain are written and rewritten, and it could help us to understand how from the multitude of connections the strangeness of a human mind might be built.

The nerve impulse races over a neuron, reaches a terminal branch. Beyond it is the gap, the cleft, and beyond that is the membrane of the next neuron. It is the adding up of the electrical force of many membranes that determines whether the torch is passed on or not passed on. The runner touches the next runner, and that touched runner runs or does not. Many must touch if he is to run. Once he starts running, he keeps on to the end of his lap. Once a nerve impulse heads down the neuron, it keeps on for the length of that neuron.

Dendrites → cell body → axon → terminal twigs.

A spray of terminal twigs breaks from the end of the axon. Each twig has an end foot (not end plate, end foot). Four hundred end feet may come down on one neuron. Up to a thousand have been counted.

As the region of an end foot is neared, there is increased chemical activity, so that even if the torch is not passed on, the general neighborhood has what could be called a neuron warmth, enough to cause other neurons of that neighborhood to be readier to speed some nearby torch.

CAJAL:
Observer at the Olympics

Any book dealing with the synapse, or the neuron, or some large or some small structure in the architecture, eventually also the chemistry inside of that architecture, might overlook a name, but not the name

Santiago Ramón y Cajal. Again and again that name. Usually it is just Cajal. If the frontispiece has his photograph, he is dark-complexioned, hair cut close to his compact head, soft tissues of his face laid sparely on the bone, body leaning forward, a tenseness that eludes no camera, definitely a gentleman, a Spaniard. If the book has engravings of his drawings of brain or spinal cord or single nerve cells, the engravings also are apt to be his. Even in the mountainous output of our time, where no stone is left unturned, where every book has illustrations, always the latest, rarely are his redrawn, rarely have they "After Cajal" underneath them, but are reproduced as he left them. References to his discoveries stack up until anyone would reflect on how many men sometimes may spend what one man earns.

The place of his birth was a somber village, a lost spot, an island of Aragon in Navarre, the year 1852. His father had to drudge to get his doctor's diploma, to become worthy to heal the sick and practice the art of medicine, as speakers say at medical commencements; then followed the years with the sick, the fees small, the doctor tired when after dark he was through with the day, then supper, then the Spanish nights, then the children to be supported, no compensation to be expected from this one, haunted by dreams of being the great painter of Spain, the modern Velázquez, and stubborn about his dreams. Two wills, father's and son's, pitted against each other, the father winning but only because he was the older and had the greater experience, a victory that had to be won over and over and never really was won. In the son's memoirs he described that contest. All one can say is that it was not as one-sided as a bullfight; the father tried him in this school, tried him in that, the passion for painting was flogged out of him, unimaginative and dour priests kept him at Latin and whatever else boys are kept at for the good of their souls. Regularly the pent-up feelings burst through in escapades punished by starvation, by imprisonment, once a literal prison with bars, for a child. Possibly he deserved the floggings. Possibly he required them. Possibly they drove into genius the discipline necessary for it to survive. What is sure—he got them. The memoirs tell how he saw a miracle in the discharge of any firearm, used always to admire his father's gun, kept out of reach. He took timber and an auger from a carpenter's job, strengthened the timber with metal, produced a cannon, hoisted that to the orchard wall. Loaded. Aimed. Fired. At a neighbor's new gate. And there fell down on the Spanish earth the wreckage, and on him the bitterness, this is a village that must save every stick and stone. He was eleven.

One reads the robust tale, sees it from the father's side, thinks the son had something of the devil and the father the good heart; then one turns the tale around, sees it from the son's side.

Came a day when the son too got his diploma. *Doctor Santiago Ramón y Cajal*. All things do come to pass. His advance through the medical school had been distinguished because of anatomy. He had his eye for shapes, could draw what he saw, the eye focused, not on the Spanish landscape or a hidalgo with a horse, as he had dreamed, but on a dead brain, or part of a microscopic field, a single nerve cell, many nerve cells. There the eye stayed. A talent tramped down at one point had pushed up its head at another. Son and father together prepared their anatomic atlas.

Besides being a student of the intricacies of that system of the body that most underlies mind, and an artist fascinated by the living machine, and an artisan who could fabricate the tools to study it, Cajal was a patriot, passionate in that as in everything. He was not obligated to enter the army but demanded they enlist him. This was at the time Spanish-American animosity was brewing. The army sent him to Cuba, where he contracted malaria, as many did, and dysentery on top of the malaria. He was shipped back to Spain. At home he rested. Then he started to work, and started to hemorrhage from his lungs, probably from the tuberculosis that frequently followed malaria. But illness does not kill that kind. He recovered his health and up the ladder of Spanish universities he practically ran, soon was professor of anatomy at Madrid, his knowledge of the nervous system amassing day and night because day and night was his pace. He invented methods, altered them, flung them aside, took up others, cut section after section, stained the sections, tracked the single neuron through the wilderness of spinal cord and brain. All present-day advances in these fields somewhere rest on him, all have a greater firmness because of the foundation he put under them. What he discovered in one animal form he verified in others—mouse, rat, rabbit, guinea pig, man, when he could get a man, a dead one. He had the shrewdness to examine embryos at that stage of development when the neuron is small and stands alone like a tree in a white winter. Saw what other men had not been able to. Saw with the ordinary light microscope. There was as yet no electron microscope. There was a method of Golgi's, the Golgi silver method, abandoned by that eminent Italian because of its difficulty; Cajal wrestled with it, forced it to produce. He took his slides to science meetings, struck fellow scientists with amazement. When someone could not see what he saw, out came his impatient pencil, and with quick strokes he indicated where to look, made the significant visible to slower eyes.

No one has brought to the nervous system more of the reverence it deserves; no one has been more humble before it, and humble before nothing else. He cleared away the weeds. Actually, there are no weeds, but there is the unexplored, and genius knows what of that can for the time be put aside. Without ever saying it, without ever being concerned with it, possibly, he was helping to build that flesh bridge which would join nervous system to behavior, to language, to thought. Farther than anyone, Cajal comprehended the minutiae of this one base of our being, and in the respect that brain relates to mind gave mind an earthiness.

His discoveries he published in a journal that he launched, virtually wrote, paid to have printed. Its text was Spanish. The world was not reading Spanish, and after various editorial evolutions the journal was French. "A bitter thought," he says in his memoirs. Drama was in his discoveries, drama in his life, and it can be claimed for the memoirs that they do not smother drama. When he had his microscopic sections ready, his drawings ready, his engravings ready, he wrote. His entire life seems at the point of pencil or pen. Every year we are more convinced that Ramón y Cajal was one of the greatest of all anatomists, Spain's greatest man of science since the Renaissance, the all-time, all-world greatest student of the details of the nervous system.

He won a Nobel Prize in 1906 but had to share it with Golgi, who probably hated him and whom he probably hated. Golgi's Nobel address was a polemic against the neuron doctrine (a nervous system built in units) that Cajal's lifelong labor had proven. Such was the atmosphere that year in Stockholm. Already as a child Cajal seemed to know that there can be no perfect moment in an extraordinary man's life. In a fool's there can be. During his last illness he spattered the wall of his sickroom with ink. Lived with a pen, died with a pen, and the subject of his year-in-year-out thesis was brain and spinal cord and neuron viewed from one end of the barrel of a microscope. The year of his death was 1934.

INHIBITION:
Handicap

The idea of excitation gives no one much trouble—something excites something. A stimulus excites. Inhibition may be more troublesome, but inhibition is common, operates everywhere in the body. All movement—the way at this moment a reader turns a page—must include a force opposite to excitation. That is inhibition. It long has been studied in the physiologist's world. In Moscow an international conference, November 1963, was given over entirely to inhibition.

Excitation. Inhibition.

Both are positive. Both operate together: in our chest to coordinate the expansion of our lungs with the movement of our vocal cords; in the circulation to coordinate the heart's beating with our breathing; in every skeletal muscle to coordinate its contraction with its antagonist muscle.

One could imagine two kinds of throttles, chemical throttles and electrical throttles, depending upon which kind for the moment one is choosing to give one's attention to. More than two kinds; at least, more than two kinds of inhibition are postulated, though hereabouts there is murkiness. In any event, the machine employs as positive forces both excitation and inhibition. "A machine that breathes, a piece of animated property," Aristotle said of a slave, making us realize that that sharp Greek intellect thought machine when it thought body; also, that intellect was sufficiently casual about slavery.

The inhibition goes on in every nook and cranny of us independently of every other nook and cranny. Without inhibition Phidias could not have lifted his mallet, Demosthenes could not have lifted his voice, no twentieth-century white rat could run a race. The white rat is a professional. He will run ten miles in twenty-four hours if encouraged—a dishonest word—and during every inch of that ten miles, which is on a treadmill, that rat is employing inhibition, slows something, reduces something to half-action, to no-action, is always inhibiting one part of the machine to keep it out of the way of another.

Crabs have frank inhibitory nerves that go to the muscles of their body wall. These muscles also require at the proper moment to be gotten out of the way of other muscles, require to be inhibited in order that the total crab will be kept in a proper crab motion. Human beings have no such frank inhibitory nerves to the body wall, but their bodies have the same need as the crab's, their contracting muscles also must not interfere or be interfered with. For this the gear shifting takes place inside the centers of the spinal cord and brain. Physiologists speak accordingly of central inhibition.

To repeat, inhibition operates hand in glove with excitation. It exercises its force in all parts of all the bodies that creep or run or fly over the globe. The bodies are always proclaiming that they want nothing so much as freedom. Nevertheless they cling to the earth. Everywhere the earthbound are checked inside themselves. Everywhere inhibition is a first principle.

REFLEX:
"Des Passions de l'Âme"

So many baby fingers have hopped away from so many hot stoves by a compounding of reflexes, accompanied by pain, accompanied by scolding, that it could seem obvious what was nature's intent for the reflex.

A neuron is a nerve cell. It has its cell body and its fibers and, as Cajal taught us, is a unit structure of the nervous system. The reflex employs at least two neurons. A signal travels to a center, is rerouted, travels to a muscle or a gland. The muscle contracts. The gland secretes. The contracting and the secreting are unit responses, and the total act is a reflex. A performance is more. At the barest, a performance is some summing of reflexes. One dapper neurologist defined life as such a summing, and what an impoverished definition that was.

Unit structure = neuron. Unit response = reflex. Unit performance = life.

Zeus coughs. Aphrodite yawns. Those are Olympian performances employing Olympian reflexes employing Olympian neurons. (No mortal can comprehend the gods entirely.) Zeus rises to speak and coughs. Aphrodite disrobes and yawns. Aphrodite is not unlike the naked beauty Praxiteles, the sculptor, yesterday morning, summer 340 B.C., hired as a model for his Aphrodite of Cnidus. Yesterday evening Praxiteles went to the stadium, watched the glistening runners, last night sat in the theater of Dionysus at a revival of Sophocles' *Oedipus Rex*, watched the actors, watched their bodies within their masks, watched the reflexes. Praxiteles is always about his business. Praxiteles is a visual animal.

A reflex can be described thus: something strikes a living body, the mechanical energy of the stroke is transformed to chemical energy, to electrical energy, signals start racing over nerves. Signal follows signal follows signal. Neuron meets neuron meets neuron. Something enters the nervous system, something comes out, the in-out being the reflex, and when reflex is integrated with reflex, that is the performance.

An engineer's definition of the reflex is terser, three items: an input, a gear shift, an output. The input could be anything, the setting down of one foot of a fly onto one hair of the flank of a sleeping dog. The fly walks along several hairs. That is too much. That in the nervous system is summation. The dog, a hound, immediately has Zeus-only-knows-what signals dispatched here and there, gears shifted, a re-alignment of his nervous machinery, a tensing of his muscles, then the output, that ancient response that in all dogs seems excessive, the scratch, the impatient rhythmic thump. Effect is out of proportion to cause: slight energy applied, the weight of a fly, and, as can happen in the nervous system, great energy released. Effect is really not out of proportion: that fly has six feet.

Everyone remembers his physician testing his reflexes. The physician taps the lower right of the four quarters of the belly wall, and under the skin a muscle leaps up like a toe under a sheet. Taps a wrist and a hand rotates. Strokes the sole of a foot. Strokes the palm of a hand. Draws a wooden applicator between closed lips. For each of these stimuli there is the mechanical response. The physician is testing the nervous system.

René Descartes of Paris wrote the first downright description of a human reflex:

> If someone quickly advances his hand toward our eyes, as if to strike, even though we know he is our friend, and that he does it only in jest . . . we have nevertheless difficulty in keeping ourselves from closing

our eyes, which proves that it is not at all the intervention of our spirit which closed them . . . but it is because the machine of our body is so built that the movement of the hand toward our eyes excited another movement in our brain, which sent animal spirits into muscles, that caused the eyelids to close. . ..

That is a literal translation from Descartes's book *Des Passions de l'Âme*, published 1649. The reflex fascinated Descartes—this result rolling off the assembly line, able to bypass us, we not able to prevent it. In the first half of this century much was done to clarify the reflex. The scientist who did most, who acquired great fame, who extracted most knowledge from the reflex, said shortly before his death forty-two years ago that the study of the reflex was exhausted. That gadget, he said, had lost its usefulness. He no doubt meant usefulness to the laboratory, because its usefulness to you and me and any dog and any cat, as to all flies, goes on day and night.

XIV

BRAIN
Cold Earth of Life

SKULL:
"Alas, poor Yorick"
—*Hamlet*

A medical student wearing a crisp white laboratory gown, a blasphemous fellow, put his finger onto a man's brain, drew away, a cold brain reeking of formaldehyde. It was given him—he could take it home to study if he liked—in the winter of his freshman year by the department of anatomy. Some months earlier he had sauntered, a novitiate, into the dissecting room, bodies wrapped in white sheets, on tables that decidedly were not couches, and when the sheets were jerked away, the naked dead lay one beside the other, and the tables were parallel and that added something. Pride was all that kept down horror. Nevertheless, a few days later this student was leaning his elbow, not on a nearby windowsill, but on Adam's rib. A callousness finally takes care of us all. Even so, it would again be a few days before he could be debonair about a brain in a jar in his bedroom next to his bed. That brain is a fact of anatomy, yes, but it has an idea attached. *He* had been in there. He is not in there now, no, but something of him was left behind when he vacated, a shirt on a nail.

To pick up a knife and cut into the formaldehyde-hardened haunt would be still another few days. Then, with a sudden viciousness, in he went. Probed the cut surfaces. Probed the depths. Turned upside down that assemblage of tissues we call a brain and underneath where the geography is obvious located the obvious.

The medical student dislikes anatomy. Dislikes the smells. They follow him into the street, percolate up through his meals, get into his beer, and under his fingernails. And that eternal memorizing.

Each night he adds new terms. Writes new lists. Copies someone else's notes.

Then, one disenchanted evening, three evenings before the final examinations, he recognizes that he cannot waste another hour. He will study this stuff systematically, get done with it.

He begins there where the brain joins the spinal cord and works his way up. Medulla. Pons. Comes out at last on the roof. Roof brain! Cerebral hemispheres! What a relief!

How can this freshman be expected to have the maturity to realize that while he is at this drudgery, he is also placing for himself the perspectives of an architecture that is grander than Florence, than Rome. Where is the professor who could inspire him with that? Where is the professor who, if he were himself bursting with that inspiration, could afford to let it escape into a classroom?

The freshman is not even thinking of the crasser gains: that this drudgery will help him better understand the controls on the living body, important to a physician. The freshman shrugs away the thought that science and common sense long have considered the brain to be the foundation of mind. That is too obvious. The brain is flesh, the dissecting room is dead body next to dead body, and one learns from the dead, and a dead brain has that which a dead foot has not.

Another student at midnight under a lamp has been trying to memorize the Latin names of the bones of the skull and the grooves and the holes where the nerves and blood vessels during life passed through. He stops. Did that skull lodge a princess? It *is* a female skull? What did she wear at her wedding? "That skull had a tongue in it, and could sing once." Gentle Shakespeare. Out at the front of the skull there are still some of the half-attached frail bones that gave shape to her face, framework to her smile, the smile with which that day she swept down the aisle of the hushed cathedral. "Let me peruse this face." Different from the two students, a paleontologist this afternoon in an ancient pit found a fragment of skull, and this evening is reconstructing the mind of a man of a clan that foraged lands and ravished women eighty thousand years before Andreas Vesalius was born. Padua was the center of anatomy of all the world in Vesalius's day, because of Vesalius.

MENINGES:
"Tie my treasure up in silken bags"
—*Pericles, Prince of Tyre*

The old morgue was at the back of General Hospital. This was across the street from the dissecting room. Recently there is a new morgue, on this side.

The pathologist, the chief, has for years cut open bodies and heads to find out why the late renters died. A nimble-tongued fellow. Nothing about him suggests postmortem. Even so, he has just turned a corpse over on its face, immediately starts the humdrum day-labor of lifting spinal cord and brain out through the corpse's back. He incises skin, fat, fascia, muscle, in that order, does it with a harsh knife. The arches of bone of the vertebral column he snips, wields a neat bone shears. At the head end of the corpse he slashes through the hairy scalp, which is bloody in life. Next he saws through the bone of the skull, two plates with a spongy lacework between. He opens the skull from behind. Spinal cord and brain in their membranes lie in that excavation in front of him. Splinters of bone and other rubble are scattered about, yet, considering the circumstances, it is a clean place.

Three membranes—three meninges—cover the entire central nervous system. It has a triple coat. Each large nerve trunk goes out for a distance with a sleeve of that coat. The nerve trunks go in pairs, one left, one right. Of the triple coat, the outermost makes one think tarpaulin, and the second and third make one think inner tube. Had there been a hemorrhage, the tarpaulin would have been a shade of brown that depended on the time between hemorrhage and death. With a scissors he then clips in succession the pairs of nerves that still pin down the system, then hoists the whole of it in one piece out of the corpse. *Requiescat in pace.*

Alternatively, the pathologist might have run his scissors only through the outer of the three membranes, in which case spinal cord and brain would have been hoisted from the wreckage with the second and third membranes still around them. The hoisted he drops into formaldehyde, and not until this moment has the bystander realized what a fortress vertebral column and skull are, and what a pliable inner stockade those membranes are.

Those membranes have Latin names: the outer, dura mater (hard mother); the second, arachnoid (cobweb); the third, pia mater (tender mother). The pia is fitted skintight and dips into every irregularity of surface of cord and brain, while the arachnoid bridges across the irregularities. Between pia and arachnoid flows, if anything so slow can be called a flow, the cerebrospinal fluid, making a shallow lake in some places, deep gutters and ravines and cisterns in others, with a fine stream of it running down through the middle of the cord, and a good quantity filling the caverns inside the brain.

The amount of the fluid? It would fill a teacup.

To tap some of it is to perform a lumbar puncture, an LP. A neurologist asks the sick man to lie sidewise near the edge of the bed, curve his back as much as he can, bend his head forward, pull his knees toward his chin. The last helps further to curve the back and to separate the vertebrae. In goes a two-inch-long needle, advances firmly toward the cord, into the space between arachnoid and pia, and out comes the fluid. The neurologist measures its pressure. He drains off a sample to send to the laboratory to study its composition. The proceeding sounds simple. A senior medical student was not so sure. He had tried it.

SPINAL CORD:
"Pillar of the world"
—*Antony and Cleopatra*

The spinal cord with those membranes around it hangs in the spinal canal. Nerves lead into it from neck, arms, chest, abdomen, legs, make connections at junction points inside, and other nerves lead out. Highways at the same time streak up and down, to and from the brain.

The canal is roughly cylindrical, is built in sections called vertebrae. Together they make the vertebral column. Joining the vertebrae are tough ligaments that help hold the column firm, and at the top end the canal opens into the skull, and in the skull is the brain.

The vertebrae themselves are light in the construction but architecturally strong. One vertebrae rocks on the next. The freedom of the movement differs in neck, back, pelvis. A thin man may bend forward to tie his shoe, a child bend backward till her hands are flat on the ground, Little Egypt bend in all directions for the education of her audience, and a dignified gentleman in the front row bend the least possible, but he manages to see all. Throughout the bending, the spinal cord and the nerves are seldom kinked or pinched. Between the vertebrae are shock-absorbing cartilages, disks, which occasionally slip, and then nerves get pinched, produce pain.

Thirty-three vertebrae in a human skeleton. Seven are cervical—*cervical* means neck. The uppermost of the seven supports the head and is fitted onto the next below in a way that permits the head to pivot, to see how the other heads live. Twelve thoracic vertebrae—*thoracic* means chest. Five lumbar—*lumbar* means loin. Five sacral—*sacer* means sacred. These five are fused and form the back wall of the pelvis. Four coccygeal—from the Greek for cuckoo. Those also are fused, and abbreviated, sad remembrance of a day when each of us still could wag his tail.

Nowadays the neurophysiologist may impale a single nerve fiber, neuron, of the cord. He does it with a microscopically fine electrode. By the electrode he picks up the electrical messages, or the reverse—he stimulates. When impaling, it may be cell body, dendrite, axon, or the surrounding of these, and he knows where it is by the recordings made via his electrode. In the old days, the experimenter simply sprinkled salt on the raw surface of the cut cord of an animal. The salt was the stimulus. Crude. Yet our glorious new days rest on those old days.

DESIGN:
"The spring, the head, the fountain"
—*Macbeth*

All our life the brain sits up there above our spinal cord except during the conventional eight hours at night when it capitulates and with our body topples to the horizontal.

That space must have the right size and the brain the right size, not be too large, the head not too large, not too small either, space enough for the great nerves and nerve pools for eyes, ears, nose, for the arriving nerves from all parts of the body, the connections between the arriving and the departing.

Any anatomist could make a blueprint of the spinal cord. A blueprint of a brain would not be that simple, though there are plenty of what amounts to blueprints in dissecting guides and other books. These are apt to show us the long-known details, and possibly some recent detail, because while the pendulum swings, the anatomist adds, a little.

If he has found a lush brain tumor, that enamel pan with the slabs would have been sent to the hospital photographer, the photograph filed, and a pennyworth added to the anatomy of disease, hardly to the design of the brain.

An architectural blueprint could instead have been based on the brain's development in the fetal skull, which would have been based on the brain's slow development in the species. That loftiest part of each of us began in a microscopic line of cells. The cells multiplied. There resulted a narrow, longish plate of cells. To either side of this the cells multiplied faster till there were two walls with a narrow, longish trench between. The trench deepened. The top was covered over with cells. Thus did the trench become a tube.

Down sank the tube. While it did, the embryo's back was covering over the tube, and over everything in its neighborhood. At the head end of the tube were the beginnings of the future brain, at the tail end the beginnings of the future spinal cord, most astonishingly odd shapes at the head end.

That head end now divided into three, with cavities connecting the three. Development was relentless. Differentiation was relentless.

The first of the three divided, the second did not, the third did. So there were five. Despite folding, despite the tight engagement of the baby's head in the mother's birth canal at the time of birth, nowhere were the cavities obstructed. Latin names were given to the five. *First ventricle. Second ventricle. Third ventricle. Aqueduct. Fourth ventricle.* Ventricle means hollow, a little belly. During life when the cerebrospinal fluid has been drained from brain and spinal cord, and air injected to replace the fluid, and X rays taken, the darker silhouettes, darker because X rays are less blocked by air, produce what could make one think of some ancient relief pieced together by an archaeologist.

Arteries and capillaries and veins of the brain have thousands of times had X-ray-opaque material injected into them, been photographed, and those photographs too have added to the design that has been aggregating through the centuries. In general, the old Latin and Italian names have resisted change.

Medulla—small and immediately above the spinal cord.
Pons—small and above the medulla.
Cerebellum—large and sticking out behind.
Midbrain—small, the smallest part, forward of the pons.
Interbrain—comparatively large, forward of the midbrain, including hypothalamus and thalamus.
Basal ganglia—various sizes of clumps of cells, some close around the thalamus, some at greater distances.
Limbic system—an encircling core at the bottom of the brain, and holding much of our history and some of our hope.
Cerebral hemispheres—tremendous, on top, outside, all around, enveloping the whole.

Lop off cerebellum and cerebral hemispheres and what is left is brain stem. It could seem a stout umbrella handle. Snap open the umbrella and that would be the spreading of the cerebral hemispheres, but not a thin fabric, of course. In man and in his evolutionarily close relatives there is more spreading, and it is all comparatively new, so the cortex that surfaces it is called neocortex. One can think, if one likes, of those spreading hemispheres as giving contour not only to the human head but to the human mind. The formaldehyde-hardened substance is not easily damaged, the living brain is. So, take up the pretty bambino with care, padre. Nature started this, nurtured this, wrapped this in membranes, boxed it in bone. Care, padre!

MEDULLA:
"He fetches breath"
—Henry IV, Part I

The medulla lies just inside the skull, just above the large hole at the bottom of it. An inch of special flesh, it has great age. Andreas Vesalius knew it well. Swallowing, vomiting, articulation of speech and song, the roguery of the tongue—these and more are entrusted to that inch, such as the control of the beat of the heart, the size of the blood vessels. A cut through the brain stem above the medulla in an animal, or an accurate guillotining above the medulla by a drunken driver, leaves the medulla continuous with the cord, and there may be no change whatever in blood pressure, though changes enough otherwise; but a cut below the medulla, and the trunk of the body and the extremities are deprived of their controls, blood pressure collapses, and the victim goes into spinal shock and may die at once. The doctor is glum when he decides his patient has bulbar poliomyelitis, bulbar palsy, bulbar anything, *bulb* being another name for medulla.

Vesalius published the *Fabrica* and the *Epitome* both in 1543. He was born in 1514. So he was done with anatomy essentially when he was twenty-nine years old. He did accurately recognize the brain as the center of human faculty.

The brain cutter, the modern one, may have reason to examine slice after slice of that inch, the medulla. Being a man who wishes to understand his trade, he may then quietly in his mind refit slice after slice, reconstruct the great amassings of neurons, the embedded cables that run up and down. Those that run up have come from scattered parts of the body toward the spinal cord, passed up the cord, passed into the skull, and now pass through the medulla on to the top of the brain, bringing information. Contrariwise are the lines that come down from the top, through the medulla, through the cord, scatter via nerve trunks to all parts of the body, delivering orders. On the front of the medulla great numbers of those down-traveling lines cross, right to left, left to right. This explains why damage on one side of the brain produces paralysis on the opposite side of the body, as was known to old Hippocrates.

On the back of the medulla lies a part of the fourth ventricle, a shallow, diamond-shaped lake roofed by a thin veil, and under the lake, in the solid substance of the medulla, nerve cells and nerve fibers of many sizes and kinds.

Some of those nerve fibers have an insulation of white fat, some not. A crisscrossing results. Generations of students of the nervous system have shown it to contain clear-cut areas that are centers of vital body control. There are hereabout similar areas less clear-cut as to work and less clear-cut as areas. They make a reticulum—called reticular formation—that we are meeting here in the medulla for the first time. We shall meet it again higher up, because it goes upward in the axis of the nervous system, and also downward into the spinal cord. This formation has received close study, and new light has been thrown onto some of the ways the nervous system works. Sleep and waking have become somewhat more comprehensible; some think mind more comprehensible.

Among those clear-cut centers of the medulla is the respiratory center. It too is built of such a reticulum, latticework of cells and fibers. Here is the central exchange, where it is determined exactly how deep the next breath shall go, exactly how fast. Galen, second century A.D., bold vivisectionist, slashed through between spinal cord and brain, paralyzed breathing, knew therefore that this center must be above where he slashed.

Italy was on all sides fascinated by structure, yet Andreas Vesalius, professor of anatomy in Padua, though he did at so many points clarify the anatomy of the living, did not know the respiratory center. That knowledge still had to wait. A time would come when anatomists and physiologists would seem with a ruler to be measuring off a plot under the floor of the fourth ventricle and saying: "There." Later they would again be less dogmatic and less diagrammatic. They would think the respiratory center reached over a longer length of brain stem. They would recognize that its action was less simple than they thought, indeed exceedingly delicate. A cloud of molecules of a gas moving via the blood through the flesh of the medulla would determine that next breath. A drink of water, a laugh, a sigh—each would modify the next breath; a thought could, a dream, especially a nightmare, as most of us have experienced. In this nervous flesh of the medulla is the control tower for the rhythm of our breathing.

PONS:
"Come you from the bridge?"
—*Henry V*

Just above the medulla is the pons, second piece of the brain stem, again one inch. Spinal cord . . . medulla . . . pons . . . We are on the way up. All of it has been ancient brain. All of it is roughly the same in all mammals—cat, monkey, man. All of it belongs to the part that has been spoken of as reptilian.

Pons means bridge. A broad band of curving flesh builds this bridge. Seen from the front with the naked eye, it reaches from one side of the brain stem to the other, the ends disappearing into the masses behind. The broad band is nerve fibers. There is a graceful rising from the one side, then a falling toward the other. The full name is *pons Varolii*. Varolius was a professor in Bologna in a brilliant period of the sixteenth century. He was physician to Pope Gregory XIII, reached thirty-two. Seems a short time to have lived, but time enough to make many studies.

Pons is next above the medulla, the next piece, and through it run those same nerve lines from high to low, low to high. On its back is that same fourth ventricle, that diamond-shaped lake, and under the floor of the lake there is still some of that center for breathing.

The chief pathologist, having with his scissors completed the snipping of the pairs of nerves that pinned down the spinal cord, continued into the cranium, snipped pairs there too, and since he was a systematic man, which a pathologist is apt to be, he began with the lowest pair and continued to the highest. Because the pairs now were in the cranium, they were called cranial nerves, twelve pairs. The lowest of the twelve is the nerve that gives orders to the tongue. Then, a nerve that goes to two of the powerful muscles of the neck. Then, a cable that distributes its branches to the throat, the windpipe, organs of the chest like the heart, organs of the abdomen like the stomach. A nerve for hearing comes out where the pons joins the medulla. A nerve that controls facial expression. A nerve for sensation in the face. For swinging the eyeballs. For blinking. For chewing. (We have left the pons behind us.) For seeing. For smelling.

When finally the pathologist had entirely unpinned the brain, and had lifted it out of the skull, those pieces that were hanging there were the amputated stumps of those twelve cranial nerves.

A dangerous place to be damaged, the pons. If a child has run into the street into a truck, and if a hemorrhage has torn its way into the pons, there are frightful signs, bizarre paralyses. From human misery, learning sometimes comes. From deliberately produced animal misery, learning sometimes comes too. The pathologist and the anatomist and the physician and many researchers have forever in their minds, with help from their hands, been fitting together the crashed or slashed nervous system. It is their business.

Despite the charm of life, a man may think he has had enough. The suicide who prefers the mouth route rests his revolver against his palate, aims straight back, blows out everything in the neighborhood of the fourth ventricle. To dissect with a revolver is not real neat. If he uses a shotgun, he can aim much more casually, just north, blow off his head.

CEREBELLUM:
"It is too little"
—*Two Gentlemen of Verona*

Kleinhirn is what the German calls the next part. The word means diminutive brain. We call it *cerebellum*, and that word too means diminutive brain. When anatomy was young, anatomists enjoyed attaching Latin names to every nook of our body.

The cerebellum is a wedge, looks like a separated piece tucked under the mushrooming large cerebrum, peeps out. Immediately in front of it is the brain stem.

In weight the cerebellum is one-eighth of the brain.

If one places one's finger high on the back of one's neck, the nape, inside there is the cerebellum. Three diminutive feet join it to the rest of the brain—peduncles, the godfathers called those feet. Through them run nerve tracks leading to and from the rest of the nervous system, which means spinal cord, brain stem centers, cerebrum. Everywhere in the nervous system there is a leading to and from. Wealth of connection makes the body what it is.

The godfathers looked over the surface of the cerebellum and saw that it was everywhere grooved, that one groove dug more deeply than those near it and therefore more definitely separated one lump from another, producing apparently independent organs, small lobes within this small brain, this *Kleinhirn*, and everywhere over that surface they sprayed names that stuck. For instance, a longish roll of brain that runs down the middle between the right and left cerebellum suggested worm, its name therefore becoming *vermis*. It must have been a satisfaction for the godfather who thought of that. To either side of the vermis are the lobes, and the lobes have gray and white coloring, the gray, nerve cells, the white, nerve fibers.

If a thin-bladed knife slices through the cerebellum, on each face of the slice one sees a decoration that suggests tree, *arbor vitae*, tree of life.

The cerebellum often has confused the medical student, because when he lops it off, hardened as it is in formaldehyde, and holds it in the palm of his hand, flips it about for a minute or two, he no longer knows top from bottom, front from back, and that might have embarrassed him in the dissecting room when the professor of anatomy quizzed him, with the other students listening. He will forget the embarrassment later when he is a physician and his gluteals are tired from driving all afternoon seeing patients, or sitting all afternoon in his office doing the same.

MIDBRAIN:
" 'Twixt which regions"
—*The Tempest*

The course upward has carried far and deep into the brain.

Medulla . . . pons . . . cerebellum . . . midbrain . . .

The midbrain is forward of the medulla. Medulla was an inch. Midbrain is less. It is the smallest division in that original design. If instead of an adult corpse the brain cutter had gotten for himself a several-month-old dead fetus, and if he had cut open the back of that skull, he would have seen four little hills at the corners of a small square, colliculi. In the fetus the four little hills loom large, but in a few months they will have sunk quite out of sight. In the adult corpse, to find the four little hills the braincutter must take hold of the brain stem and pull it away from the brain above. There they are, four. The two below are junction points for hearing—we hear a sound and automatically turn eyes and head that way. The two hills above are junction points for seeing—a sight is half-seen and we automatically turn eyes and head in order to see all. The recent evidence is that these colliculi are concerned more exactly with changes in the *rate* of movement.

Ask now that the dissector jab into the solid parts of the midbrain. He will come on an island of nerve cells reddish in tint, called nucleus rubor. Another is blackish, substantia nigra. Here again is the reticular formation—in the midbrain now. Special work is assigned to a variety of aggregations of cells of the midbrain: to wit, the control of movements of the eyeballs when we look from far away to near, the control of the movements inside of the eyes when we look from a black cloth to a glaring incandescent lamp.

The dissector cuts into a narrow canal. That canal is the aqueduct, and through it the cerebrospinal fluid creeps from third ventricle to fourth. Vesalius knew this. We do not know all that he must have known, besides the anatomy, about the function of the parts of the brain. He with Paré, the great surgeon, was at the bedside of Henry II of France when that gentlemanly king died of a wound inflicted by a lance broken during an encounter at a tourney, the metal passing through the eye and into the brain. Henry II had spells of consciousness and unconsciousness during the ten days that he survived, and the surgeon and anatomist both would have been talking, and how interesting it would have been to listen to that talk.

INTERBRAIN:
"And underneath that consecrated roof"
—*Twelfth Night*

It was said of Agassiz that he gave a fish to a student, did not tell him what fish or what waters it came from; the student just had to find out everything with his native powers, and away with the fish went the student, and being a Harvard student returned with a description that surprised the fish. The reader of these pages has not been given even a photograph, not any pictorial help for a pictorial subject. If he still is reading, it can only mean that he is possessed of a patience Agassiz would have admired. The reader has read himself up the brain stem.

The brain stem, for all its importance to you and me, is small. It looks unexpectedly inconspicuous if one has the chance, rare, to view it toward the bottom of some living human skull during surgery.

Just inside the skull, just above and out of the hole in the vertebral column, was the medulla. The pons was next above. The midbrain was forward of that. The cerebellum was tucked in behind. And now it is the interbrain. Spinal cord . . . medulla . . . pons . . . cerebellum . . . midbrain . . . interbrain . . .

The interbrain is at the top of the brain stem. In it are the controls for most of the body, controls for some of the head. Two of the interbrain's parts should get attention: hypothalamus and thalamus. Also, the basal ganglia hover close. Also, these parts are more and less anatomically fused.

Hypothalamus. In our day the hypothalamus has been actively studied and some new importance found with such regularity that one has ceased to be surprised. The reader, to get his bearings, might do worse than fancy himself a surgeon who had picked his way along the outside of a human brain—a child's with a fast-moving tumor—and come down toward the bottom of it. There he sees the midbrain. He is not concerned with that now, not with those four little hills, but is concerned with two little breasts, mammillary bodies. Mammillary → mamma → breast. The mammillary bodies are part of the hypothalamus, and the hypothalamus is one of the more manifest parts of the interbrain.

So far as the relation of the hypothalamus to skull goes, it is situated immediately above the middle bottom. Forward of it is the crossing of the two great nerves for the two eyes. And between that crossing, in front, are the mammillary bodies. Weight of the hypothalamus? Approximately one three-hundredth of the total brain.

A surgeon who is operating in this neighborhood, under local anesthesia, presses lightly, with cotton, but it is still too heavy, and his patient loses consciousness. A breath may affect the living brain, so sensitive it is, yet let death come and the pathologist will promptly proceed with his rough work. "Why does he suffer this rude knave now to knock him about the sconce with a dirty shovel, and will not tell him of his action of battery?" Duller men than Hamlet have thought similar thoughts when they saw a brain knocked about on the scrubbed morgue table. Instead of the knáve's dirty shovel, the pathologist uses a clean knife. Down through the dead brain he goes and, at the middle bottom, goes through the hypothalamus.

The thalamus sits on the hypothalamus. In the classic world, from where that word *thalamus* was plucked, it was the inner chamber of the house. In the brain it is not a chamber but is inner. Two thalami, a right and a left. The two are grossly fused for part of their length, and fused also with the brain flesh around. Nerve fibers enter the thalamus from every part of the body. On their way they side-branch into the reticular formation, that core of the brain stem.

Each thalamus has two sides, has a back, a front, a bottom, and a smooth and rounded top. That top of the brain stem is the floor of the first and second of the four ventricles, those caverns filled with fluid. During life, the clear cerebrospinal fluid fills the caverns, and in it are strewn a seaweedlike material richly supplied with capillaries. Harvey Cushing, eminent brain surgeon, actually saw fluid oozing from those capillaries while he was operating.

From this waterway through our brain we might get yet another idea of the brain's geography, as a visitor might sail along the Thames and get an idea of London. Spinal cord . . . medulla . . . pons . . . cerebellum . . . midbrain . . . hypothalamus . . . thalamus . . . basal ganglia . . .

Those last, the basal ganglia, are scattered about the top of the brain stem, but also down in it, islands of nerve cells, lumps of gray, sometimes interlaced with white. What lumps to include among the basal ganglia has been debated. Perplexing to locate. Perplexing to explain. Perplexing names. Caudate. Putamen. Globus pallidus. Amygdala. Claustrum. Like names on Welsh railroad stations, except these are Latin.

HEMISPHERE:
"Grey vault of heaven"
—*Henry IV, Part II*

The roof brain. We have climbed there now. Top of the house. Human cerebral hemispheres with their cortex. These have waxed from mammal to mammal and culminate in man. They have long been regarded not only as each human being's personal crown, but as the crown of evolution. To be sure, it was the human mind that did so regard them.

In the embryo the hemispheres began as head-end buds. The buds grew. Luxuriantly. For a brief time in embryo and fetus, those four little hills of the midbrain still peeped out behind, but then they were covered over entirely, the cerebral hemispheres with their plating of cortex overgrowing everything, and while they were growing, they were keeping close to the inside of the skull, which was growing also. Those hemispheres were building a dome. They were completing with a dome the edifice of the brain. What portion of brain to designate cerebral hemispheres is roughly agreed upon. What the immensity of those hemispheres can be in the higher creatures, as against lower creatures, is recognized. And what the immensity may mean is in some respects recognized also. Gray on the outside—nerve cells. White underneath—nerve fibers.

When a hemisphere is surgically removed because of disease, one-half of the barn of a man's skull vacated, whoever sees that sight never forgets it. Louis Pasteur suffered his noted stroke, lost most of one hemisphere, but appeared, judging by the great work he accomplished, not to have lost any mind. Possibly it would be allowable to speak of the two hemispheres as the vault of heaven. They did invent that other heaven, gave it its shape, the earth child's vault. Kepler, great seventeenth-century astronomer, believed that it was simply because of the hemispheric hollow at the back of our own eyes that we see the sky as a hemispheric hollow. The heaven, that "vision of fulfilled desire," was a hemisphere up there because of an item in the geometry of our anatomy down here.

LOBES:
"Heaven's vault should crack"
—*King Lear*

Like a monkey, a man will crack open anything, find out what is inside; crack open a skull, and there is heaven's vault, the brain. He has studied the grooves on the exposed side of it, established landmarks among the grooves, and these have aided him in parceling the brain into lobes, and then over years he has investigated into the lobes to learn what each does for us.

The first of the grooves appeared late in the life of the fetus. In the fourth month of fetal life, that cortex still had had a smooth unreality, but then came the grooves, shallow ones, deep ones. Say that the seven days of the Creation are too amazing to believe, but no more amazing than and rather like what is taking place in every sleeping fetal head. Is it sleeping? Is that sleep? Waiting to be awaked?

A deep ditch—*longitudinal sulcus*—runs from the front to the back of the brain, dividing it into the two hemispheres. A rugged furrow—*sylvian fissure*—cuts in on each side at about where sideburns begin. A decidedly less rugged furrow—*central fissure*—takes off midway from the top, the crown, runs diagonally downward and forward. So, boundaries dividing left from right, dividing high from low, dividing front from back. There are others, but those are the large ones. It is they that have parceled man's brain into lobes.

Frontal lobe. Parietal lobe. Temporal lobe. Occipital lobe.

One of each of these is on each side of the billions of heads. In late years there have been other regions of study. There has been talk of the limbic lobe, also of the limbic system, supplying a cortex over that old reptile's brain, and still down there in ours. It lies middle, low. Some of what is considered limbic by one scientist is not by another. It is still a dividing to encourage us to keep climbing, keep trying. Much has been learned. The dead have and do and will enlighten the living.

Frontal lobe—everything in front of the central fissure. Parietal—everything above the sylvian fissure and back as far as a vertical line drawn an inch and a half forward of the brain's rear tip. Occipital—everything behind that line. Temporal—everything below the sylvian. Limbic—the ring encircling the brain stem, for twenty-five years thought to deal with the emotions, with things ancient in the mind of the creature.

Such scholastic cracking of heaven's vault ought never to let anyone forget that the brain is one, that the closest we ever come to one world is one mind, and underneath one mind somehow is one brain.

Franz Joseph Gall (1758–1828) was the first brain surveyor to make surveying fashionable. In the century and a half since his death, that surveying never has flagged. Phrenology was started by Gall. Phrenology claimed that the bumps and shallows on the outside of the skull could predict the strengths and weaknesses of the Johnny inside. An observing barber should have been good at this. Phrenology was sure to sell well, because it was easy to understand; the customer for his money had something his fingers could touch. It was ABC truth. Gall was driven from Vienna as a quack, honored in Germany, died rich in Paris. Goethe was impressed by him but laughed at him. Flourens had no faith in phrenology but admitted he never truly looked at a brain till he saw Gall dissect one, and Flourens was a great neurologist. So, Gall could dissect. Quackery mixes with excellence often. Grooves on the surface of this flesh had significance to Gall and to Flourens, would get more and more significance, and latterly, chemistry would join in.

ARCHITECTONICS:
"A rich embroider'd canopy"
—*Henry VI, Part III*

Vesalius was dead two hundred years when an Italian medical student noticed something that previously had gone unnoticed. This was the year when across the Atlantic gentlemen were affixing their names to a declaration, the year 1776. What the medical student noticed was a white line near the tip at the back of a freshly removed brain. Even today when one knows beforehand that that line is there, it may be difficult to see. It is white against gray, fibers against cells. A landmark. The general region of brain is where arriving nerve impulses some-how give to a mind sight. Sixty-four more years, and a French psychi-atrist reported that the white line was two white lines, with a dark line between. Later the microscope was brought in to rove over the region and over the entire cortex, that rich embroider'd canopy. It was proven now to be teeming with detail, all previously unnoticed. The cortex of the brain had layers. There was a layer with few cells . . . A layer closely packed with small cells . . . A layer of medium-sized cells . . . A layer . . . A layer . . . Whole lives would be devoted to scrutinizing that frescoed vault, no doubt sometimes with the secret hope of finding what in the structure of brain might be the basis of mind, that mystery, find what in anatomy might be transmuted into thought and feeling, in order that we might perchance more securely grasp what is at the foundations of us, find the lost language on the buried stone.

In a single year three independent publications divided the cortex of the total brain into small plots based on local cell and fiber arrange-ments. Maps began to appear. Cartographers opened shop in many countries. Areas of the frescoed vault were given numbers. Area 4. Area 6. Area 44. One cartographer said there ought to be twenty. Another said fifty-two. Another, more than two hundred. Under the microscope these areas were different, and that was very likely because they had different work to do. Some cartographers were beginning to say that other cartographers had divided too much. Too many maps. Too many small plots on each map.

The labor had probably taken the anatomist as far as anatomy could. This quintessence of architecture had probably reached the limits of what it could reveal. The secret of the vault must henceforth be sought in other ways.

XV

SENSE
Message from the World

ANALYZER:
City Desk

Monday morning.

Sounds, sights, smells tumble down from that second floor. She slams the icebox door twenty times an hour. And the January wind rattles the windows. Drafts climb in over your shoe tops no matter where you sit. Everybody knows Monday morning. Stomach stretched with Sunday dinner. Monday's socks still sensing Sunday's newspapers all over the floor. Muscles numbed from no exercise. Day of rest? In-laws just dropped in—dropped. If only spring and baseball and the violets would come. Damned wind, a beat in it, since eleven o'clock last night. On a Monday your heels jar your head, and you do pay the five senses a wry respect. Praised be Jehovah that not any more gets into us than does, that each sense has its limits, its walls, its thresholds.

But our walls we do have. We are mathematically walled. Touch, heat, cold, pain, taste, sound, sight, smell—each has a minimum before the attack from that sense begins, each a maximum where the attack shades off into sense-nothingness. Were it not so, this frail construction that each of us is would long since have been banged to death. Alerted body and head must be, perpetually, peace-less in order to have peace.

The stimulus?

Size—must be big enough. Intensity—must be strong enough. Duration—must last long enough. Rate of change—must increase or decrease fast enough.

A curious fact—each sense is able to adapt. I am permitted to tighten my belt to the third notch, for a few seconds feel the pressure, then the pressure adapts, and I tighten again, feel again, adapt again, and if

I am fool enough to keep that up to the point of pain, the system poorly adapts to pain. I am not permitted to become indifferent to pain.

Receptor?

Anything that receives a stimulus. A receptor is any living instrument, chemical or physical, that picks up any kind of news, from outside the body, from inside the body. It is an antenna. Myriads pick up news. Large populations are packed close together in places like ear, eye, nose. What the receptor does is to invite in one kind of energy and keep out others.

Analyzer?

Anything that analyzes the sentient world. A young neurophysiologist with a microelectrode in his hand may shiver at that definition. He knows how inexact. Admittedly *analyzer* is less exact, more general, more inclusive than *receptor*, more nervous system involved. Besides receptor and the news wire that carries the message to spinal cord or brain, there is the often baffling wiring that handles the news inside cord or brain.

The term connotes also why we creatures lug about with us this equipment of radios, transistors, cameras, to sense a universe that must for the purposes of body and head be ceaselessly analyzed, each sense in its manner, with its limitations, its ends. The higher the animal, the more comprehensive the analysis. Sheer survival may be the whole evolutionary intent of this built-in machinery.

SKIN, SENSE ORGAN:
At the News Front

Our skin, that outer barricade, is crowded with receptors.

An ordinary microscope easily sees these sense organs.

A thin slice from the tip of a man's amputated finger is stained, and the sense organs are observed. One looks as if it might register weight, and does. Another reports heat. Another cold. A still more definite one, large, lies almost at the center of the illuminated round field. Appears to contain a jelly. Sunk in the jelly is a nerve fiber, the jelly probably protecting the nerve fiber's membrane in such a way that it may safely be stretched, and, consequence of the stretching, a rise of electrical potential and nerve impulses starting conventionally toward headquarters.

At almost four o'clock of the illuminated field there is a Pacinian corpuscle, famous, named for Filippo Pacini, who first described it. It is enormous, two millimeters long, visible without a microscope, has been widely experimented with because of its size and its comparative isolation from other sense organs. There is space around it. It is found where pressure must be appraised. That would be particularly toetip, fingertip, joint, tendon. A tendon moves, presses, the Pacini is stimulated, the news telegraphed.

Many different shapes of receptors are in the skin of that amputated finger. Bulbs. Disks. Cylinders. Beaded nerve endings. Some are more bizarre than others, some merge into others, some become unrecognizable by the merging, dissimilar similarities everywhere. The specialist freely admits that he often is not sure just what he is seeing, let alone what is being reported, and whether the report from one part of the body by the same type of receptor is the same as the report from another part. He states what he sees, and waits.

The skin of an abdomen may have only a few hairs. However, each hair is sunk in a moat, and the moat is crowded with receptors. Whiskers around the mouth of a rat bring in news from distances twice the diameter of the rat's head. Deprive that rat of whiskers, it dies. It needs its whiskers.

The message must get from the person's skin to headquarters, to the central nervous system, to the brain. An energy—from near or far—meets the skin.

Touch paths, heat paths, cold paths—all of the skin paths have a length that depends on the body's geography. A path may start farther away than a man's ankle, travel in a trunk line up his lower leg, upper leg, pelvis, spinal cord, cross to the opposite side (can cross higher), thence through the brain stem, reach its destination on the outside of the brain, the cortex. The sensory cortex. That destination is a point in a strip of that cortex, of what might rightly be considered a most marvelously frescoed surface, the strip being about midway between the front and the back of the brain. Maps of the sensory cortex have been produced in three ways: (1) In human beings, successive points have been stimulated with electricity during surgery, the patient under local anesthesia and able to state what and where he is feeling. (2) In an animal, a small area has been damaged, the animal not able to state, but observations made on what it does that it did not do before. (3) In an animal again, a monkey, the skin has been touched at successive points with a pin, or with a feather, the electrical consequence picked up from the strip of the cortex from the point that corresponds, the point recorded. A map made in that manner has taken twenty hours, the experimenter needing to keep hustling because at the end of the twenty hours the monkey was dead. The maps produced in these various ways indicate that the maps produced in man correspond to those produced in animals.

An experimenter studied the cortex in a pig and found that the snout does the message sending. The four chubby short legs, the barrel belly, immense though these are, had no space allotted, or none was discovered. Farmers verify this in their own way, claim the snout is the most intelligent part of a pig, that the pig digs, explores, eats with the snout. Also, the pig is sensitive in the skin of the snout, and farmers apparently know where to deal with a pig if it gets piggish.

Our skin, our sometimes tired yet usually willing skin, that barricade with its sentinels, is with us early and late, and mostly we pay no attention to it. A warm bath caresses the mind with the smooth water and the friendly temperature; then without warning there is an absentminded turning of the wrong faucet, and the soul is slapped with an arctic blast. What dozing skin receptors, dozing skin paths, dozing analyzers flew into action at that instant. A warm bedcover lies innocently upon us, our skin receptors adapted, hardly reporting the touch and pressure, we soothed toward sleep, when a crumb of last night's snack uncouthly awakens a crowd of the sleeping sentinels, instantly whole analyzers in action. Everywhere coded messages are traveling from the outposts of a belligerent who is defensive when he is forced to be, offensive the rest of the time.

PAIN:
Complaints

Why do we have pain? Why do we have this sense forever waiting? Why? Is it only Nature's determination for her own purposes to protect the living creature that can take only so much damage? The creature must be forewarned. Pain is preventing destruction from going too far too fast by making a quick announcement whenever anything has begun threatening the flesh, and sometimes almost always is.

An Irishman, a man who perhaps lived not wisely but well, would not pay his pain so much as the courtesy of recognition. His doctor believed the pain frightful, shook his head. "A patient man." His wife objected to that. "Why shouldn't he be patient when he has brought the whole thing on himself?" She was German. His mother-in-law endured the pain of her cancer in a silence that accused all those around her of their good health. She was German too.

Pain is forgotten once it is over. That often strikes us as remarkable in ourselves, though commonly it does not strike us at all. We just forget. A mother screamed while the head of her baby was pushing open the exit from her birth canal, which women say is a special pain, yet the next day she remembered clearly only that she had had a bad time. Some kind of thrifty economy of our mind requires that.

A tall woman with syphilis of the spinal cord had the pain of her disease for twenty years. Drugs did not relieve her, nor surgery. Poor skin-and-bones, pain had used her up, and she groaned, moaned, squirmed, feebly. She would have been worse off had the pain been present for twenty years and she had not moaned, been muted by it, as happens. A man cried from his open window into the street when a kidney stone passed through his ureter, and though his ureter was the same as anyone's, and the cause of the pain the same as anyone's, namely drastic distension of a living tube, yet even that pain, that choice variety, as every doctor knows, will in some definable respect be that man's own. It has taken on his character. A big-bodied man let out all the vowel sounds in the English language while a gallstone crept through his gall duct, but even to an ordinary ear his vowels were his vowels. Each of us does think of his pain as his own, nurses its privacy, may exaggerate in speaking of it, may minimize.

Every doctor accepts how roundabout he may have to be to find the facts of a pain. It is part of his routine. It may be talent. Does the pain throb? Does it burn? Does aspirin stop it? Where is it? Is it? Is it pain or is it only unpleasantness? Is it nothing at all? Is it not on some other subject? What is the subject?

Headache, that old hag, has done its part to dampen joy, discourage Pollyannas, defeat psychiatrists, dent sobriety. It may start from teeth,

sinuses, eyes, muscles of the neck, membranes of the brain, hemorrhage in or around the brain, harassed blood vessels inside or outside
the skull, many flesh causes. Then there are the mind causes, the mind
as elsewhere employing anatomy and physiology.

Trouble in a man's life may be the cause of a specific headache,
migraine, that is nauseous, usually quick in onset, usually accompanied by what briefly might seem serious neurological signs, usually on
one side of the head, attacks recurring over years; then he has gotten
old and everything has lessened, and he may not even know when the
migraine finally departed. A few physicians still insist that all migraine
is invariably an allergy.

From the skin inward to headquarters, the path for pain is much
like other skin paths. Two questions arise. Does the pain path of itself
tell us the location of the pain, and does the pain path explain referred
pain, where the cause of the pain is in one place and the sensation in
another?

Beaded naked nerve endings, no insulation of fat, are the pain
receptors in the skin. Soon the nerves lead into treacherous-looking
nerve nets that overlap and underlap. There are two sizes of nerve
fibers. Large fibers carry the bad news swiftly, are responsible for the
brilliant burning of pain. Small fibers carry the news slowly, are responsible for keeping the pain hot. Together they sometimes make a
Dante's Hell. The slow news reaches the top of the nervous system
instants later than the swift.

When the cause of a pain is below the head, the path leads toward
the spinal cord; when in the head, it leads into the brain. The paths
cross, right to left, left to right, then continue centrally into the nervous
system: medulla, pons, midbrain, thalamus.

Part of our bodies are less equipped to experience pain than other parts, fingertips less, but fingertips are more equipped to experience touch. Arms and legs at their junction with the trunk have
four times more pain points than touch points, and, whatever the
explanation, the creature would then acutely know when destruction was starting in the direction of its vital organs. The surface of
the eye, the cornea in the middle, as every person has experienced,
is highly sensitive to pain, and in that middle are found the naked
nerve endings. Recently, by testing the cornea with nylon threads of

different coarseness, it has been possible to excite touch separately from pain, so there must be touch receptors there. Jets of air of varying temperatures directed toward this area have shown it sensitive to temperature.

So, the cornea senses pain, touch, heat, cold, but has only naked nerve endings. Naked nerve endings must respond to several kinds of stimuli. Furthermore, in the skin, pain points and touch points and other points appear to change their localities from day to day.

Pain paths exist that will not be called upon once in a lifetime, nevertheless stay vigilant.

What explains referred pain? We all experience it sooner or later. The heart is ill, but the left arm aches. Physicians every day make diagnoses with the help of pain that has been referred to the surface from organs inside chest or abdomen. A theory is that the paths from the damaged organ and from the healthy skin reach the same junction in the nervous system; this junction is barraged by stimuli from the organ, so that the mildest stimulus from the skin, which ordinarily would cause no sensation, now does. Brain and mind have had much experience with the skin. The species gained by that. An animal was guided by pain in the skin to pluck out a thorn from exactly where the thorn was. The species would probably not have gained had the animal been able with equal exactness to locate pain in an inner organ; there would have been little use in distinguishing an abscess of the liver from an abscess of the pancreas. Vague pain might have reduced the animal's activity, and that might have been useful, whereas a geographic locating of pain in a deep-lying organ might only have inspired the animal to lick and gnaw and bite and increase the damage. We know how animals bite at themselves with the greatest enthusiasm. Sex often has excited them to a wild activity in spite of painful destruction of the flesh. As for the top of the top, the cortex, possibly it is not as much concerned with pain as has sometimes been thought. Assuredly, however, it has been concerned with teaching the human being how to alleviate pain, how to write a prescription, how to extract and concoct and administer morphine. The extracting of roots, seeds, leaves, and flowers must have been among man's earliest skills, calling on his highest nervous-system capacities.

SMELL:
Gossip Column

On Easter morning, moving in the direction of Saint Patrick's Cathedral on the other side of Fifth Avenue, there went a small bouquet of violets. When the smell particles had crossed the avenue, they were so diluted that not even the micromethods of the modern chemist-physiologist could have proven that anything noseworthy had walked that way to the Easter service, but it had. The blind man at the corner knew perfectly.

Methyl mercaptan, when diluted in the proportion of one molecule in fifty thousand of breathed air, registers garlic. From that it follows that if anyone dislikes garlic, it is always spoiling his life, whereas an admirer can tell whether it was three times or only twice that the chef waved the clove over the salad bowl. A delicate sense, the sense of olfaction, truly. Artificial musk registers when the concentration is four one-hundred-thousandths of one gram in a quart of air. Skatole, putrefaction molecule dancing in human residue, reports at 0.4×10 to the—6 milligram liters. Oh, well. Animals often have saved their lives, successfully battled through brute history, because they were specialists in the manufacture, the recognition, the study, of excrement. They are dedicated students. They and we share the one world together, sty and laboratory.

To the smells that are ours, many, lumped this way by this specialist, lumped that way by that specialist, add the four primary tastes—sweet, salt, sour, bitter—to give the flavors, and the possible sum of that comes to such a richness that it is a complete mystery why cooking can be so bad. In themselves the tastes are dull things. One would check them off as mere background were it not that they had other importances, help our automatic selection, lure us to substances our body requires—table salt, calcium, vitamins.

Taste buds are the receptors, distributed over the mouth, throat, tongue. As we get older, they atrophy, but they do it so gradually that an old man has to be taken aside and told quietly that, really, he is not tasting much of anything anymore. He thinks about that for two minutes. Then he agrees. It might hurt his feelings if he were told too suddenly that he also was not smelling much of anything anymore. (That last has advantages.) On the tongue the taste buds make a pattern.

Salt and sweet are mostly at the tip, bitter at the back, sour at the sides. The taste pattern travels inward over taste paths; meanwhile the smell patterns are traveling over the smell paths.

Insects and vertebrates were evolved over widely different routes, yet have taste talents and odor talents that sometimes are quite near, or apparently. Their sense world is forever closing in upon them too. The minimum of sugar that can be tasted is nearly the same for bee, butterfly, man. Bees have been used by man to investigate the so-called stereochemical theory of odor, and have been claimed able to recognize the same primary shapes of molecules as we do.

The smell receptors are in the olfactory membrane, a sheet spread out about a square inch in size in the high attic of each nostril, protected there from bacteria and dirt and cold. A few odorous particles rise up, warm as they rise, dissolve in the fluid that bathes the receptors, the mind gets a nudge, the nose gives a sniff, many particles accordingly rise, the mind is unmistakably informed. The color of the membrane appears of some significance. Negroes have a dark membrane and great sensitivity to smells. Albino animals sometimes are poisoned because their smell is poor and they eat what they should not.

For the rabbit, smell substances that are soluble in oil are said to stimulate the back of the olfactory membrane; substances soluble in water, the front; some degree of solution in oil or water always deemed necessary. The rabbit is frequently used for experiments in olfaction because its smell system is easily approachable surgically, the overlying bone chipped off the system, and the parts of it more easily separable than in other animals. All questions can be studied more singly. Also, much learned from rabbits appears true for man.

Blunt naked nerve ends equipped with five or six hairs poke through the olfactory membrane. Each nerve has two parts, two poles, one directed toward the cavity of the nose, one toward the brain. Thousands and thousands of such make a bundle, the bundle passes through a perforation in the skull, entering a longish bulb, named the olfactory bulb. In smell-talented animals this is enormous, has two principal types of cells. Is a bloody area. Is actually a small brain. Cajal studied it microscopically, and in late years it has been studied electronically and with other stimulating energies. A rhythm comes off it.

Up into the nostril sneaks a smell, and that starts something electrical. One rhythm might consequently be playing into another rhythm, and smell pattern after smell pattern advance into the brain, and a man be seen to look around him and say he has a haunting thought and does not know where he got it. The thought departs. The smell adapts. The adaptation could occur either in the receptors in the nose or in the depths of the brain.

Dulled for skatole, the sense of smell still would be fresh for Chanel No. 5, then would adapt, would dull for Chanel No. 5. Who has not started into a roomful of friends, been knocked back, continued on in because he was polite, soon found himself happily panting with everyone else.

Far far back in time and in the old old world these senses were evolved in the seas and rivers where the one-celled as now was lured by food. When this lure acted from a distance, that was probably the ancestor of smell, and when from nearby was probably the ancestor of taste. Their dinner floated past those hungry ones, and they partook.

It is stated that if a wad soaked with meat juice is held to one side of a catfish, it snaps to that side, but if the pertinent nerves are cut, it no longer snaps, nevertheless continues to distinguish salt from acid, and this would suggest that the catfish employs, let us hope also enjoys, both smell and taste, therefore enjoys flavors, as we do. Some insects walk in their food, have receptors on their feet, that is to say, smell with their feet, or taste with their feet. A fly has taste receptors on its feet, and if a left foot were not to pick up a strong enough impression, the right foot would reinforce the impression. Some insects have taste receptors on feet, on antennae, in the mouth, and their world should be rich. To a male eggar moth, a female eggar moth is recognizable at two miles. He presumably thinks of her in smells. A male dog communicates with a female dog in smells. To a hound and a bitch sitting together in the sun on the porch, the fresh earth of spring is a Haydn trio, quartet, symphony, whatever their word for it is. She slightly turns her head, he turns his, both heads in the best position for the smell music.

A human being has a nose not as acquisitive as a dog's, or apparently a moth's, a shark's, nevertheless quite adequate to convey the quality of a steak, a spray of jasmine. Also adequate to lure a small boy to a peanut stand, his father to a house he meant not again to visit, his grandfather at the beginning of the century to a midnight bakery, coal-hot ovens from a cellar that filled with odors that city street, right by the side of a stagnant canal.

Neurologists are concerned with odors because they lead sometimes to diagnosis, a frontal-lobe tumor sometimes detected because of a changed acuity of smell. Brain surgeons say that the odors awakened by stimulating the brain usually are bad. The family doctor is concerned with odors. The garbage man is. Artists are. Maupassant enriched many a page with an odor. Baudelaire did. Possibly the air around those Frenchmen was olfactorily more enticing than that around us, but possibly their noses were more sensitive, more disciplined also, though neither as sensitive nor as disciplined as the noses of their dogs digging in the dank French soil of April.

WAVE MOTION:
News Flash—Tree Fell in the Forest

The stimulus for hearing?

A stir in the air, a wave motion, a wave that repeats, and for any animal society to pay attention to it, it must fall within that society's hearing span. It is molecules of air swept together by some force, a shout, a sob, then immediately, because of their elasticity, springing apart. Swept together. Springing apart. Hence a wave. Textbooks speak of condensation and rarefaction of air, speak of increase and decrease of atmospheric pressure, so-and-so many times per second.

A musician strikes a tuning fork. The prong of the fork moves in one direction, compresses the air ahead of it, leaves an emptiness behind, air rushing into the emptiness. Then the prong moves in the other direction and everything is reversed, that back-and-forth repeating until the energy put into the fork has been spent and this locale is at tonal peace once more. Airwaves are not like water waves, cannot be seen, but the movement of the tuning fork can easily be recorded. One back-and-forth is a cycle, and the number of cycles per second is the fork's frequency. Air's elasticity allows frequencies up to 100,000. Middle C on the piano is 256. The lowest frequency to which the human ear can respond is about 16, lower than the low of a bass violin and heard more as a beat than a sound. Highest in adults is 20,000. In children 40,000. The total span is about ten octaves. Music falls mostly within seven and one-half octaves. Supersonic is any frequency above the creature's range. Dogs hear higher than men. Insects higher than dogs. Bats guide their flying by frequencies between 50,000 and 100,000, first produce the frequency, then listen for it as it is reflected from objects around, a radar that the bat uses with an efficiency comparable to our vision.

Male canaries have singing contests. Male grasshoppers do. It has been reported that if a male grasshopper goes to a man-built telephone and a female grasshopper thirty miles away hears his voice, she stays at the other end of the line to harken to his wave motions for as long as he is in the mood. Parrots hear less intricately than we, but have excellent analytic capacity inside our speech range, and that is partly why a parrot can be so clever about imitating his madam. The ordinary voices of her regular customers the parrot distinguishes as correctly as she.

Where man differs from the other creatures is that he can if he chooses stand off from frequencies, despite the fact that he is always a prisoner within them, make observations upon them, draw science and philosophy and often error from them. When that tree fell in the forest one thousand million years ago, there were the stir, the cycles per second, the spread, the dying out, but no living instrument to receive them. How careless of somebody.

EAR:
Receiving the News

The instrument that picked up the pressure variations in the air, and did it with high efficiency, was of course the ear. Its receptors, in the case of our human ear, are set in a cavern in bone, this communicating with the outside via complicated corridors, and the bone is the hardest of the skull.

Three divisions: external ear, middle ear, internal ear.

External Ear. The pinna is that part of it that we see, that sticks out from the side of the head, has a canal that runs down in, and this we ordinarily do not see. A man cleans his a bit when no one is looking, and a dog cleans his when everybody is looking. A dog has only a blunt paw to do it with, a few fingernails, no effective fingers. The monkey and his relatives have the effective fingers. Add soap and washcloth to the effective fingers, and that is culture.

Cartilage gives the pinna rigidity. Pinnas differ from person to person, may be small, may be large, may be rolled, unrolled, objects of charm, gargoyles. Every woman who uncovers hers knows why, and knows why if she doesn't. A person whose pinnas stick out too far has an early seriousness thrust on him. If he lacks one, was born without it, the chance is he will develop into a sentimental altruist, a superman, or some other kind of psychological cripple.

Man does not especially need his pinnas for hearing, but he needs his ears, and he is better off when he has two. Two help in establishing the direction of a sound. Two ears are useful, too, because they enable him to sleep on one and have the other free to bring him bad news.

Middle Ear. None of it do we see from the outside. It is part of that cavern hollowed out of the bone, is filled with air, has three ossicles inside the hollow, three small bones, and it is separated from the external ear by the eardrum. Two hundred and fifty eardrums stacked one on top of the other would make a stack an inch high, so each must be thin but tough, presents a large surface for the airwaves to pound against or lap against, is convex toward the middle of the head. It is sensitive to anything vibrating in its neighborhood, vibrates with it, provided the vibrations are in its range. Students of the ear have wondered how our mind is spared the annoyance of hearing the flow of blood in our head. That might result from some mechanical block in the flow, some block in the nervous system, some block in the mind, and might also be merely a matter of cues, because only the cues that are cues are cues, to keep up the doggerel. Our mind responds to what it wishes, as does, with some redefinition of the word *wishes*, the nervous system under it. What we can be sure of is that no more than a fraction of the life-related energies of world and universe reaches our nervous system, and only a fraction of that fraction reaches our mind.

The pressure of the air pocketed in that cavern of the middle ear is kept close to that of the outside air by a tube connecting with the throat. When an airplane rises or a train dips down under Hudson River, passengers swallow, open the throat end of the tube, so equalize the pressures on the two sides of the eardrum, making those heads more comfortable again.

The three ossicles, small bones—a hammer, an anvil, a stirrup. The handle of the hammer leans against the inside of the drum, and when the drum moves the hammer moves, and the anvil and the stirrup move because they are latched to the hammer by joints. The stirrup has what is called a footplate. That fits into an oval window in the wall between the middle ear and the inner ear. Those three ossicles act like a lever, and when the mechanical advantages accruing from lever and drum et cetera are calculated and added, the force of the wave motion that arrived at the drum will have been multiplied thirty times at the footplate, or, by the calculation of the most noted living student of the ear, twenty-two times.

Some say that the footplate moves like a door on a hinge, some say like a piston, fast in either case. For the comparatively low frequency of middle C the piston moves in and out 256 times in a second. Two tiny muscles are attached to the ossicles, and these by their graded contraction reduce the blow delivered by too-loud tones, especially by low loud tones, giving quick, reflex protection to the system. If the tones are explosive, the muscles may not be quick enough and the ear is damaged.

Internal Ear. It also is a cavern in bone but filled not with air, with fluid. It is called cochlea. Cochlea = snail shell. A conical snail shell spirals two and one-half turns around a column of bone. It is intricate. It is a jewel. Each turn is divided into an upper and a lower gallery by a spiraling membrane, the upper gallery again divided by a spiraling membrane; so it is three galleries filled by fluid placed deep in the bone on each side of our head. Fluid is in the internal ear, air in the middle ear, and that means air is pushing against fluid. That is not mechanically the most effective. Fluid is incompressible, and when the footplate pushes from the middle ear toward the internal ear, something must yield. Something does. A membrane. It fits into another window, the round window, which yields in the direction of the middle ear.

Thus has the scene been set for action. If now a tree falls in the forest, a firecracker fires, a two-ton truck snorts, a bell rings, or my beloved blows her nose, two instruments are waiting to dispatch the variety of the news to my mind.

BASILAR MEMBRANE:
Sifting the News

The membrane that divided the two and one-half turns of the cochlea into an upper and lower gallery was the basilar membrane. Highly sensitive. Highly specialized. Hermann Helmholtz studied it, studied ears, studied sound, studied music. He developed a theory—the harp theory—that has stood up against ninety years. Helmholtz was one of the greatest physicists who ever lived, and one of the noblest of men.

He was born in 1821, at fifteen had chosen science for his life work, by twenty-seven, independently of Julius Robert von Mayer, formulated the law of the conservation of energy (the energy of the universe being constant), gave it mathematical statement, and it was one of the gigantic achievements of the human mind.

In his harp theory, successive groups of fibers of the basilar membrane corresponded to successive strings of a harp or piano. The living strings ran side by side, each had a frequency, as do the dead strings, and vibrated sympathetically whenever that frequency was set up in the neighborhood. Later the basilar membrane was shown to operate more complicatedly. At low frequencies it did respond frequency for frequency. At higher frequencies groups of impulses went into the brain. At still higher there was something quite different. A fluid wave moved along the two and one-half turns of the internal ear, and this put a shearing force on that other nearby membrane, exciting several kinds of electrical play; nerve endings were stimulated, and impulses went into the brain. It was the crest of that fluid wave that had excited the dominant frequency. Brilliant studies had established this later observation and interpretation, and for it Georg von Bekesy won the Nobel Prize in medicine in 1961. Helmholtz's overall idea meanwhile continued to stand much where it stood.

In a human ear the basilar membrane is somewhat more than an inch long, mounted on flesh pillars, tiny of course, two rows of them, leaning toward each other, and on either side of the pillars are cells with hairs, these the receptors, and above the hairs is the membrane. As the fluid wave moves along, the hairs are alternately squeezed and released, squeezed hardest in the neighborhood of the crest of the wave, electricity generated, nerve impulses traveling. To either side of the crest other nerve impulses also traveled, inhibitory impulses, which dampened, pedaled the tones to either side of the crest, leaving the dominant tone less interfered with.

The nerve impulses that result travel inward over the great auditory nerve. It has twenty-five thousand fibers.

At its one end, at the peak of the snail shell, the basilar membrane, with another membrane above it and sensitive cells between, is broad; at its other end, at the base of the snail shell, it is narrow. The narrow end, like the harp, vibrates to high frequencies, the mind hearing high tones; the broad end to low frequencies. Boilermaker's disease produces victims deaf to high tones. At postmortem the damage is where expected. All of the facts fit. All is mechanical. All is inevitable.

Helmholtz had decided on medicine as a career because he could embark upon it without having money in the bank, the government paying the expense of his schooling. It did mean barracks. His pre-medicine he took in Potsdam, and he included the study of Latin, Greek, French, English, Italian, Hebrew, and Arabic. He wanted a piano in his room. He insisted. He got it. Much of our knowledge of acoustics and of the physical basis of music began, accordingly, in a barracks.

Nowadays, motion pictures can be taken of the basilar membrane and the structures attached to it. A glossy falseness may lie over those pictures, as over most motion pictures, but safer to have the pictures than to leave everything to the imagination. And here the imagination might be too modest. We might, for instance, imagine the basilar membrane as a stiff prim parchment on each side of our head. It is the uttermost reverse. With proper stroboscopic lighting and a magnification of a hundred times or more, that membrane is mobile past imagining, its surface shimmering in the lighting, and when the biggest fluid waves rise because some crash has beaten the air, one's mind sees the surface of the sea. If that last strikes someone as exaggeration, he is wrong, but let him think instead of a symphony hall, the strings of the string instruments and the drumheads of the drums and the air columns of the woodwinds and the organ, all dutifully vibrating in their totals and their fractions, setting the loose molecules in the space of the symphony hall into a corresponding vibrating, while in front of the orchestra, built into his civilized anthropoid head, are the two highly cultivated analyzers of the conductor, periodically annoying themselves with picking out a single sour note in a single viola, while behind him in the dark glow of the hall two to four thousand pairs of basilar membranes are swinging in a sinuous harmonic motion, the energy of the swinging resulting in electrical messages that travel inward to a final destination in the hearing cortex in the temporal lobe of the brain. Each ear connects via way stations with both temporal lobes, and that is why serious damage may be done to one side of the brain with no resultant hearing loss in either ear.

In each of our skulls there is a strip of cortex that has a point-for-point correspondence with the basilar membrane. The high frequencies are forward, the low backward, and every two millimeters means another octave. Upon that final keyboard of cortex, the final news is played. Each side of the brain has, in fact, two keyboards. Even three. A second and a third have been mapped in animals, and the order of frequencies of the second found reversed, low forward, high backward.

No one understands all of the mechanics but in general understands well enough until the attempt is made to pass from mechanics to mind, when everything is suddenly blurred, and thin theory takes the place of hard fact.

SOUND:
Newscast

Textbooks say sound has three qualities: pitch, intensity, timbre.

Pitch places a sound in a scale. E-flat. Pitch depends on frequency, but there is the ear, that instrument, and there is the mind, forever intruding on everything. Not surprising, therefore, reasons enough, that sound should be wrenched somewhat from the place allotted to it by its frequency. High sounds, for instance, are heard higher, low lower. High sounds are not as masked by a streetful of high as by an engine room of low. *Intensity* is volume, loudness, depends on wave size, and that depends on the energy put into the wave. *Timbre* particularizes sound, separates a human voice singing on the beach from the foghorn moaning at the end of the pier, from a cricket who is on the scene as an interloper. Timbre depends on wave form, complicated often, repeats for as long as the source produces. Our ears break up the wave form, but fuse it also, and our mind bats back and forth with all of it. Timbre lets us distinguish a clarinet from a flute, a single flute from six flutes, each flute belonging to the species flute but obligated also to its flutish self to be true.

Pitch, intensity, timbre—these allow us for the sake of the defense of our body and for the sake of the slow evolution of our mind to isolate the peculiarities inside a sound. They allow the physician with his stethoscope to know if the fetus in the womb has its pump still pumping, that therefore it probably will arrive alive, add that one more to the directory. Allow the physician also to hear inside a sick man's skull or neck a dangerous rush of blood.

A plucked violin string vibrates through its length and produces the fundamental, the base tone. Any ordinary ear without training hears that. The halves of the string vibrate independently and produce the octave. Any ordinary ear hears that too, hears fundamental and octave as individual tones, at the same time hears their fusion. Then there are the halves of the halves, and the thirds, and so on, all the overtones that the ordinary ear has long since stopped hearing individually, but does hear more or less of the richness that comes of their fusion, the mind in consequence continuing its evolving, perhaps exceedingly slowly now.

The ear itself, the instrument, sifts frequencies and intensities with an efficiency that no one stops to consider as he reads the cold notes off the page of piano music and bangs the keys. From these siftings, whatever gets past the ear and on into the nervous system, more is lost there, and that which finally is left for the mind is the sound.

A tone that repeats rhythmically and is agreeable to the mind is music. Bring together legions of such. Bring in changing frequencies, changing intensities, and different wave forms. Bring in, that is, trumpets and the rest. Bring in accelerandos, retardandos. Bring in quavers and semiquavers and arpeggios and minor scales and major scales, and modulations from one key to another. Bring in a new theme, or a rearrangement of an old theme, let it fade, emerge, fade, emerge.

Noise is different. Noise also is sound. Noise also is a passage from frequencies through a physiological machine to the parallel-lying mind. Noise is disagreeable to the mind usually, unwelcome to the old music if not to the new, excited by tones that do not repeat rhythmically, or that leap from here to there with inconsiderate speed, or a waxing and waning of intensities that produce a beat that scratches the soul. The sources? A wagon spring. A pencil against slate. A falling dishpan. A village orchestra stumbled into misfortune. A violin whose string snapped. A bad line in a bad poem that stayed bad because among other reasons the poet did not have the patient ear of Keats. The out-of-rhythm bogus bass of the boss. Gumchewing Lucy's hi-fi through the hotel wall at 2:00 A.M.

Mind is warned by noises. Informed by noises. Inflamed by noises. Enraged by noises. Driven to despair by noises. There are noise meters that give numerical values to noise, and satisfaction to meter readers. A psychiatrist was of the opinion that noise sometimes has been the straw that broke the psyche's back. Even when not that destructive, it still is morally destructive to have music fed fitfully between clicking typewriters into a bank in order to sweat more typing out of mind-deprived Lucy.

AUDIOMETER:
Applause Meter

It was the frequency of the waves that gave the tone its pitch. The size gave intensity, loudness. Speech therapists and some doctors use an instrument that relates pitch and intensity. The audiometer. It produces pure tones electrically. The therapist needs only to turn one knob for pitch, another for intensity. Right and left ears are tested separately. If an ear does not hear a certain frequency at a certain intensity, the therapist turns up the intensity knob. If then the ear hears, a mark is made on a printed chart, the successive marks are connected by a line, and the chart is examined and filed. It is a rough but useful description of that human ear's capacity. Normally, the two ears of a person test much the same, and audiograms of members of a family are apt to be similar.

Alexander Graham Bell invented the telephone, and in his honor the intensity unit is called the *bel*. One-tenth of a bel is a *decibel*. The pressure of the vibrating air molecules might itself be measured, but the measurement is technically difficult. Instead, the tones are graded. A tone close to the lower limit of hearing is the reference tone. If another tone is ten times the reference tone, that is 10 decibels. If a hundred times, 20 decibels. Ten million million times, 130 decibels. (It is the logarithmic scale, 1 for 10, 2 for 100, 3 for 1,000.) A 130-decibel tone is almost more than the ear can bear, even though the difference in actual impact on the eardrum between 10 and 130 is not great. Our breathing sounds are around 10 decibels.

The ear has an indescribable sensitivity. It is far more sensitive than the eye, possibly because in the primeval forest night was more dangerous than day. At any rate, it is. A good ear can eavesdrop on anyone's gossip.

DEAF:
News Flash—Void Revisited

A human being may have lost his hearing for no reason that anyone knows, or for tumor, or for infection. He may also have had some abnormality of genes, or something went wrong in utero, either way a silent creature there in that silent womb. He then arrived in the outer world, deaf.

As many as ten million deaf have been credited to this nation, and a more enthusiastic figure would put it at 80 percent of the population, the figure to be brushed aside because it surely ignores that each of us is rushing toward old age and getting there, and getting deaf. We may manage both tantalizingly gradually.

An eccentric said he was pleased he could no longer hear the world, but more commonly a man insists rather that he is not deaf, not very, that what you are noticing is absentmindedness. He has other things besides your voice to listen to. His liberal neighbors make a praiseworthy effort not to hold it against him that his body has a flaw; and he looks right through his neighbors. That small button in his ear, they say, no one would ever notice if he were natural about it. The button would give nothing to his hearing if the damage was in his hearing nerves. It must be somewhere on the way to the nerves, somewhere in the transmitting system; then a hearing aid helps. His neighbors are right; his aid has been so cleverly placed that no one would see it except for that self-consciousness of his. One kind of aid amplifies all frequencies, practically puts a loudspeaker into the ear, another kind amplifies only the impaired frequencies.

That word—*deaf*—is soaked with mood. It is gray or black. The limiting walls have moved in closer.

Difference in disposition is brought out by deafness. One man behaves as if he were constantly thinking of his lost antennae, thinking how his power to grab has been reduced, everybody else's power increased. So he just sits, throws into the wastebasket the good-advice letters from his sister, who hears. Deafness may produce a wasp, or a pillar of sadness, or a gusher, or an honest philosopher, and which depends on what the persons were to begin with.

Rarely, but it does happen, a person is born both deaf and blind, from the beginning heard no evil, saw no evil, thought no evil. It is as apt to be, heard nothing, saw nothing, thought nothing. It is quiet inside his cave. One might conclude that he had been left far back on the road of the living. But it is not so, or may not be. The person was born with the intelligence and the force to circumvent anything. The latter was stepped up by his loss. The force may be excessive, as if the mind correcting for a loss did not know how far to correct, when to stop correcting. He may have persuaded another to join her fate to his. That may have helped. The community may have made a community project of him. On the contrary, he may have made a project of society, where he reaches farther into it than it into him. Society might instead have been scrupulously honest with him, but that is too much to ask. Even in the presence of affliction, society prefers the afflicted to behave as it does, accordingly flatters the afflicted to go about as if its senses were his, as if he actually saw a Cézanne, heard Debussy, thus still further unsettling for him the boundary between fancy and fact. This man could give us aspects of the world that belong uniquely to him, that reach him through senses superbly sensitized, powerfully disciplined. Touch. Pressure. Pain. Taste. Smell. Even temperature. However, we are all too tired to break through our own numbness so as to assist him to assist us.

So, here we have a human being whose two dominant senses communicate nothing, and yet his mind remains a mind. It is a most sane mind. It has brightness. It has comparative selflessness. Instead, it might have had maniacal selfishness, maniacal imperiousness. The point is that this mind has not died, has not grown dull, and never may.

LIGHT:
"And There Was Light"

Vision brings us the evening star. Hearing could not do that. No other sense could. The word *vision* has as its principal meaning that common one, sight. He has vision—he sees. She lacks it—she is blind.

Light excites vision, and light is two: (1) dead radiations, (2) a living receiver. "And there was light." This could not have been written had there not been the two, what arrived from outside, what waited inside.

Even a creature of one cell, the entire creature one cell, may live and be directed by light. The sun's radiations arrive, cause a change in the creature's chemistry, it turns toward the light, or turns away. A beetle flies out of the night toward a street lamp, crashes, has used those radiations to die. Some invertebrates are light-sensitive over half their bodies. One nocturnal species buries itself in sand, has a tube to siphon in air, shoves out that tube, and if the radiations of the sun strike it, back in it goes, has received the message, buries itself safely again. Life has obeyed light. Other species defend themselves or are offensive, withdraw or attack, because in dim illumination they are able to concentrate light, to see, though better not say see, better say receive. Other species have the shape of a hollow sphere with an inside light-sensitive lining, and a pinhole opening to the outside. Others have light-sensitive units fitted one against the next, as the dragonfly. Many inventions. A spider's compound visual system produces a mosaic of blotches, which becomes a signal, a language, and the spider says in translation, "There approacheth a fly." What the spider is seeing we cannot know, but the way it runs from a moving shadow, which is a warning, assures us that the spider has something that makes use of dark and light, and that is worth its having. Two human beings may have a surge of human feeling because both can see, then a surge of melancholy because what the one is seeing the other is not. Rays come from taper or oil lamp or sun and flash in all directions, and somewhere someone has the proper receptors to receive them, to add them to its chemistry, to be notified ahead of time about something that is still at a distance.

EYE:
A Camera

The eye is the instrument.

The eye is the camera. That has been said a thousand times and is correct enough, on the surface. Both eye and camera have a talent for light. (And for dark.) We creatures were born with the talent, and its carrying out depends upon facts and rules that have been and will be studied.

The foremost student of the chemistry of the retina was one of three recipients of a 1967 Nobel Prize; all three recipients were discoverers in vision. Those living test tubes there in the retina are two kinds of cells known to all intelligent grade school students: the rods and the cones. Their shapes give them their names. They are set in the hollow on the inside of the back of the eyeball. Whole books have been written about that sensitive, many-layered inside, the retina. Man's has 125,000,000 rods and 4,000,000 to 7,000,000 cones. At the fovea, which is a pit in the macula, which is a circular patch near the center of the hollow, there are 120,000 cones to every square millimeter. So, at the fovea, cones; at the macula, cones and rods; at the outer edge of the hollow, rods. To risk another figure, the eye is a silent lantern hung at the front of the brain, waiting to be lit from beyond itself. The two eyes, two lanterns paralleling each other in construction and in work.

Had that comparison of eye with camera been with a motion-picture camera, it might come closer, still not very close. To be honest we ought to say that the nature-built has possibilities that could not be dreamed in the man-built.

Why is the white face of the clown, which is brought by the eye's lens to the retina as an upside-down tiny image, seen in the mind right side up? Why? And seen actual size? Actual is of course only a word. But the facts can be bewildering. What happens in the visual centers in the depths of the brain? It would be no answer to lay everything to experience, because this would be to make the whole problem one of learning, and that problem has similar ambiguities, and, anyway, we are born with our visual system, and the evidence is that it begins to work at once. To be honest we ought, furthermore, to explain, if we can, why the white face stays out there in the circus ring. The image is in our own heads, is it not? Why does it stay out there then? In fact, why do we see at all?

An old and famous and simple experiment encouraged support for the comparison of eye with camera. College students performed it. They took an eyeball from a freshly slaughtered cow, thinned the rear wall of the eyeball with a knife, fitted it, front side frontward, into one end of a mailing tube. Then they pointed this rude instrument toward, say, a tree across the street, and on the thinned rear wall as on the ground glass of a camera there immediately appeared an upside-down picture of the tree. Tiny picture. Delightful! Surely we must be near to the solution of the mystery! Those who look for comparisons of eye with camera would snatch onto that ground-glass picture. They would say that the retina there was the camera's film.

Another experiment, old too, discouraged that famous comparison. The experimenter put a patch over one of his eyes and over the other fitted an inversion lens, so that images on the retina now were not upside down. Whatever traveled back to the receiving parts of the brain should be the reverse of normal. Notwithstanding, in eight days the experimenter's mind had completely made the adjustment. We cannot help being astonished at how adjustable our sensations are, how pliable our learning, how detachable and reattachable our mind to our brain.

The front of the eye has its transparent living glass with the pain receptors and the touch receptors. Rays of light reach this curved glass, are bent, and focusing begins. The glass is bathed behind and in front by a salt water that keeps it clean and keeps it shrunk, the salt water only slightly salty, steadily manufactured, steadily drained away, manufacture and draining steadily balanced. That is necessary. Any sharp rise of pressure inside an eyeball could cause sudden blindness. From the cornea the rays continue inward, strike other transparencies, the most important being those of the lens. If a lens is lifted out of a cow's eye, it still bends rays as a glass lens would. Helmholtz is reported as saying that had his lens grinder ground a lens as poor as Nature's, he would have discharged the man. Helmholtz probably was speaking only of the ray-bending power of the living lens and of its optical errors, not of that miracle that grows in the embryo from apparently nothing, soon is expert at its performance, serves in all illuminations from high noon to midnight. Being living, it consumes oxygen, gives off carbon dioxide, must receive nourishment, must have its wastes removed, and all of this depends on flowing blood, yet blood vessels are not permitted to cross a lens and cast shadows, so the chemical transactions must all be via that salty fluid.

The curtain that we all know, the one in front of our own lens, has a round hole in it, the iris. An iris may be a heavenly blue in the young but usually grows darker. Its inside surface is continuous with the retina but has no rods or cones. The iris as we look from outside is dark dark-red in brunettes, light dark-red in blonds, chocolate in the Negro. The tint is a chemical, melanin, and the corresponding tint in the camera is black paint. In both cameras, accordingly, reflections are reduced and blurring is reduced. As for the total eyeball, that rolls harmoniously with its partner in a fat-protected orbit, the harmony faltering in sleep when one eyeball may wander, swing upward and outward, make grandpa dozing in his chair even older than he already is to his five-year-old grandson. An eyeball is about an inch in diameter. From it the optic nerve fibers, which carry the messages, pass diagonally backward along the middle bottom of the brain, meet the nerve fibers of the opposite nerve, the inner fibers of both nerves crossing and continuing via the way stations to the back of the brain. If the right half of the brain is destroyed in an apoplectic stroke, the opposite half of the world is still there but is not reporting. In spite of this, brain and mind will soon have adjusted, each of the two eyeballs will henceforth swing somewhat farther and somewhat faster.

That person with the stroke may be quite convinced that he is seeing all there is to see. Every one of us is willing to be convinced that he is seeing all there is to see, visually and otherwise. A quite blind person may insistently claim that he sees.

PHOTOPIC:
Day Editor

Sir Isaac Newton—Shakespeare's contemporary and countryman with his so different quality of mind—directed a beam of white light toward a glass prism, and out the other side of the prism as out of the spray of a garden hose there spread a band of color. White light had the stimuli of color in it, the glass prism had broken the beams into its component radiations, these were bent by the cornea and lens and other media of the eye, were focused on the retina, and the mind saw red, orange, yellow, green, blue, and indigo, with the hues between. Recombine all of these and there would be white again. Various other combinings, as red and green, any so-called complementaries, also gave white. Amazing forever to the two naive minds left in the Western Hemisphere that a red light and a green light in right proportions will give a white light.

Black is color. That point was debated to apparent conclusion through a century. We see black. We require, that is, a retina for there to be black. Where there is no retina, nothing. Black is not nothing. Black is the apparent result of all color stimulation being withdrawn from the retina. The no-stimulus becomes the stimulus, this transmitting in and out of the communication pools of the brain, and a mind is in possession of black.

Your dog sitting quietly at the edge of the universe and looking out sees black and white and shades of gray, nothing else, it is believed. A parrot sees color. A lizard sees color. Bony fishes do. Crabs do. Turtles do. Rats do not. The cat does not. We have proven this for the cat, we believe, but no cat will commit herself in writing, and it is possible that she cheats during the experiments, but the fact that impresses us is the chill casualness with which Nature has dropped so useful a talent as color-vision here and there in the ranks of the living. Some birds see almost the same colors as we do, and have notches chipped out of their rainbow as we do.

A human eye discriminates a hundred distinct colors, fewer or more, and if all mixtures are taken into account, a thousand. The sun strikes a sweater, and the dye in the sweater absorbs some wavelengths, reflects others, and what is absorbed the eye does not see, and what is reflected the eye sees as the color of the sweater. If the wavelengths for the greens and blues have been absorbed, then the sixteen-year-old next door has come out of her house this morning in her prettiest red sweater.

The color-blind are classified on the three-color theory, are blind or enfeebled for one or two or three of the primary colors, therefore need to experience their world in what is left. Color blindness is inherited, and women are ten to twenty times less subject to it than men. A completely color-blind person would be rare, one who saw his world in shades of black and white, and very likely he would not be as discriminating for black and whites and grays as his dog. A color-blind painter might not necessarily know he had a deficiency, would paint a gray sea with gray waves and gray clouds, his painting to the rest of us would look either wrong or piquant or highly original.

The purpose of color vision?

Please do not say it evolved for survival. It seemed to. Do not say either that it was to get the best mate, discover the best prey, play dead in one's own color against a background of concealing color. It seemed to. If you could not resist putting it so, then at least admit that the human mind was driven also by a wish to reach out, sometimes climbed to the highest state of wonderment that it could, and color sometimes played a part in this. A face or a countryside had not only planes and lines but had color to start dream and cogitation.

SCOTOPIC:
Night Editor

Night changes all of this. Night changes what is seen and changes the eye that sees. Night prepares the eye for the affairs of night. An Indian walks in the moonless forest and never stubs his toe or bumps into a tree. An infantryman can kill in a foxhole in Okinawa or Vietnam, or a citizen murder a citizen in the springtime in Paris and wish only that the Parisian night had been less embarrassingly bright.

The murderer would have been fitter for his business had he first spent some time in the dark, gotten his eyes ready, because eyes do get ready. This they do rapidly the first five minutes, slowly the next thirty when dark adaptation is near to complete. Thirty minutes is also the time from sunset to night, suggesting that our eyes were evolved to get us ready for night in the same time it takes night to draw its curtain across the sky. The process continues triflingly for another hour, vanishingly triflingly for twenty-four. The adaptation in those five minutes was in the cones, the daylight receivers of the retina; the remainder was in the rods, the night receptors.

Some persons are night-blind, cannot adapt to night, some are day-blind. The chicken's eye has no rods, hence no night vision, and poor twilight vision, which is why a chicken decides wisely to go to roost early.

The stimulus for night vision as for day vision comes finally from sun and stars, from light energy packeted in quanta. The size of the quanta varies with wavelength. One quantum decomposes one molecule of the eye's photochemical. Two or three quanta are enough for a sensation.

When we go from broad daylight into the movie, we see nothing, bruise our shins, or somebody's, then after a while see everything. Later we come out, and the long immersion in the darkness has allowed large amounts of photochemical to rebuild; the daylight attacks it, a fierce decomposing, a fierce rushing of messages, a fierce brilliance, not the happiest sensation, almost pain. People in the far north exposed to the cutting reflections of sun on ice or fishermen exposed to the sunlit ocean may seriously injure their night-vision machinery, and it is a habit among them to shade their eyes whenever their hard life allows. Then comes that other part of the year, the months and months of dark.

Visual purple is the photochemical in the rods. Light bleaches it there in the rods much as if it were in actual bottles. Green or blue-green in dim illumination are more easily seen than the other colors. On a summer evening, as day dies down and one color after another is turned off for the night, the greens of evening grow strong and have to an unusual degree the power to cast a spell. That last, no doubt, is because of past summer evenings associated with contentment.

The visual-purple molecule is large, a protein with a pigmented something attached; when light breaks this attachment, a number of compounds form, visual yellow and visual white among them. Vitamin A is required for rebuilding the visual purple. If one's vitamin A intake is insufficient, one can eat liver, which has it. Oils of the liver of fish long have been employed in the treatment of night blindness.

An unpleasant dramatic experiment dates back to the last century. A rabbit was kept in the dark, then for an instant one of its eyes was exposed to an old-fashioned daylighted laboratory window with bars; the next instant the rabbit was killed, an eye removed, its retina immersed in a solution of alum, as a photographer immerses a film in a developer, and there on the retina was the window, dark bars and the bright blotches, the unbleached and the bleached.

What scenes are spread on our retina and quickly wiped out! One thinks of the night pilot the last moments before he crashes into the mountain, how exasperatingly calm the words coming from the control tower two hundred miles away, how calm the electric lamps over his instrument board, how calm the chemistry in his retinas. The large molecules are methodically dividing, the coded messages marching over nerves into the depths of the brain, the kaleidoscopic record rolling, then the final frame, the camera smashed. In the future a chemist will figure out how to reach chemically into and beyond the eye of the mangled for data on the crash.

Sometimes four hundred rods connect with one nerve fiber, and that is to say that an area instead of a point, four hundred rods instead of one rod, report over one wire. This also inclines night reports to be vague. The beauty of night and the mystery have the vague besides the dark. It is all different from day, and the glory of the difference is due to that chemical, visual purple, and to those four hundred messages over one wire, the clarity blurred, softened, even before the messages start into the depths of the brain, where somehow night interpretations attach themselves to night sense-data. The glory is somehow still something else. *Scoto* means darkness, and scoot-pic vision is night vision. As for the light source, though heaven's great candle is out, and its small candles out too, somewhere some wax candle is going up a stairs past a window to bed, somewhere some shimmer slips from beyond the black shutters or the black clouds. The human eye at its minimum in the night responds to one ten-billionth of its maximum in the day.

FORM:
Layout

Already a century ago it was understood that two stars would be seen as two (sense-registered, brain-handled, mind-interpreted) provided the lines that came from them made an angle of one minute at the eye. The eye will be able then, as one says, to resolve them. Even a century earlier it was recognized that "a spot of the moon's body" must make an angle of at least one-half minute for it to be seen on earth. Some eyes do better, some fantastically better. If, instead of two stars, a thin thread is rightly lit, rightly set in a contrasting background, has a right reflecting texture, the angle need not be one minute, or one-half minute, but one-half second. At the retina that would be less than the diameter of a single cone. That cone is able to "see" that thread. This explains why the tailor at his bench is free to think of other things, of his new wife, while he works. The thread gives his eyes no trouble. Some birds of prey do better than the tailor.

When an eye is not able to resolve a sequence of points, they fuse into a line. A line has length and direction. Add genius, and the line may produce the famous one-line cat in the drawing of Hokusai. Hokusai wetted his brush in black Japanese ink and by a single sweep left us the cat.

When an eye fails to resolve close-lying lines, they fuse into a surface, and a surface may have many qualities. A black-and-white reproduction of Hokusai's self-portrait would have both the sequences of points to fuse into lines, and then the close-lying lines to fuse into surfaces, and the resulting play of blacknesses against white passes on to us how that greatest of Japanese artists saw himself.

Recall the eye chart in the doctor's office. There are rows of letters of successive decrease of size. The breadth in any stroke of any letter in any row or the space between strokes is such that the eye can resolve stroke or space at a distance from the chart that depends on the letter's size. The person being fitted for eyeglasses sits 20 feet from the chart. An average eye reads the 20 type at 20 feet, has 20/20 vision. Stroke or space there is making an angle of one minute at the retina. When the eye can read only the 60 type at 20 feet it has 20/60 vision. The fitting for eyeglasses becomes an investigation into this human being's two-dimensional vision.

Form may be three-dimensional. Form may have depth besides length and breadth. Depth can be seen with one eye, better with two. So there is monocular and binocular depth vision. Illumination expectedly helps. A spherical surface looked at head-on—with one eye—is spherical because the light reflected from it diminishes progressively toward the edges of that spherical surface. Shadow helps. Contrast helps. Whoever has picked black fuzz from black cloth knows it would have been easier had it been white cloth. Parallax helps. Parallax is the apparent movement of objects relative to one another. When looking out the window of a taxiing airplane, one recognizes that the moving faraway distance, the middle distance, and the near distance have different speeds and directions. Our mind, long before it jerked itself up and paid attention to such speeds and directions, knew about them, and from them learned something more about depth.

Perspective helps, of course. Two lines approaching each other in a drawing give us a street running off into the distance, and if those lines are felicitous, the drawing may require not another detail. Tanyu, the Japanese painter, painted an arc on a rectangular *kakemono*, just the arc, placed it low on the *kakemono*, did it with perfect knowledge so that to anyone's vision that arc is an evening sun sinking into the sea. Interference helps. When an oak gets in the way of a maple, the eye sees the maple as farther.

Two-eyed, binocular, vision is better than one-eyed. The right and the left eyes are not looking at exactly the same scene, the right is reaching farther around to the right, the left to the left. Textbooks instruct us to lift a finger in front of our noses, then alternately to open and close right and left eyes, the finger then alternately jumping from side to side, proving that each eye singly was not looking at exactly the same view of that finger.

The developmental psychologist has said to us often that the various senses help each other. We rub our fingers along the baby's crib at the same time that we see the crib. We touch the baby's face. The baby touches our face. Hand teaches eyes. Eye teaches hand. Whoever has watched children bending over their drawings knows how the entire body helps, how then after a time the formless takes on form, the two-dimensional becomes the three-dimensional, and it is always striking how this occurs suddenly, overnight, one thinks. Latterly a Puerto Rican has been teaching the blind of his country to draw. In human beings deprived of their dominant sense, a teacher has taught the hands without benefit of vision to put form into minds. Our artists teach us. Our painters and sculptors see shadow where the ordinary eye sees none—may avoid seeing in deference to an earlier tradition of painting and sculpture. The face of a Japanese geisha in a Japanese print has little or no shadow because through generations she has been painted so; then one day we come into the presence of the flesh-and-bone geisha and powdered white face appears round and shadowless, though her anatomic face is of course not round and not shadowless.

Shadow and substance signal from the void, and signal from an ink drawing, from a watercolor, from an oil, signal from a mouse that flees along the side of a restaurant kitchen and into a black hole next a pipe. That mouse is a light source, stimulates a human retina, and a human mind thinks mouse thoughts. "Wee, sleekit, cow'rin' tim'rous beastie . . ." And a mind wishes it were not in this restaurant.

BODY IMAGE:
News Flash—Man Loses Limb

Each of us knows he has a body. When did he find that out? Usually the question never occurs to him. It may at that horrible moment when he is introduced to someone and reaches to shake hands, and the person has no arm. He may be sure at that moment that his senses teach him everything. In any case he knows that he has a body and the woman has no arm. He knows that he is in possession of an infinitude of masses and shapes and lines and textures and moistures, all apparently reporting to one another. These apparently do somehow give him his image of himself. The image has two legs, two arms, a trunk, a neck, a head.

Until a few decades ago no one realized there was anything remarkable in this. Then it was discovered that there were and must always have been persons who did not with any of their senses sense a body, so did not have it, or did not have some part of it. A case was reported of finger agnosia, no knowledge of fingers, these lost to the mind, the person's own fingers and everybody else's. The person could not name a finger, not point to the correct one if named, had a blank spot in the mind. Nevertheless, the sensations when tested proved normal. Skin receptors were normal. Vision normal. Smell normal. News came from the world as it should, but not news of fingers.

It appeared on study that the damage might be in the flesh of the nervous system, or it might be in the mind. Nervous-system damage. Mind damage. The person might be totally unaware that anything was wrong, or he might be suspicious, might be tormented.

A man washed the right side of his body, laughed miserably when it was pointed out what he was doing, and he proceeded to wash the left side too. He had overlooked the left side because it did not exist, to him. This case was reported as nervous-system damage, demonstrable blighting of mind because of demonstrable blighting of flesh. The case was brain damage.

A young girl lost her body, the whole of it. This occurred suddenly. Earlier in her life she had had a hopeless reduction of force and feeling in one leg and some reduction elsewhere. This was indeed nervous-system damage. She had gone along well with her trouble, was brave, as one says, but that had been because she had an imaginative family, and had a close friend her own age. The friend fell in love—the patient lost her body. She could talk calmly about both losses, the presumed loss of her friend, the loss of her own body. The psychiatrist gave her a Rorschach test. Everybody has heard of the Rorschach. The subject tells what he or she sees in some carefully worked out ink blotches. This subject saw only animals. Blotches that nearly everyone sees as the whole or the parts of a human body or face she saw as animal body or face. She has wiped off the human body. There were to be no more human bodies in the world—because she had such an unhappy one. It struck one immediately as being her mind's way of protecting itself from dwelling on a disability with which she was doomed to live. The case was mind damage.

When persons who have losses in body image make a drawing of a body, they produce what could be mistaken for the work of a child, though of course the psychiatrist does not make that mistake. To him every detail of that drawing has adult meaning. If one listened to the psychiatrist talk (or the psychologist), one could soon think that every line in any drawing had meaning, any detail, that everything has mind meaning.

When then do we find out? How do we come originally to know we have a body? Perhaps we learn it. Perhaps the developmental psychologist is correct when he suggests that our senses are our first and last and only teachers. I touch my hand. I hear my one hand strike my other hand. I can tell in the dark where my hand is. And those facts are connected in the nervous system and mind. That teaches me that I have a hand, and similarly that I have a body. The knowledge keeps accumulating in every hour of every day. To the knowledge of my own body and its parts I add knowledge of other bodies and their parts. The baby looks, looks, looks at another baby, also looks, looks, looks at himself. Somewhere in the course of this he comes into possession of that other baby, that embodiment into a single of what are already many parts, and possibly with his eyes closed he now can see that other baby.

Yet the whole of my knowledge of body image cannot be a mere addition of the teachings of my senses. It had to have been built on something I was born with. There is excellent evidence for an inherited location in the brain for such uniting into totals. At postmortem in persons with loss of body image, damage repeatedly has been found in a definite location, parietal lobe, one side, toward the back.

A patient was asked to look at her left arm, which was absent to her mind, then to put her right hand on her left shoulder, even though that might be difficult because that arm was absent, then move her right hand from that shoulder down that arm. She saw clearly what she did but adhered to the logical opinion that a person must believe her intuitions and not her eyes. Another patient. He had his arm hanging off the side of a hospital bed. The doctor asked: "Whose arm?" The man answered: "Yours." The doctor lifted the man's arm in front of the man's face, asked again, whereupon the man suggested, yes, it was conceivable that that arm was his own arm.

For centuries there was the notion, less common in this less superstitious century, that each of us at a time and place enters his body. Each of us might therefore under unusual circumstances, or even under usual ones, leave his body again. Indeed, if a mind exists that does not know it has a body, or anyone has, would not that make the disembodied a fact? One gets a new respect for such a simple after-dinner performance as taking one's body out for a walk.

XVI

THE HEAD
The Rest Exists for This

MIND'S BIRTH:
From the Wings

Imagine a father in a distant village. Let it be Ecclefechan, Scotland. It might be New York, United States. He in Scotland is looking with God-fearing fervor into a crib, into the face of his newborn son, Thomas. He is looking and seeing there Thomas's mind, right in the pink flesh, or somewhere near, and he is not in the least distracted by the still unformed putty of Thomas's nose or by the two arms that pump and pump as mechanically as oars, or the legs that equally mechanically kick the bottom of the crib. To that father this newborn son might still be having some trouble breaking into the world, might be held back somehow by the flesh, appear shy, but the person exists. The mind exists. If he were asked to say something more about that, where for instance it was, he would not say anything about genes, does not know about genes, but might unperturbedly answer, "Oh, somewhere around that small head." At the least he is mistaken about the size of the head. It is not small. It is one-quarter as long as Thomas's body.

Imagine another place, an infant developmental diagnostic laboratory. Yale for many years had the famous one. Imagine a psychological tester and Thomas in an interview. A secretary has been writing down everything the tester said. A cinema camera has been photographing everything Thomas did, and tape has been recording the coos and screams. To the tester Thomas has exactly five days' worth of mind—because he is five days old. Five days since birth—five days' worth of personality. Mind is building, has been building from the beginning, womb to now, nothing to riches, and will go on building week after week, month after month, year after year.

Plainly, there might be two ways of imagining the birth of mind, and no doubt others.

Of the two the first is closer to the Church, the second is closer to science. Mind to that psychological tester sprouted and grew from no mind, personality from no personality. There had been nothing. Now there was something. It branched. It was gene-directed, but it emerged always as a product of the back-and-forth between this machine and a world, between a small lump of molecules and a large lump of universe. It followed: change the world, change the universe, change the mind, change the personality, and that is the only way mind and personality can change, in fact could be born in the first place, could appear. There is nothing strange, mystical, inexplicable, even if some items are not yet explicable.

Quite the contrary, to Thomas's father, had he troubled to think about it, had he been able to think about it, mind was—well, it was there in that crib. It not only already existed, it always existed. As the astronomer said of the astronomical, it always was, it always would be. The back-and-forth that in the psychologist's view was producing the first dawn of it, in Thomas's father's view was setting it free, was allowing it for a while to go about on the earth. It arrived at 4:20 A.M., with an IQ—no matter what the IQ, the mind was busy, if tentatively, establishing its contacts with mother, father, siblings, neighbors. The first wisps of a capacity to give attention had occurred in the first hours, by the kind of evidence the scientist properly accepts, the most eminent of child psychologists having produced it.

Thomas's father's view was the view of the centuries before Copernicus, before Darwin, before Freud. Those three changed the view. Thomas's father's view must strike a computer age to be as distant as Ecclefechan, but it is just possible that one ought not to forget that Ecclefechan still is right there in Scotland. Also, though you are a robust citizen of one of the colossal clans of the last of the twentieth century, it might be wise, or rather it would be courteous, not to dismiss Thomas's father's view too entirely and too easily. Call it the senile view, or, if you are feeling charitable, call it the old man's view, and show it the respect sometimes shown the old.

MIND'S FIRST STEPS:
Downstage Center

Often the small face looked illumined, the light not cast on it but somehow coming from it. Healthy infants glow that way. A pale doomed one may. However the newborn seemed on arrival, now there was mind unmistakably. Even at its pudgiest, this baby never was what one pediatrician said a baby is, a bag to stuff with canned apple sauce, canned meat, correct milk formula. The infant now turned its head at its mother's voice, was discovering her to be more than something to suckle, and she was discovering it to be more than something that suckles. They stroked each other. The behaviorists say that that stroking is first love, as restraint, any restraining, is first hate, or first rage.

Days went by. Weeks went by.

The infant tumbled less awkwardly. Its parts looked less like those of a toy put together, or blubber. Its actions looked less automatic. True, suddenly the eyes still might stare as if mind were somewhere else or nowhere, stare for minutes. Then suddenly mind was in the eyes again. Then there was that special moment when the eyes discovered other eyes. The effect of this was astonishing, because intelligence appeared to flash from eyes that had none. Moving objects were watched more continuously. The restless was being separated from the quiet. Persons were the most restless. The world had been all one with this one—there was no outside. But now something was being pushed off, gradually, gradually. Something was being separated off. Psychologists speak of "subject" separating from "object." The world was lost, found again, lost, found. At moments one thought that one could actually observe how sensations from different senses were floating together, sight and smell and touch and hearing. At moments they seemed mixed, two of them, three of them. Then there came a time when one knew that something knew that a wall of persons had closed in around its crib. A person in there was recognizing persons out there, then forgetting. Over weeks and months the eyes were succeeding in breaking that wall into mother, father, sister, brother, the child next door.

Abruptly the infant heart breakingly wept, stopped. Some weep their way out of the womb and go on weeping through life and into old age. Weeping in an infant seems not much touched with mind.

The cramp in the belly, to judge by the face and by the tears, was not a pain in the mind or a pain in the belly but a pain in space.

Not so with laughter. When no one was expecting it, a smile, meaning attached, slipped shyly out, slipped shyly back, next time might stay out longer. Mind was emerging with hesitation but emerging. A laugh on the right side of the face pulled the left side along, then was all over the face, then died. A harsh noise from the street had killed it. More and more, something generally comprehensible attached itself to laughter, and those around the crib were watching and were pleased to see how the solitary island was beginning to join the mainland. Mother, father, sister, brother, each at a time was a piece with the laughter. Weeping stays off alone.

At six to eight weeks it was evident to anybody that the infant was seeing increasingly with something more than eyes. Call it mind. Call it a forerunner of mind. If your definition of mind requires language, your definition is wrong; there is mind without language.

Anyway, the infant was noticing. Parent and pediatrician and expert of the infant developmental diagnostic laboratory noticed the noticing, and, to those who notice, noticing is credited with mind. (Pavlov would have credited his dogs with no more than the exploratory reflex.) Whether this infant arrived on earth with mind or with no mind, now there was mind, and now no need of a definition. Nevertheless, without warning the mind still would blank into the nowhere. A pup may do that, concentrate on you, blank, concentrate. A macaw may. Until yesterday the eyes appeared always to swing in an unnatural, parallel like two beacons, today they rolled inward, focused, converged, or converged more often, and not yet upon a letter of the alphabet or a decimal point but upon an artificial butterfly that was attached by a string to the crib.

How tinged with mind that converging of the eyes made that face. Who would not wish to fathom what the butterfly was reporting? What was going on in him, in his personality, definitely a personality now, when through his eyes he was seeing butterfly? Was butterfly attaching itself to his own plump hand, to which his eyes also shifted, on which they also converged for long periods. Is this what a mind does, brings one thing into itself, and having brought one, brings another?

The infant began to play. The mother called it play. What is play? Why is there play? Is it the machine oiling its operations in order to be ready for the living of its life? It may be. But is he—in there—not living his life now? He is of course. And maybe he will not be much less an infant—in there—when he is forty. What we can say of this immediate act is that that infant was doing something that undoubtedly looked like play, that his mother joined in, and that both laughed.

Faster and faster the two eyes went in whatever direction the world called. At twelve weeks, the pediatrician could demonstrate to onlookers that the eyes focused on him. At six to eight months, if there were two like the pediatrician, the eyes selected one, deserted the one, went to the other, came back to the one. Mind was rampantly on its course. At ten months, the eyes selected one from three. At a year, one from a roomful. Another half-year, if the eyes were seeing one, and another got in the way, that other was bodily pushed aside, hands joining eyes to accomplish it. The infant now could decide to push.

Ears, too, were emerging from the cotton, and these ears from the start had been sensitive ears. A heard world was needling and weaving into the fabric of a seen world. A child's psychology still was taking its first steps. Earlier the child had wailed each time it was startled by a sound; now it listened instead. That was a large advance. To tones of lower and lower intensity it listened. The face was more and more a listening face. Mind was receiving water from many sources, many rivers. Arms and hands were continuing to help in whatever way they could. They reached, groped, clutched, let go. Sometimes the hand reached where the eyes had reached previously. Hand had joined eyes. Sometimes the left arm and hand operated independently of the right arm and hand, a further independence of body parts that went along with a further independence of mind. Skin sense and muscle sense both were woven into the fabric. Sounds in the laughter also were changing, and at sixteen weeks were unmistakably helping this person to express what he had to express, which was much. Much more had been going on—in there—than was indicated in any movement. But movement did indicate. For another year he remained unpredictable and impoverished as to sounds, but every month reduced the poverty, and by the time he was five years old all infant characteristics would have been erased, parents and the neighborhood willing. Mind was enveloping the persons as the ocean a reef.

Unexpected gains accrued. It was natural that a bright young mother should aim to have her bright young son housebroken ahead of the Joneses'. It would convince the Joneses of the value of good stock. Triumphant days were followed by days when the young mother despaired of anything ever bringing order into the ghetto habits of her genius. Then he fooled her and performed expertly. Lapsed. Performed. Observations like these were bound to inspire research. One of two identical twins was placed on the chamber every hour, and the responses to the "chamber situation" were scientifically recorded; the other poor twin was not given the advantages of his brother. He was scientifically neglected. Notwithstanding, without benefit of training—imagine everybody's surprise—he performed as expertly as his brother. Training before the nervous-system mechanisms are matured is useless, perhaps may even be harmful, may take away a freshness from the mechanisms when they do mature. It was a Saturday morning after many mornings when the young man, no, the two young men approached their mother spontaneously and well ahead of the physiological act, whispered each into one of her ears, conveyed to her their thought, accompanied her down the hall, disappeared, reappeared, had chalked up two more neat victories.

DOG'S:
In an Ounce

In 1892 Friedrich Leopold Goltz, Strassburg, Germany, reported on three dogs whose cerebral hemispheres he had removed, both hemispheres, a formidable surgery, especially at the time. It was the removal of the latest, highest portion of the nervous system, with which is associated, or so it was and mostly is thought, the latest, highest portion of the animal, his mind.

In a dog, the cerebral hemispheres are a large fraction of its brain. In us, a much larger fraction. Hemorrhage was Goltz's problem. His dogs bled to death on the operating table. Again and again he failed before his historic successes.

The first of the three survivors lived fifty-seven days, the second ninety-two days, the third eighteen months and would have longer except that Goltz wanted a postmortem to make sure of what he had removed. He operated on that third in stages: June 27, 1889, a portion of the right hemisphere; November 13, the rest of the right; June 17, the following year, at a swoop, the left.

So, June 17 much or most of the presumed organ of that dog's mind had been erased.

Another dog, a bitch, Goltz allowed to keep her hemispheres, removed nothing of her brain, but she had to give up her spinal cord. The consequence was a dog paralyzed from her neck down, a spinal animal. She had her brain but what could she do with it? Goltz arranged for her to conceive and to carry a litter to birth. A macabre triumph.

Der Hund ohme Grosshirn (The Dog without Cerebral Hemispheres) was the title of Goltz's classical monograph. He reported those three "brainless dogs" in language a tired market woman could have understood, their behavior, as we say today, and the fact that he does not mention mind somehow calls attention to it. One of those dogs might stand up, start walking, was apt to walk in a circle like a numbed prisoner or a numbed inmate of an asylum. On a rough floor it walked securely enough, on a smooth, slipped. If it had not been fed for a time its pace increased, or it might stop the circling, rise to its hind feet, put its forefeet on the bank that surrounded the cage, stare out. Meat had to be placed right under the dog's nose if it was to eat, which it did like a stoker. Drank the same way. Emptied its bowels with that put-on, it always appears, absentmindedness of a normal dog, began with the ritualistic spiraling, which Goltz described as having been "very lively and rapid."

Sleep descended on the mutilated as on the unmutilated, and Goltz reported the quiet breathing, the closed eyes, the motionlessness. (Mind gone deeper in?) The dog slept too much. (Bored?) Loud noises were required to rouse it, the surgery having blunted its hearing or blunted its attention, probably both. Most effective, in Goltz's words, was "an instrument that is recommended to bicyclers to warn innocent wanderers that are coming along the road."

At the assault on its hearing, the dog "twitched with its ears, as if it wished to be rid of something unpleasant, then finally stood up." It would push a paw toward the ear nearer the sound, and that action, again for some reason, not clear, suggests mind, a worried mind. Hunger waked the dog. Stroking might. Rough handling did. If the dog was lifted from its cage, a spell of rage took possession of it; it growled, bit the air, had lost aim. Lifted back into the cage, the rage dropped off like a cloak. It left upon one an impression of a de-individualized animal, a member of a species, René Descartes's machine, no longer a dog with the dignities and privacies that to a dog's degree add mind to the world. Goltz's own cerebral hemispheres were substituting for the dog's, brought the necessities to the dog, protected it against injury, gave defense to a body that would otherwise have stumbled into death.

Experimental brain surgery might be anatomically more exact after this, might be technically cleaner, parallel human brain surgery, might be more revealing neuro-physiologically, but never again was it as ground-breaking, or more haunting. It was bound to excite the continuing effort to understand the back-and-forth between brain and mind. Experimentally, it was an achievement of the first magnitude, and Goltz shines through it not only as an unusual man but one who in spite of appearances was concerned about his dogs. He knew them. He recorded that Moog UU was a "stupid dog." He knew their characters. Between him and them there was concern. "The human being," Goltz wrote, "can pretend to have feelings that he does not have. The unlearned speech of animals may be a much more dependable expression. If I step on the foot of a dog, I will not doubt that I have hurt him if he cries and howls." Simple words that one may make too much of; one remembers them.

MAN'S:
In Three Pounds

Thousands of years ago men already believed that the human body and the human mind met inside the skull.

It is understandable that Nature built such a stockade around that precious flesh, bone that is light but strong and shaped into a form that has elicited the admiration of architect and engineer. Neolithic people bored holes into that bone, sawed out windows. The Incas did. With them it was religious rite—shoo out evil spirits. It might have been sheer bravado. Sometimes this could not have been bravado, was a physician trying to save a life after some fool had got a skull fracture in a brawl. Fracture lines can be distinctly seen in an occasional ancient skull. Often it has been perplexing that even with large penetrations there were no signs of infection, meaning either that the victim died promptly or we are not understanding something. Hippocrates in patients with sudden blindness bored holes to relieve pressure. Galen six hundred years later bored holes. Physicians in the Middle Ages did in cases of head injury. We do.

Some editions of *Guilliver's Travels* have a picture of two heads with their tops sawed off, so as to interchange the half-brain of a leader of one party with the half-brain of a leader of another party, to achieve peace. Swift explained.

> The method is this: you take a hundred leaders of each party, you dispose them into couples of such whose heads are nearest of a size. . . . it seems, indeed, to be a work that requireth some exactness, but the professor assured us that, if it were dexterously performed, the cure would be infallible . . . the two half-brains being left to debate the matter between themselves within the space of one skull, would soon . . . produce that moderation . . . so much wished for in the heads of those who imagine they come into the world only to watch and govern.

This was Jonathan Swift's cleansing irony, was not meant to give any view of mind, had not anything to do with the problem of brain-mind, and yet the old picture did kindle the thoughts of a child. It was a quite satisfactory picture. It was the one of all remembered years afterward, the first inkling, it could be, of the awesome intimacy of brain and mind, of body and head.

THE OPERATION:
Exploring for It

No modern surgeon enters a skull to find mind. If he is entering a baby's, he is not trying to find the beginning of a personality. It is to cure. We may not be interested in his opinion of mind—depends on the surgeon—but we may still be interested in the vantage point at which he stands. There is glamour there. It may be only the glamour surrounding the mountain climber, for someone below, but even to work one's way from station to station around that conical mountain, see up toward its peak, never seeing clearly, has some of the daring and doubt of exploration.

A brain surgeon's operating room should vibrate differently from other operating rooms—dental, rectal. Like any operating room, it was built to be kept clean. Whoever enters wears a sterilized costume; there is a laundry for this on the tenth floor. (A New York hospital.) All knives, scissors, chisels, suturing needles, catgut, silk, gauze, cotton, and towels are sterilized. Because human beings exhale bacteria with their stinking breaths, mouths and noses are covered by gauze masks, double gauze if anyone has a cold in the head, in which event he also avoids talking expulsively. Instead of gauze it might be copper screen. A cap covers every head. Shoes are grounded so that no spark of electricity may set off some tank of inflammable gas. Each person in the room has scrubbed his hands and scrubbed his hands and scrubbed his hands, then plunged them into sterilized gloves. That is not enough. Even an informed hand may stray and spread infection, on which account the chief surgeon and all the others of the hierarchy make a routine of keeping their gloved hands lifted in front of them until the work begins. They look, those green-gowned, green-capped, hand-lifted ladies and gentlemen of various heights and girths, as if about to be initiated into a secret order. It is. It has as a mostly unrecognized incidental project the searching in that bony container for one of the secrets of the world.

Whoever watches TV knows most of this.

For one operation the surgeon works with his patient sitting, for the next has him stretched out, or lying on one side, head lifted, or head dropped, or face down. In that last position a fixture spans the forehead. The table itself may in the still-voiced middle of an operation be tilted to some other angle from that at the start, orders gone out from the chief surgeon. A few pioneers, Harvey Cushing notably, were responsible for much of the technique of this surgery.

The chief surgeon has precisely in his mind the steps he is planning to take. So has his assistant. So has his assistant assistant. So have the nurses. Each is skilled at his own task.

All equipment is now ready on stands, and not only have the items been scrupulously anticipated, but every item is in its place, because, at the moment needed, life or death could conceivably follow from whether it was or was not in that place. An ultraviolet machine may be spraying the air in expectation of killing bacteria, the ultraviolet directed away from any cut tissues.

The patient is wheeled in. Drapes have been thrown over him, a space left in the drapes where the green-gowned chief surgeon will with his tools penetrate that presumed sanctuary of mind. Quickly, he stitches into the skin the towels that edge the space, to prevent their slipping. So there is now a neat frame around what is spoken of as the operative field, an area of naked skull. The entire skull has been shaved and scoured, and the scalp sterilized. Often nowadays the operation is under a local anesthetic, which must be injected by a succession of fine punctures like mosquito pricks, not really painful in skilled hands. The patient is of course groggy, has been made so. A nurse sits under the drapes in a kind of tent, and there she can undisturbedly check pulse, breathing, blood pressure, writes everything on a chart, leaves a record of the patient's shifting body states. Should the patient at a point seem not to be doing well, she promptly but with professional quiet notifies the chief above. He may regularly ask her. A formal dignified dialogue may run between them, yet everybody at the same time, more in one operating room than another, tries to be as lighthearted as the circumstances allow.

Bottled blood is ready for transfusion, and a needle to deliver the blood is in an arm vein, the needle held in place by adhesive, the bottle raised on a stand.

Mind hovers around that place. There would be the chief surgeon's mind. The assistant's. The assistant assistant's. And, again and always, the patient's, that would be somehow somewhere in the vicinity of that substance where the work now begins.

By quick knife cuts the surgeon establishes definitely the boundaries of the operative field, a line of scarlet following his knife and serving as a guide for deepening of the cuts. Down through the scalp to the bone he goes. The patient feels the cuts but no pain, the anesthesia successfully blotting pain. Rules for establishing those boundaries have been laid down by a long experience. The assistant surgeon has kept just behind the chief surgeon's cuts, mopped the blood, pinched off and tied the bleeding vessels. The patient feels the mopping but no pain. The scalp is stripped back, the bone along three sides of the four-sided field scraped. Procedures differ, but burr holes now are drilled into the bone at the four corners of the field. This is apt to be with a motor-driven trephine, but if hand-driven it would look like any auger that was cutting out buttons of bone, the touch of the surgeon keeping him informed of the depth his auger is going, the substance of the brain also mechanically safeguarded by the construction of the trephine. So, four buttons. It takes some prying to get them out. A flexible strip of grooved metal with the groove facing upward is carefully fed into one buttonhole and brought out the next, a wire saw slid along the groove, handles fitted to the two ends of the saw, and the surgeon saws, up and down, up and down. The bystander holds his breath, thinking that the saw may snap. It sometimes does, but no harm because the substance of the brain is now protected by the metal strip. Three sides of the four-sided field are thus sawed, the fourth side left unsawed, and the surgeon slowly and tenderly lifts at the slab of bone, then, this appearing sudden and rough, cracks through that fourth side. He folds the slab back along that side and away from the brain. A door has swung open. Long forceps are clamped across the cracked fourth side so as to squeeze off leaking blood, and the door, with the attached scalp and wrapped in gauze, turned out of the field.

What one is now seeing before one is deeper into that lair of the mind, but not yet seeing the brain because the membranes still cover it. The outermost is the dura. Should the pressure inside the skull have risen abnormally high, as happens, and the dura bulge through the doorway, the surgeon pushes in a trocar, a sharp-pointed hollow tube that he advances into a ventricle where the cerebrospinal fluid is, drains off some of it, and the bulge recedes. Next he cuts a doorway into the dura corresponding to the doorway in the bone, and now at last there is the brain itself. It is heaving gently each time this patient takes a breath.

The character of the disease, which often has mind signs, dictates what happens next. If a tumor, the surgeon tries to determine the extent, whether there will be more loss than gain to the patient in removing it, or removing part, and this depends too upon the kind of tumor, the kind of cells, the area of the brain, how critical the area, whether it might be better not to do anything, just relieve pressure, close up. Instead, he may be needing only to trim out an old scar. Open an abscess. Explore for hemorrhage. Tie or clip an abnormal and dangerous blister on a blood vessel, an aneurysm. He works when he is talented with what could well be described as courage, must strike boldly if he is not to damage the delicate living thing inside, because any faltering may forever silence that mind, or worse, keep a maimed mind roaming and alive. A brain surgeon moving with his fingers in that so-called organ of the mind can strike one as an archaeologist digging in a buried city, small but ancient city, asking questions of the stones.

The critical part of the surgery complete, blood vessels are tied where necessary, the operative field flooded with salt solution that is siphoned off, the field scrutinized, flooded again, siphoned again, scrutinized again. Is any small vessel oozing? It must not. All bleeding must be stopped, this according to principles that studies in blood clotting have made routine. The dura is stitched back into place, the bone wired back, the scalp sewn back, this according to principles that studies in wound healing have made routine. Then he or she, head wrapped round and round with gauze fashioned into the shape of a great white turban, is wheeled back to his or her private room, H219, looking rather jaunty.

HEMISPHERECTOMY:
Throwing Away Half of It

An American surgeon was the first to remove an entire cerebral hemisphere from Homo sapiens, half a brain. The other half stayed where it belonged in the skull, else there would have been no mind. The hemisphere that is removed must be the right in the right-handed, because if it is the left the man will have become what people for some reason like to call a vegetable. (Recently there have been several cases reported where the left in the right-handed was nevertheless removed, with some recovery of mind, some recovery of speech, the risk taken because of malignancy that could only go on to death.) This species of surgery is not experimentation; at first thought one might think it is, but then one remembers again the patient, the person, how everything living, including man, wants to live, keep on living, however absurd. It is forty years since one-half of a brain was first removed from a human skull and hemispherectomy had become an accomplished fact. Having performed one, that surgeon performed a series. In the case of an elderly colored man, the total flesh cut out including the tumor

weighed 584 grams. A healthy adult human brain weighs on the average 1,350 grams. The elderly gentleman was brought to the Johns
Hopkins Hospital in coma, was operated on, lived three and one-half
years, and died because of a recurrence. Three and one-half years is
not a negligible span in which to observe and reflect.

Human hemispherectomies were adding, and in the future might be
anticipated to add, to our understanding of mind, not much, though
that might depend on who and how, and they would keep awake the
controversy around the brain-mind.

Hemispherectomies began to be reported both in this country and
in Europe, oftenest performed on children born with a damaged hemisphere, the damaged interfering with the healthy, so, better the sick
be removed. Another elderly patient who survived a hemispherectomy
for more than a decade was reported puttering and hobbling and losing his balance and enjoying his garden and leading a generally contented if bodily enfeebled life. A decade would be a still less negligible
span in which to observe and reflect.

Walter Dandy, the American who first risked the operation, left us
snatches of dialogue with a patient whose right hemisphere he had
removed, a man fated to live only a few days, and who had a fever
while talking.

> PATIENT: They could get a lot of things around here to build!
> QUESTIONER: What things?
> PATIENT: Grass and things.
> QUESTIONER: What is this place?
> PATIENT: Johns Hopkins Hospital.
> QUESTIONER: What would they use the grass for?
> PATIENT: To build a nest for the young ones.
> QUESTIONER: Who would?
> PATIENT: The mice.
> QUESTIONER: I didn't know that mice used grass for nests.
> PATIENT: Yes, ma'am, and feathers and cord and things.
> QUESTIONER: What made you think of mice?
> PATIENT: I don't know, I just thought of them.

A human mind associated with a half-brain, or dwelling in a half-
brain, or even being a half-brain, as some scientists would consider

it, was still able, judging from this dialogue, to poke about in what interested it, a half-brain apparently space enough for that, or whatever way to regard it. Presently the conversation went elsewhere. Then $3.00 was mentioned without provocation.

QUESTIONER: What do you want $3.00 for?
PATIENT: To get to Cambridge.
QUESTIONER: Why do you want to go there?
PATIENT: My wife wants to go.
QUESTIONER: Why?
PATIENT: Because her mother is there.
QUESTIONER: Does it take $3.00 to go there?
PATIENT: NO, $1.65.
QUESTIONER: Don't you have that much?
PATIENT: Oh, yes'm.
QUESTIONER: If it is only $1.65, why do you need $3.00?
PATIENT: You see, my wife's got to go and I've got to go.

That last was spoken, it sounds in the last line, with some annoyance. Realize, it was a dialogue between a patient and a nurse, or, to be needlessly literal, between one cerebral hemisphere attached to a brain stem in one skull and two cerebral hemispheres attached to another brain stem in another skull. Lately, animal brains have been surgically sliced through much of their length, right side cut from left, and, after recovery, each side taught to do something independently of the other, in the experimental psychologist's sense of taught, and what the one side had learned the other side did not know. Guesses can be made as to the meaning of this, but no firm conclusions can be quickly drawn. The old large questions did insistently recur: could the objective methods of science be applied to the subjective questions of mind?

SHERRINGTON:
Not One but Two

Instead of a whole brain there might be the scant nervous flesh involved in a reflex. It might be of the skeletal system. It might be of the visceral system. That might be another station from which to view the problem of brain-mind.

The scientist who first saw the reflex in its full variety was Charles Scott Sherrington. He came to it slowly, came to it with great talent, soon was completely absorbed by it, and not for weeks but for two generations; then in his life he reached a conclusion about mind.

There was a cholera epidemic in Spain, and Sherrington, a young Englishman, went to help, no doubt to help himself, was striking out to find himself, as young men have before and since. He was a physician inclined to research, the direction of it not settled. He met Cajal. Years later Sherrington would say of the Spaniard that he had "introduced a new conception of the nervous system as a whole, and no man who ever lived did this to his degree." It was an accurate statement.

Sherrington's research, once he had settled the direction, was the scrupulous study of the so-called simple reflex. Dust blows into your eye and you wink. The doctor taps under your kneecap and your leg kicks out like a jack-in-the-box. But Sherrington over the years had demonstrated in detail how the reflex was not simple. He derived concepts from it. He pursued it with persistence, with intensity, with some combination of artist and artisan. Usually his experimentation kept away from the terrifying complexity of the top of the nervous system, the brain, but by no means always; he was fascinated by and worked with that complexity; but usually he stayed down in the spine, where the technical problems were more manageable. He did stimulate the motor cortex in primates, and his mapping of it was regarded as correct and standard for years; it was in all of the textbooks still at a much later date. Once the primate was a gorilla. On the day of the gorilla an assistant had a revolver ready, because a gorilla with a point of view has the stamina to carry it out. For the spinal studies he used mostly cats.

In the course of the prolonged experimentation, he gathered the data which, with his and later interpreting, gave a more fundamental comprehension than any before of how the body stands and moves, and, that firmly set, he was able in English-gentleman fashion to worry the scientific community with his carefully thought-out statement of what was the nature of mind, or rather what was *not* its nature.

So he can be said to have made the reflex, somewhat indirectly, one station around Fuji.

Sherrington's career branched originally from that of Friedrich Leopold Goltz, the German who in 1881 had made the trip from Strassburg to London to the International Medical Congress to report on the dogs whose cerebral hemispheres he had removed. Goltz had one of the dogs with him. A glittering congress. Charcot was in the audience, French neurologist and psychiatrist at the peak of his power. Ferrier was there, British physiologist, who demonstrated a monkey paralyzed in one limb, consequence of a sliver trimmed with jeweler accuracy off the outer surface of its brain. Charcot saw the monkey, burst out: "C'est un patient!" Whether Sherrington heard that remark is not recorded, but he also was there, might be considered the younger generation that comes knocking at the door. Later he would be Sir Charles Scott Sherrington, O.M. For him the importance of the congress was the opportunity to join in doing a postmortem on the nervous system of that dog "which was exhibited by Prof. Goltz at the International Medical Congress of 1881." Those were Sherrington's words. Goltz had killed the dog in front of the audience and taken out the brain and spinal cord, and Sherrington's teacher, Langley, got that material, and together teacher and pupil studied it microscopically. What were the changes in the low parts of the system when the high had previously been destroyed? Sherrington's first article dealt with that—the young man's first published article. The importance of the congress for the world was that it turned Sherrington finally into a field that he would plow and plant and grow and weed and from it glean not only a new neurophysiology but also, for whoever cared, another view of brain-mind.

Like Cajal, Sherrington shared a Nobel Prize, Cajal getting his early, Sherrington late. Graduate students began coming to him from near and far, many of them Americans. Looked at historically, his effect on neurophysiology because of his students became even more worldwide. A generation moved by, another, another. He experienced lapses of memory, told another man's story to the man who had told it to him, and in 1952 at ninety-four years of age he died.

He left a heritage that was like Cajal's in this: it would never cease being drawn on. He had worked with the reflex as methodically as a general of the army who isolates one regiment of the enemy, decimates it, isolates the next. Ever since Descartes, three hundred years earlier, the reflex had been a unit mechanism of the machine of the body. What Sherrington did with his imaginative thinking, with his careful experimentation, with the techniques of one and two generations ago, was to make that mechanism quantitative.

Since the body in action could be regarded as an unrelenting hierarchy of reflexes, Sherrington did add his part to the century's fatalism. In the laboratory he was cool. He built a guillotine that chopped off the top of an anesthetized animal's head together with a proportionate amount of its brain and threw the animal into the state that Sherrington labeled decerebrate rigidity, ghastly to look at, painless to the animal, which last fact Sherrington always took seriously. He vigorously supported legislation against laboratory cruelty or indifference.

The decerebrate state was seen sometimes in human beings after an automobile crash, after an industrial accident (a workman struck by a moving crane became a two-legged decerebrate), was also seen as a transient phase in the fading of motor controls during the closing hours of any life, but never before Sherrington satisfactorily explained. At the half-point of his career Sherrington wrote *The Integrative Action of the Nervous System*, a book which sixty years later still appeared assured of its kind of immortality. It made the machine of the body even more machine than Descartes had intended, and Sherrington had not intended that either; it screamed from his work.

Liverpool had him as a citizen for his eighteen happiest years, but probably had little notion of the distinction. Liverpool was one of the world's great ports, hence rats, hence cats, and whenever Sherrington needed a cat, he had only to send a boy down to the docks. Unusual man though he was, he apparently let himself be caught in the usual feuds of higher institutions of learning. Between his department, physiology, and the neighboring department, anatomy, there was a door, and the two professors each kept that door bolted on his side. A poor lecturer, he would have fits of absentmindedness, would turn away from the class, make some calculation of his own at a corner of the blackboard, turn again and continue the lecture, on another subject.

His work with the reflex completed, as he saw it, he was ready to turn his mind toward brain-mind.

His fame now was truly international, truly deserved. Oxford offered him its chair of physiology. Honorary degrees began to cover the walls. Invitations to memberships in foreign societies arrived like the first of the month.

Meanwhile he kept producing more and more nonphysiological writing, sought to clarify for himself and for the reader his view of life and the world, attained a vexatious style, and his ruminating book *Man on His Nature* has exasperated many a reader in and out of science.

In contrast, his biography of the sixteenth-century physician Jean Fernel, who defined physiology and pathology, had the scholarliness of the trained historian.

Anyone who has intelligently and scrupulously year after year worked with the nervous system, let alone worked with genius as this man did, must inevitably and frequently have reflected on how mind relates to brain. Of the gross functioning of that unique flesh he said: "As to its mechanism, perhaps the point of chief import for us here is that those who are closest students of it still regard it as a mechanism." (A Sherringtonian sentence.) What was mind? Where? How? In many a twisted phrase he argued that the nervous system was everywhere the same, bottom to top, everywhere input-output, only tremendously more input at the head end because there were the eyes, the ears, the nose, yet not anywhere in that tier-on-tier could he find mind. And he did not find it among Cajal's neurons, not in the mammoth assemblage of them at the head end, not in any multiplication of his long-studied reflex. Someone might comment that he was unwise to look in those places, but the someone would have to remember how sustainedly this gifted man looked, how shrewdly, how he had disciplined himself, how prepared he was by temperament to find his way through that orderly jungle the brain. What he said *must* be listened to. The brain he described as an organ of liaison that made connection with mind, was associated with it, but in what manner he disavowed understanding. To him we were two. We were brain and mind. "As for me, the little I know of the how of the one, does not, speaking personally, even begin to help me toward the how of the other." The dualism never worried him. "That our being should consist of two fundamental elements offers, I suppose, no greater inherent improbability than that it should rest on the one only." Was mind energy? A twentieth-century scientist might be surprised that a twentieth-century scientist should ask that. Sherrington thought it unanswered, and unanswerable.

COMPUTER:
In Metal

Sherrington would have rejected as nonsense any computer-is-a-brain analogy.

Two decades ago Sherrington—a young man of ninety in a wheelchair in a nursing home, his long study of the reflex praised throughout the world—labeled the reflex an exhausted gadget. Two decades ago, Cajal's neuron and the uses to which it could be put had been available for a third of a century. By that time the computer had comfortably settled itself among us. It was everywhere—up with the astronauts, below with their earthbound controllers. It was in medicine. In physiology. So-called special-purpose machines were part of nervous-system experimentation, and general-purpose machines were serving the computer-is-a-brain analogy. Computer models were replacing one another like automobile models.

At the computer's earliest appearances, someone implied it was a mind. "In principle . . . I do not see why it should not . . ." From mathematicians and psychologists and other scientists came the phrases. "No reason a computer should not be as original as a mind, just a matter of space enough and money enough; with space enough and money enough . . ."

Everyone agrees that a computer can receive data touching the living if the data are mathematically expressible, and that the data can be symbols. In our society what cannot serve as a symbol can serve as a model, and a treachery of models is that they may start as an aid to thinking and end by seeming to be the thought. For example, the computer began as a model for a brain, became a model for a mind, then computer was mind.

World War II gave a push to such ideas. Three developments touched the life sciences. One, the atomic bomb. Two, the computer. Three, any device that employed feedback.

A thermostat was at the time Example No. 1 of feedback, as Norbert Wiener explained in his *Cybernetics*, which was responsible for the first great thrust of the mind-computer analogy. The temperature in our house rises, the thermostat receives the information, sends a report to the furnace, which turns itself down; temperature falls, furnace turns itself up.

An antiaircraft gun was the happy-warrior example during World War II. Example No. 2. The gun swings toward its target, swings too far, corrects, overcorrects, corrects back, keeps that up, at the same time corrects for target speed, wind, rotation of the earth, the corrections smaller and smaller until missile and aircraft meet at X with the desired dead soldiers. Our muscles by innumerable corrections attain the smoothness of the dance floor, and if driven by genius attain the manual skills of Michelangelo. The human ear corrects the human voice, the voice the ear, back and forth, till Pagliacci hurls his tragic, perfectly pitched lament.

A second engineering development of the war was the computer. Though it associated with physiologists and psychologists, it still was merely the great-grandson of the adding machine. Then it matriculated at MIT, matriculated at Harvard, entered government service. The clerics have come in too. Several years ago a Church of Scotland clergyman using his computer had a bitter quarrel with a Massachusetts Episcopal clergyman using his computer as to the authorship of the epistles of Saint Paul. Was Saint Paul six authors? The conclusion seemed to be that he was one. In Italy five computers commandeered by a cleric, if the press releases could be depended upon, interpreted the shades of meaning in the thirteen million words of Saint Thomas Aquinas, wherefore thanks to five computers each of us might help to arrive on schedule in the House of God.

Throughout, the computer goes on doggedly solving problems in arithmetic. These must be presented as the human mind presents: input, order, output. *Program* is the term for this.

To dub a human mind a computer is good or it is bad. Good where it encourages anyone in any field to use this tool to minimize human drudgery. Good where some special-purpose computer is expediting some special job, such as exploring the activity of any living cell. Good wherever human life and the human mind have produced hills of arithmetic in places where there were only anthills and the hills need to be leveled to the plains. Good—but now we must be careful—where the idea computer-is-brain-is-mind pushes the thinking of someone who honestly wants to perceive, even more than to understand, the ways that brain and mind may conceivably relate. Say that the someone is a researcher, capable of keeping separate fact from fiction, talented in the building of computers. He collects data. Constructs curves. Goes to a meeting. Returns to his drawing board. Builds a better computer. Compares this to the brain. It compares well but not well enough. Builds another. Compares.

Bad to dub a human mind a computer when it introduces more machine thoughts into a society already bulging with them, and the computer is a suffocating machine thought. Is there first thing in the morning in the telephone. In the fingertips of the girl at the airline office when she informs you in forty-five seconds of the reservation you may or may not have between Fairbanks and Nome next January 30. Every week our lives and minds are more enmeshed.

But should that person, who from long back dreaded the computer that he thought he was destined to be, sense now a computer breathing down his neck, let him not despair. A noted cyberneticist-psychologist-specialist a few years ago told us that the metal brains up to then had attained no more than the developmental stage of the head ganglion of an earthworm. Ought to cheer the earthworm too.

In New York City at 590 Madison Avenue, the then world headquarters of International Business Machines, IBM, there dwelt for four years, 1948 to 1952, a computer visited by thousands. It was second in a line of great computers. The first, Mark I, was at Harvard. That at 590 Madison appeared to be a room that one walked into and looked around. Rows of vacuum tubes. They covered walls, everything busy counting, averaging, comparing, gathering information, storing information, all this information releasing other information. Toward the back, punched tape kept moving along. In cabinets, cards were being punched, being read. Confessedly, one did not immediately think of a brain. "Of course, of course, but. . . . In principle . . ." The 590 was fabulously fast, its descendants fabulously faster, the upcoming faster. "And what will he do when he gets there?" said the Chinaman watching the racer.

The 590 had judgment. When a crossroad in a thought process had been reached and the computation had to continue north or south, the machine decided which. This judgment was self-determined by the arithmetic whether so many billions or so many. The 590 had sex, female, said the engineer; when a group of professors from Harvard visited, up she perked, performed phenomenally.

Name any aspect of mind, the 590 had it. Norbert Wiener years ago had already developed a psychopathology, a lexicon of mind diseases, for computers. A computer could have a neurosis. Could have a psychosis. Norbert Wiener should be considered the Hippocrates of computer physicians.

But could a computer have the capacity of the living to learn? The engineer thought it could. On one occasion the 590 solved an old problem faster than a new, so it must have learned. Someone suggested that the machine parts for the new might not have been used for a time, that dust might have accumulated on unused contacts. Which would make absence of dust, learning. Presence of dust often has been.

René Descartes's machine of the body has, then, in our time been pushed up into the skull. A cocktail party where machines walked about or sit on stools. Ingenious machines. They dance in the Bolshoi Ballet. Talk of Sartre and Ionesco and Beckett. Perpetuate themselves in small replicas that grow. One wears a summer tuxedo. One wears bifocals. One wears contact lenses, which he takes out and loses. One has a tonsured pate. One invented the lever. One wrote the Song of Solomon. In Atlantic City at a national meeting each wears a white ticket so that everybody may know which machine he is. A hundred fiddle in the symphony. Art, science, equations issue from that vent each has at its head end. They call theirs the machine age. Not entirely to agree with this, or something equivalent, marks a man as an unsocial machine. One died on a cross.

CONDITIONING:
It Learns

Returning to flesh, getting away from metal and plastic, getting away also from the simple reflex that Sherrington studied, would bring us to the only less simple reflex that Pavlov studied, the conditioned.

The conditioned reflex.

A dog stands alone in a room, a trained dog, a small room. It could be that the dog is tired of these daily experiments, but is it not a human being performing them? The experiments must be sublime. Any vivisectionist who took advantage of that trust in the dog would be a low beast, wouldn't he? Pavlov said of the dog: "This friendly and faithful representative of the animal world." Pavlov spread that feeling through his physiological laboratories; gentle places they were, and physiologically lively.

The dog stands facing a wall of the room. A dark room. No sight. No sound. No smell, so far as a man's nose knows. No stimulus. No change of energy. The intention is that for this dog the world will change only when and where the experimenter chooses to change it. For the duration of the experiment the room will be, sense-speaking, unvarying except as something pleasant or unpleasant or neutral is introduced by this meddling member of another species, the researcher. Russia saw the first such room almost three-quarters of a century ago, but it has since been duplicated in every country where science thrives, modified of course.

The following is a typical, and also the original, conditioned-reflex experiment.

A button is pressed. That rings a bell, or sounds a buzzer, or lights a geometric figure in the wall the dog is facing. Almost immediately another button is pressed and the dog is served an exact portion of meat powder in a cup, the dog needing only to bend its head to eat, like breakfast in bed. The portion is small. A hungry dog stays more alert, has better intelligence, will do a better experiment; we are the same, a bit of starvation exciting our intellects. Loose straps are slung around the dog's legs to prevent it from wandering, prevent it from investigating, exploring, as is said in the laboratory.

With the eating of that meat powder the first experiment ended. It will be repeated any number of times on this day and on subsequent days, each repetition spoken of as a reinforcement, meaning that what has happened will have been etched deeper into that dog's brain.

The experimenter throughout the experiment has let the dog be alone in the small room, not able to see him and be aroused, wag his tail, have his mind stimulated. That last the researcher would never have said. Nervous system stimulated, yes.

The researcher is watching the dog via periscope or one-way glass.

How can he be sure that in the course of the repetitions of the experiment something is building up inside that dog? How can he prove it? How can he know that something is being etched deeper into the flesh of that brain? Simple. He counts the drops of saliva that drool from one of the dog's salivary glands.

In the classic experiment, a minor surgical operation was performed in advance; a slit was made through the dog's cheek, and the duct of the parotid gland, which is a conveniently close salivary gland, was brought out through the slit, saliva from then on trickling, not inside the dog's mouth, but outside on its cheek. During the classic experiment a funnel was sealed over the slit to catch the drops. The first time the button is pressed and the geometric figure is lighted there are no drops, because the figure does not yet speak meat to that dog's brain. But from then on, the figure does speak meat.

The effect of the stimulus is first generalized in that dog's brain. Generalized is the inventor's term. This means that the stimulus reaches anywhere, and, possibly, from what we today know of the interconnections in the brain, everywhere, signifying nothing specific to that brain; but with repetition of the stimulus it does signify something specific. It has become particularized, concentrated, means meat. It reaches one exact point in that brain, the point isolated from the rest. This was the inventor's explanation.

To emphasize another side of the matter, a more everyday body side of it, when the stimulus is particularized, reaches the one point of brain, the dog's physiology now neither wastes saliva nor does it fail to provide it. Saliva flows for meat or the promise of meat, and the amount of saliva is the right amount for the job. That brief lighting of the figure was the stimulus. It sent out notice that the meat was coming, and the brain now responds accordingly.

To emphasize still another aspect, a more obscure aspect, more difficult to establish, there is no emotion when there is no assurance of meat, but when there is assurance there is emotion. The geometric figure in its overall effect, emotional and otherwise, is like the dinner gong that starts saliva, and on a rolling ship starts nausea.

When the experimenter on a first occasion places meat in a pup's mouth, saliva flows. This saliva does not depend on previous conditions in that pup's life, is unconditioned. It is an unconditioned stimulus, and the reflex that provides the saliva is an unconditioned reflex. On the contrary, when the geometric figure is lighted and saliva flows, that saliva does depend on previous conditions, the figure is a conditioned stimulus, the reflex a conditioned reflex. That dog's brain has learned.

A second type of experiment goes still farther into the dog's thoughts, if one can say thoughts, a word that would have thrown the inventor of the reflex into a fury, a pink flush over his pink-white Russian face. In this second type, the lighted figure again is presented numerous times, with meat, the dog's brain learning, then presented numerous times with no meat. And what the brain does now is un-learn. It sends no-meat signals to its salivary glands. These secrete less and less. For this the inventor's word was *extinction*.

Curious what did in fact happen. As the repetitions with no meat continued, the spot of brain concerned was pushed down not merely to zero but below zero. It now had negative force. It had the power to resist excitation—had the power of inhibition. And this inhibition could accomplish things. It could lessen force as a brake can. It could, as the inventor understood it, reduce something taking place in its neighborhood, could separate off and make exact some spot of excita-tion by digging around it a moat of inhibition, which to the inventor was how the brain separates: for example, one blue dress from all the other blue dresses in *Harper's Bazaar*. The separation can be exceed-ingly fine. For the dog it is not shades of blue but shades of gray. All shades except that which signifies meat are inhibited. All pitches of sound except that which signifies meat are inhibited, and a dog thus develops a hearing as sharp as Toscanini's, but not for the glory of God, for meat. All smells also, and indescribably acute is a dog's sense of smell. The inventor's phrase for this inhibition was *internal inhibition*. He worked with it widely. He got everybody in his laboratories to work with it. Each year the techniques were modified to take advantage of developments in other sciences. Many scientists at the same time were denying its existence, or denying the interpretation, or scorning both, saying both were nonsense, all a lie—in the polite terms scientists call one another liars.

Later, electrodes were implanted in the brain along the path of the conditioned reflex, and attempts were made to follow millimeter by millimeter what occurred. This was the beginning of the busy im-planting of electrodes for many purposes in many parts of the nervous systems of many animals, and in man. It is going on somewhere at this hour.

Meanwhile the discoverer of the conditioned reflex often is doubt-ed, often reinterpreted, often misinterpreted, something like hated by some who borrowed from him. The pivotal position of his reflex in experimental physiology may be forgotten or buried under respectful acknowledgment, but mostly all over the world his reflex is not forgot-ten, and in the Soviet Union, where there are no gods, he has become a god. His forward-leaping imagination saw his techniques as applica-ble to mathematics, politics, language, literature, all aspects of mind, psychology, psychiatry. He was full of plans, but he died.

PAVLOV:
Not Two but One

He was even more Russian than the name Ivan Petrovich, by which everybody in Leningrad knew him.

Pavlov himself not only never used the word *mind*, he forbade his students to. What could they know of an animal's mind? They had trouble enough knowing their own. Instead of mind they should say higher nervous activities. If they did not they were fined, real money, kopecks.

Pavlov had his opinions. He had his convictions. He knew black from white. He had his vehemence. He had his passion for work. He had his scientific imagination. And he kept something of the child. Aristotle did too. Genius always does probably. Whatever happened to man or beast Pavlov squeezed into his conception of the conditioned. He spoke of it to the world—and for everything in life he spoke with his earthy statement. Even dedicated salesmen have second thoughts. Not he. During the thirty-five years that followed his first description of his reflex, he used it to study hearing, vision, neurosis, sleep.

He was born four hours by train from Moscow in the out-of-the-way city of Ryazan, had a church-school education, as a young man left for Saint Petersburg, continued his education in the capital. First it was general science. Then it was medicine. Then it was research, initially on the control of the heart, later on digestion, finally on those higher nervous activities. He was in Saint Petersburg when it was renamed Leningrad, lived through the violent events associat-ed with that turnover, went on living in leningrad for the rest of his life. Married a schoolteacher, had a warmly successful marriage, and more and more people began to know him. Soon everybody everywhere in the capital knew him, knew his birthday, and the entire country and much of the world knew his deathday. That was in February, 1936, when he was approaching his eighty-seventh year. He had said he would live to be one hundred and fifteen, arrived at that figure by mathematics, calculated the number of years it would take to complete the experiments he was at. Notwithstanding, he gave up the ghost in February, and probably it was because his son had given up his the previous November. Pavlov had invested heavily in that son, his youngest son, one of four children, three sons and a daughter. That son was planning—actually it was planned for him—that he should carry on the work after his father's death. Pavlov would that way be getting two lives, cheating Homer's gods, but the son died, and the doctors said cancer, and the father died two months later, and the doctors said pneumonia. Men sometimes die when their principal motive is thwarted, or when their arrangements are thwarted. Possibly Pavlov did not die of pneumonia. Possibly they misdiagnosed his illness—should have been "death of a son."

Every morning and every evening he walked the same Leningrad streets, flat streets by the side of the River Neva. People saw him, were instructed that he was a great man; if he looked as if he were reflecting on something, they must not disturb him. The energy in that walk could be seen at a city block, exaggerated by a limp from an imperfectly aligned fracture ten years earlier. No "life is settled" was in that walk. Youth flashed through old age. A free body. A free head. Fire in his eyes though their color pale. Love of physiology was love. Some in Leningrad said his was the only free life in the Soviet Union of Stalin's day. Pavlov had described a reflex of freedom. He had described a reflex of slavery. Dogs revealed both. Every day in those years Pavlov added to the territory of the conditioned. There were scientists who at the same time were thinking that he distorted the only and true and everlasting meaning of reflex. Cajal would not have thought so. Sherrington either.

Pavlov had a white mustache, a white beard, white thatch for eyebrows, reminded people of George Bernard Shaw, and this may explain why Shaw said that the Russian was a mere vivisectionist who had uncovered nothing about the dog's mind that was not self-evident to anyone who owned a dog.

Pavlov's father had been a priest of burials, his grandfather almost a muzhik, a serf. The grandson stepped off from all of that, as afterward from czarist Saint Petersburg, communist Leningrad, took a world place in digestion physiology, in nervous-system physiology, first place in Russian science. He won his Nobel Prize in 1904, when Nobel Prizes were hard to win because the committees still had long lists to draw from, when unmistakable achievement could be the sole criterion. Pavlov's prize was for the work on digestion. Winners of Nobel Prizes are apt not to gamble with what they have won, risk only variations on the theme. Not Pavlov. He was well into his fifties. He dropped digestion, never returned to it, year after year struck at those higher nervous activities, struck at the nature of mind, as he would see it.

In digestion he was impressed, and impressed the world, with the orderliness. The successive juices did not flow willy-nilly from countless glands, but were precise in composition, precise in amount, precise in timing, flowed for what was there to do, and when it was time, or just before, for the smell of the steak, the sounds from the kitchen, the sight of the cook. So he came to use saliva to study those higher nervous activities. Since something in the mind could determine this saliva, might that be turned around? Might drops of saliva, counted, tell what was transpiring in a mind? Could spit do that? Pavlov liked to recall how he asked himself this question for four years, by that time had convinced himself that the answer was yes, plunged. Before he was finished, he had the whole world paying attention, physiologists, behaviorists, psychiatrists, psychologists. To condition became as vulgar as to repress. In America there was an avoidance conditioning, an instrumental conditioning, an operant conditioning. Classical conditioning versus operant conditioning became Type-I versus Type-II conditioning. During the same time there was developing a behavioral approach to psychology and psychiatry and, at least it is arguable, a winding road to an existentialist philosophy. Overnight, one realizes in retrospect, he had excited international experimentation and had himself achieved global size.

He believed that the conditioned reflex could be applied at every level of nervous activity, and had anyone doubted that mind was a level, doubted that it belonged in the physiologist's domain, Pavlov would have blasted him with Asiatic scorn. He did not deny the subjective but fused it with the objective, psychology with physiology. The two were one. "What could have happened to Sherrington?" he burst out angrily; an intellect like Sherrington's accept dualism? As for himself, it pleased him to have done his part to place mind where it was, *in* the flesh. His was a satisfied monism. What right had we to impute to dog or man anything that could not be measured with a yardstick? Dog mind and dog brain were the same, and man was an evolutionarily later dog. No worker in the Leningrad laboratories would have dared let slip a phrase suggesting the psychic—the dog suffers, the dog is remembering, the dog is having a dream—on that February day when Ivan Petrovich died, when eighty trained workers came at 9:00 A.M., left at 5:00 P.M., as punctually as workers in a factory.

This scientist had no secret ambition to be a writer, put off even the writing of a scientific article, was always late for the publishers. He was late nowhere else.

Despite his priest-of-burials father, Ivan Petrovich admitted nothing left over from the other side of death, as who would expect him to; never anything of mind floating up to the skies. The idea of a soul was absurd. When he fell ill and a cold in the head became pneumonia, he called a neurologist, wanted to discuss with an expert his own deteriorating mental processes until these would no longer discuss. Somewhere in the course of the discussion he commented that mortification was setting in, his brain was mortifying. So it did, of course. Sturdy, direct, stubborn as a peasant, sure of himself, sure of science, he anticipated a day when mind and body would be expressed in a single equation. That gave him satisfaction. Like the physicist, he would like a mathematical equation to take care of everything—body, mind, planet. One equation. His talent was so powerful that it was bound to reach into the future, was bound even to affect the cold war. That last credit would have disgusted him. He expected others to speak as literally as he, had neither capacity nor patience for the indirect. To make his physiology a plinth under Soviet ideology, as has been done, was as dishonest as it may have been politically useful. The two were uncompromisingly materialistic but otherwise merely happened to grow up together, in time and space. Pavlov never thought of equating them.

That he helped Russian science and Russians to believe in Russia is fact, and good for Russia.

Once he visited America, gave lectures in his rough tongue, a translator translating. The translator was his eldest son, the professor of physics at the University of Leningrad. While his train waited in a railroad station, someone stole his wallet. He was glad when that visit came to an end, relieved to see the shores of America recede, which is the way any of us may feel when we leave the other country; but there was the wallet.

ELECTROENCEPHALOGRAM:
Spirit Writing

In 1875 Richard Caton spoke of the "feeble currents of the brain," but not until 1929 did the dramatic news come that electrical waves had been recorded from the outside of a human skull. Hans Berger had done it, and he gave what for the time was an extraordinary summary. Two principal electrical rhythms registered, one when the mind was attentive, the other when it was inattentive. There were other rhythms. They changed with changing states of consciousness.

Berger was a psychiatrist in Jena, Germany, had crude equipment but nevertheless made this starting discovery, that if electrodes were properly placed on the human scalp, brain waves would be written. They had a rate that varied, so-and-so many per second, and a voltage that varied. Berger worked in secrecy for five years. His electrodes were metal plates fastened one on the forehead and the other on the back of the head. Subsequently he stuck needle electrodes into the scalp. He found that if his subject's eyes were closed, his mind relaxed, nothing wrong with his brain, the average rate was ten per second. Berger called this the alpha rhythm. It was best obtained from the back of the skull over the vision areas.

Another rhythm is faster than the alpha, eighteen to thirty. Another is slower, four to seven. The rate from the cerebellum may be three hundred per second.

Berger named the recording instrument the electroencephalograph, and the record the electroencephalogram. This later was abbreviated in our clinics and laboratories as EEG.

Radio tubes amplify by ten million the feeble currents of Caton, and a wave writer, an oscillograph, faithfully follows the instant-to-instant rise and fall. These can instead be displayed on the face of a cathode tube, and photographed, but usually are written on moving paper by a magnet-driven ink pen. Voltages and rates are read off directly, as are the reports from the heart.

Sixteen pens reporting the brain at the same time became common, from sixteen scattered points the pens bringing together one story, one view.

Even Berger's ordinary alpha rhythm remains largely a mystery.

Each brain cell generates its small electricity, which can be made to discharge; therefore, it would be natural to conclude that some summing of the electricities of groups of such cells might account for the brain waves. However, the impulse over the frog nerve is many times one up-and-down alpha, a tremendous difference, so brain waves could not be produced in that direct way. A wave as understood today might relate to a combination of factors: to the membranes of the nerve cells, to the chemistry inside and outside them with resulting hyperpolarization and depolarization, to some electrical spread between cells, to local circuits that might reverberate among cells, to some intermittent bombardment of a huge population of cells. Great numbers of cells would need to be drawn into unison, and that is why peacemakers were postulated but they were not found.

Sheets of dendrites of neurons are directed toward the curved surface of the hemispheres. Myriads of those microscopic threads are close together and parallel and all directed outward. That represents an enormous membrane. When the axon of a nerve discharges, is depolarized, the depolarization invades those sheets of dendrites, is followed by a slow hyperpolarization, so the subsequent depolarization could not occur quickly. This alternation may pace the waves of the electroencephalogram.

When a man or his dog is idle, is not paying attention, is in a state possibly for loosely remembering, for floatingly thinking, the ten-per-second, the alpha, is apt to be dominant over a large of the skull, but particularly over the back. One can keep up with that fluctuation on the oscilloscope face.

The ten-per-second is far-reaching among animals. Many besides men have it. Brains for some reason have required the ten-per-second behind all their hell-heaven-earth activity. A water beetle has a vision ganglion useful for experimentation, and when this beetle's eye is illuminated, all of its light-sensitive cells evenly stimulated, after they have discharged once, one-tenth of a second must elapse before they discharge again. It takes that long for the rebuilding of that particular chemistry, and something similar probably explains the alpha.

One student surgically undercut a thin slab of the surface of a cat's brain without disturbing its circulation, dug a moat around that slab, but left an isthmus connecting it to the mainland of surrounding cortex. The electroencephalograph proved that that island and the mainland were beating in unison. They were connected. If now the isthmus was cut through and only the island remained, they no longer beat in unison. If, however, the slab had been thick instead of thin, and the isthmus cut through, that island would continue to beat. This suggests that something beneath the surface is feeding upward, that in your brain and mine the ten-per-second may be paced from beneath.

In the last years, the computer is increasingly brought into the study of brain waves, to compute averages, the brain operating on averages, operating probabilistically, it being more and more evident that with as many units as the brain has, it could not possibly operate in any other way. For this there are special-purpose computers. The operator is able on the oscillograph face to watch the constantly changing reports from the constantly changing brain, can stop the instrument at any moment and dwell on that moment. We are steadily if slowly knowing more what the brain is. Via the computer it is revealing an electrical flexibility that we know it evidently must have, its present always having to meet its past.

What haunts us and what mainly interests us is the simple fact of the beat, the ebb and flow, the relation to its background noise, our brain being electrically an exceedingly noisy place.

The electroencephalogram has long been recognized as varying with our thinking, our feeling, anesthesia, action, age, open eyes, closed eyes; an encephalogram is apparently unchanging if circumstances are unchanging. Into a rhythm of our brain, the affairs of our lives throw another rhythm. Those affairs may be a story. May be a sight. A soft memory. A sharp hate. An impersonal hitting the mind through an ear.

A man's character has been thought readable in his electroenceph-alogram, has a pattern recognizable if the writing is examined careful-ly enough, a pattern that consistently stays with him Monday through Sunday.

Even before birth, activity can be picked up from the forward part of the fetal skull. Four months after birth definite activity begins at the back of the skull, may have been there earlier but masked. The rate is slow. Gradually the rate increases. By the time the child is thirteen, the characteristic adult rhythm is settled and remains so until illness or old age wears it down and death wipes it out. Hans Berger took his original recordings from the brain of his young son. Berger believed us to be as individual in our brain waves as in our fingerprints. If so, it would make us ask once more, and not expect an answer: what does all the individualizing mean—the high cheekbones, the wrinkling nose, the straight gait, and these computerized electroencephalograms? Identical twins have similar electroencephalograms. Is it all just an incidental mathematics? Is it all psychologically irrelevant? It surely is not philosophically irrelevant. But is it philosophically discouraging? Is it just replicating genes? Is it just creature?

A healthy medical student, eyes closed, lies on a couch with elec-trodes pasted to his scalp. They look like curlers for the night. Wires lead from them to the electroencephalograph, where yards of ink-writ-ten paper roll off. The alpha repeats. The student opens his eyes and daylight strikes his retinas, impulses travel through his brain, and even an inexperienced observer sees that the writing has changed. Is it fast-er? It is irregular. The student closes his eyes; the alpha returns. He starts talking with eyes closed; the alpha goes. The observer is half amused that not only has the ten-per-second gone from this brain, but some equivalent change has occurred in the brains of all those watch-ing. Our brain is more sensitive than the surface of water. So are these fluctuating electrical signs of our minds. The student stops talking. His alpha returns. He is asked not to keep his eyes closed but to think hard of a dachshund. He does. The alpha goes. He is seeing in his mind a dachshund. A lapse. The alpha returns. He opens his eyes. He is asked to look at a uniformly lighted surface that flickers. The brain waves follow the flicker, up to twenty-five per second. The student's eyes now are uniformly illuminated, and there is the alpha, which stays when he closes his eyes. He is relaxed now. Begins to doze. Keeps dozing.

His eyes have been closed for fifteen minutes, but not until now could it be said confidently that he has left the waking world—high voltage and slow rate in the brain waves, every now and then a spurt of fast waves called spindles. He snores—large high-voltage waves that look random. The watchers nudge each other. If the student claims afterward that he heard everything that was said, he lies. He opens his eyes. The alpha goes. He closes his eyes. It returns. Someone hints at an escapade last Saturday night. Peremptorily, the alpha goes. Everybody laughs and the record skips and hops. The student will soon be an M.D., so he recovers his poise but not his alpha. Plainly it is appropriate to call the alpha the rhythm of inattention, and about last Saturday night this student is helplessly attentive, so has lost his alpha.

Has any of this anything to do with what interests us more than anything else during our long-short sojourn under the moon?

EPILEPSY:
Divinity Hints at It

Every day in hospitals and clinics and even in basicscience laboratories there may be talk between electroencephalographers and epileptologists. Experts consult each other. It is their problem to understand the disturbance in body, brain, mind of persons suffering seizures, those unwelcome visitations that have been adding to human knowledge from far back in history, medical history, just history. The Gospel according to Mark described them. Hippocrates objected to their being called divine.

The first epileptic attack one ever witnessed one did not forget. A playmate in school. An old man living up the street. One previously thought him merely disreputable. An overproper woman who sat straight up in the bus and the next minute was on the floor. (A distinguished neurologist related a case where sun flickered through trees into a bus and started a first seizure.) One had been told about fits, but being told was not enough; the woman shook so, her face was so twisted, and what must be going on in her mind one could not dismiss from one's own mind. Maybe nothing was. But one decided that her embarrassment at the spectacle she was creating would have made anyone hate the whole world. She had lost consciousness. Though she did afterward sit straight up, it was in apathy. Everybody in the bus knew. She knew. Her friends knew. They all always had known.

Epilepsy has put footnotes into the biographies of some of the greatest—Napoleon, Baudelaire, Mozart, Cellini, Caesar. They each had experienced this, one of the more dramatic unpredictablenesses of human body and mind. If it was a facet of mind, it was a blemished facet.

Drugs might have helped that woman. Drugs have made epileptic attacks rarer in schoolyard, polling booth, town meeting. Most seizures are controlled by drugs. Physicians the world over have learned about the drugs and how to administer them. Physicians have explained the disease as far as they could, and could quite far. They have added to our compassion and to our realism, have lessened the lurid, lessened the mysterious. It would be wrong to draw the conclusion that because the mysterious had been removed from seizures, it had been removed from the universe. The universe keeps its frame of mystery, but we do with urbanity accept what in Caesar's Rome would not have been possible to accept, that the epileptic has a flaw in his physiology, in his chemistry, his anatomy, damage directly or indirectly in the flesh of his brain.

A seizure may start always in the same part of the brain, a part in some respect different, the person born with the difference, or got it after birth, a visible difference sometimes, a molecule difference sometimes, a still utterly unaccountable difference sometimes. It may all be on the outer surface of one hemisphere or it may be deep. Electrical energy piles up at a point, and when enough has, there is the discharge, and with it the agitation of body or mind, or of both. That is how the attack has been explained. The cause might also have been some control cut off, so that that which was controlled was released, hence the wildness of the muscles. That rising electrical energy has been likened to a gathering storm, abruptly the thunder and the lightning, and the body shaking and the mind getting queer.

If such a brain comes to surgery, there may be a scar from an old head injury, may be a benign or a malignant tumor, a clot of blood, a wasting. In these cases the epilepsy is spoken of as symptomatic, the person showing the symptoms of brain irritation. Probably all epilepsy should be spoken of as symptomatic, and the causes of the irritation not necessarily so gross, merely a rearrangement of molecules, a metabolism altered. If no flesh evidence can be found but the probable neighborhood of the source of the symptoms is surgically cut out, the attacks may cease. When this happens, it must be believed that the tissue had something hidden, some abnormal pattern of nerve fibers, lack of enzyme, focus of trouble. When the focus is definite, is found at surgery or postmortem or otherwise can be diagnostically pinned down, there is still another designation—focal epilepsy.

The electrical storm may be small, brew a time, a rumble, occur most infrequently, one attack a year. The storm may be large. A hundred attacks may pass over body and mind in a single day, the misery so unrelenting as to make another moonlight, another dawn, another glorious sunset not worth waiting for. Sometimes the gathering storm is not quite adequate to induce an attack. Sleep may so have depressed the high region of brain, the cortex, or disconnected it from the low, that the high does not get through to the low, the body does not display the signs, and the man never knows.

A shaking thumb may have been the first evidence of an attack; the shaking may advance from thumb to hand, that shakes, to arm and shoulder, that shake, to the whole body. That focus was in the highest area for the thumb, the attack triggered from there, marched from there. (*Marched* is the verb used for the advance in the brain of the sweep that began in the focus.) The attack may have begun and ended in the focus. The thumb shook, no march, and the mind watched the thumb as it shook, could do nothing about it, might have been something like fascinated by it. An attack may begin at the angle of the mouth. In an eyeball. In a toe. Anything sudden may be the provoking stimulus, a door opened, slammed shut. Through the attack one may make out the lit-up architecture of the nervous system, some area of it. One sees the anatomy. One sees how mind-tinctured our mind has made our brain, sees what the physiology, if it can for the moment be considered physiology, has done for anatomy.

Epilepsy has in this manner often cast streaks of light into the architecture. Even a casual observer may think he is learning in furtive flashes something additional and important about the living. Even the specialist may pick up qualities that no drawing, no dissection, no photograph, no word could give him. The word would be best. Writers on epilepsy have sometimes had strong literary talent, and the epileptic has sometimes been an excellent literary subject, has beckoned some of the shrewdest pens in medicine, Hippocrates and Hughlings Jackson, to mention two 2,000 years apart. But a quite ordinary physician might listen with interest to his epileptic patient, keep on the heels of his story, the anatomic story also, this possibly interrupted by the clearcut statement of the misery of this person's daily life, which might have meant most of all to the physician, and been most important for the patient to have revealed.

An event outside the person may trigger the attack. A street scene. The smell of a barber shop. Music. Musicogenic epilepsy is a kind. In a well-known case it was the music of one composer, Tchaikovsky. The mind of the dead composer reached into the mind of the living citizen, mind to mind, or, for whoever prefers the other emphasis, brain to brain, touched a spot, started an electric turmoil. Recently in a widely publicized case of musicogenic epilepsy, rock 'n' roll and jazz had no effect but soft music brought on an attack, and a tedious long-term treatment with the techniques of Pavlov's conditioning cured the patient, the extinction one after another of the complex of conditioned reflexes that were excited by the soft music.

Epilepsies for decades have been considered to be of three kinds. There were other dividings. Some believe that dividings have always been justified and are inevitable, others believe that when the smoke has blown away there will be no dividing and each epileptic will be a single patient and a single person with a single mind. The three classic kinds are *grand mal, petit mal,* and *psychomotor seizure.* In grand mal there is a stiffening of the entire body, an intense tone that spreads over it, this phase called tonic, and grafted upon that is the rhythmic shaking, this phase called clonic. A tonic-clonic convulsion would be grand mal. A frothing at the mouth might go with the attack, an epileptic cry, a biting of the tongue, a blueing of the skin, a helpless urinating, and to conclude this drama that Egypt and Greece and the Bible and the Middle Ages knew, stupor and sleep. The episode might come and go in seconds. It probably is ushered in by a warning, called aura, and the aura probably mindtinctured, a wisp of sensation, a smell, a bad smell frequently, voices, a face, a crowd, an action, a strong memory, a story.

In petit mal there is no convulsion. The person briefly loses consciousness. That may be the whole of it. The mind goes off. Petit mal has been called an absence. The French have called it that. A blank gaze, a blinking, a nodding of the head, then in a moment the eyes again see the actual world around and the young girl is without any recollection of what has occurred. If her head did nod, it lifts. The eyelids that twitched are quiet. The work that she halted she takes up again. Petit mal is an illness of the young, commonest between the sixth and twelfth years. The mind usually is not damaged. Furthermore, month after month this illness may reach into this person's life, and he may never know it, never notice this fleeting trickery, and, more astonishing, his family may never notice. (Eddies in the stream of mind, whirls, may flow right past us and we pay no attention.) As to the span of the attack, that can have been five or six seconds. Dozens of attacks can have occurred during a single day.

In psychomotor seizure the person performs some unusual act. There is some abnormal behavior. It is a trancelike state that may come and go in seconds. But the seizure also may be more prolonged, suggest less of mind, less of body, or more of mind, more of body. A New Yorker finds himself in San Francisco, does not recognize the name of the driver's license in his own pocket, and the newspaper reports that an amnesic was picked up last night at the docks. The New Yorker's mind and the San Franciscan's mind were the same, presumably, but what separate compartments there sometimes can seem to be in a mind! Here each of the two compartments was unknown to the other, though both were associated with one brain and one body. It is the commonest of focal epilepsies. And the focus is the temporal lobe or its neighborhoods, the hippocampus, the diencephalon.

Helping to characterize the three classic kinds are their three electroencephalograms. Berger already knew that abnormal electroencephalograms were produced by epileptic heads. (Berger was of course the first to record any electroencephalogram from any human head.) The three epilepsies and the three electroencephalograms often correlate so closely that an expert may appear to believe that the record rolling off his instrument is itself shooting off these Roman candles.

Each of the three epilepsies and some less categoric but related disturbances of brain-mind do have strikingly individual electric characteristics. The pens may write a sharp spike. The pens may write slow waves that quickly wax larger and faster, get irregular, get unpleasantly regular, get unnaturally slow, because the man who was in tonic-clonic convulsion now is in stupor, finally write sleep waves, because he is asleep. The pens may write a sharp spike followed by a rounded dome, a monotony of that, spike and dome, spike and dome, three per second, but even in this repetitious electroencephalogram of petit mal there may somewhere be individuality. The focus is in the depths of the brain. It has been thought by some that the thalamus and cortex here are playing a game of echo, a burst of electricity in the thalamus following a burst of electricity in the cortex, each writing its part of the record. If one part fails or is too feeble, the push for the other is lacking, the attack blocked.

An epilepsy draws some of its electrical character from the part of the brain that excites it. Attacks may start apparently from anywhere in the brain, mind appear close, but the relation between brain and mind as always elusive, and the mind as always slipping away from any but literary or philosophical definition. From the scalp of a person who never has had an attack it may be possible to tell by the electrical signs whether he belongs to the part of the population that might expect to have an attack. The epilepsy is latent. But, also, there may be epilepsy and no evidence in an abnormal electroencephalogram.

Hippocrates two thousand years ago did object to epilepsy being called divine, saw no divinity in it, took it for granted that a damaged brain was responsible, taught that any convulsion must be regarded as serious until proven otherwise. He was a careful doctor. He believed that epilepsy like other sicknesses had an exact cause, which must be sought, and a treatment found. The seeking continues. The other seeking is less common but continues also, to discover in these minds something concerning the nature of all mind.

BRAIN STIMULATION:
Summoned

The boldness of brain surgery when it has added to it the boldness of calling forth with electricity occasional flashes of mind, as can be accomplished notably in the speech-associated regions of some disturbed brains, must stand among the medical successes of the century. Dog, cat, monkey, chimpanzee, dolphin, and human being—the brains of all have been stimulated by the hand of the researcher or the surgeon.

If it is the brain of the human being, and if the reason is epilepsy, the prerequisite steps for the surgery are those for any brain surgery:

the patient's story, the X rays of his skull, X rays of the arteries of the
neck and brain, electroencephalograms, echograms, anything else, all
the tests. There follows the diagnosis, the decision to operate, the con-
sent of the patient, the informing of the family, operating room and
anesthesiologist scheduled, surgeon's and associates' hands scrubbed,
and scrubbed, in order to dig from the deeper layers of the skin as
many as possible of the bacteria that always dwell snugly there, then
the powdered rubber gloves, the green gowns, and the play once more
is on.

Augering and sawing through the plates of the skull is routine, as
is the exposing of the brain. Which side, right or left, which lobe or
lobes, would have been established beforehand, yet the detailed localization, which may need to be most precise for this brink-of-disaster
surgery, might be helped to be if words could be made to come out of
that mouth when that brain is stimulated electrically. The anesthetic
is a local.

As for the substance, the flesh, even to the experienced eye that
may appear quite normal, also under the microscope, the abnormality
something not visible, something chemical. The surgeon's knowledge
loads on him the worry that the smallest bit of brain can have life-and-
death importance. It is in fact paramount that this brain be stimulated
directly, the epileptic focus located with the greatest possible precision.
The voltage that he now uses, the shape of the electrical pulse, and the
rate of stimulation are the harvest of years of trial and error.

He stimulates. Point. Point. Point. His stimulating is a nudging by
the vulgar outside of this most intimate inside. Nudge at one spot—a
single muscle off somewhere contracts. Nudge at another—the patient
speaks out, says that something tingled. Another—clear-cut mind. It
shines out. Anyone there at the operating table recognizes it.

For an instant, mind shone out. It was not something inborn in this
patient. It was something he experienced. He acquired it. The scrap-
piest scrap it was, but it was his, and it was mind, and it was forced out
mechanically; an electrode did it.

Scars in the brains of soldiers struck by shrapnel in World War I
became epileptic foci and were cut out in numbers by a pioneering
German surgeon. Scars in the brains of soldiers of World War II, and
in the brains of civilians abnormal from many causes, were cut out in
even greater numbers by a Canadian surgeon. Both those surgeons

were in their time known around the world. The Canadian's patients, since they were being operated on under local anesthesia, could hear, could answer questions.

Part of the Canadian's staff is stationed in a gallery behind glass. A microphone lets conversation go back and forth between him and them. Each point of brain stimulated is labeled with a ticket dropped onto the pia, that web of membrane that lies over the substance and has the nourishing small blood vessels. The tickets are numbered—1, 2, 3. A mirror fitted diagonally over the open skull allows the operative field to be photographed through a window in one of the walls. The photographs could make one imagine one was looking at a planted garden, the tickets marking the plants—geranium, jonquil, marigold. Meanwhile on sterile paper with sterile pencil the surgeon every now and then is jotting a note, afterward will be left with two kinds of on-the-spot evidence.

A small girl's epilepsy, cured by surgery, is talked of over years in interested circles. While an infant she was given an anesthetic and for some unknown reason, and not recognized at the time, had a small hemorrhage under the dura of her brain and suffered a passing paralysis. Later, when she was seven years old, she was walking in the tall grass of a meadow, her brothers in front of her, and a man came from behind and said: "How would you like to get into this bag with the snakes?" Terrified, she started running. Her brothers saw the man, and her mother remembered every detail of the ugly experience. Later the girl began to have nightmares, and in them relived the experience. At eleven she had an epileptic seizure and in it also relived that experience. More seizures. The experience was apparently her dominant memory. Her mind kept returning to it. And in that manner it was borne into her brain. At fourteen they brought her to the Canadian. He studied her case, decided there was hope, exposed her right temporal lobe, found an old scar, stimulated at many points in the neighborhood of the scar. No pain. The brain itself is not pain-sensitive, and the tissues around had been anesthetized. He stimulated, stimulated. Abruptly the girl cried out. "Oh, I can see something coming at me!" Realize that her skull was open, her brain exposed, she talking, and feeling that she was about to have a seizure, knew the aura. Her mind had gone on to that long-past experience, but she was also present in the operating room, and knew she was. "She seemed to be thinking with two minds," wrote the surgeon. She was in a way two persons. The one was continuing to relive that experience in the tall grass that somehow had latched itself to the flesh damaged at the time of the anesthesia in infancy, became more latched with each nightmare and seizure. Our brain probably does the same with any repeated memory, latches it deeper into the substance (not surprising that there is so much we cannot forget), only in the small girl there was the damaged flesh to somehow help, to somehow sensitize. The surgeon believed his electrode built up a local electrical storm similar to that probably built up for any usual epileptic attack.

TEMPORAL LOBE:
Peers from beneath Sideburns

This part of the brain lies inward of the temple, lies inward of that flatness, on the side of the skull, hollowed somewhat, attractive in a gentleman if he has a distinguished cranium in a narrow tall head.

The Canadian surgeon won his most dramatic successes in the temporal lobe and its vicinity. He was seeking to cure epileptics, and those seizures that caused him and would cause anyone to reflect on the nature and the meaning of mind were related to disease in this vicinity.

The small girl frightened by the man who offered her a place in the bag with the snakes suffered temporal-lobe epilepsy. After the surgeon excised the bad flesh, the epilepsy was gone, but she still had the memory of the experience. So that must have been cross-filed, deposited somewhere else in the brain, presumably in the temporal lobe on the other side. At least, something mental acquired within a lifetime of an individual seemed deposited in more than one place, and could be called forth by an epileptic seizure, or by an electrode, or by a mind's decision to call it forth.

The Canadian was a scholar given to the formulation of hypotheses, and he conferred upon this region of the brain the term *interpretive*, which may have been assuming too much, but it shows how he felt about that called forth, the quality of it, and it was based upon a surgery done upon many human brains. Mephistophelian success he seemed to have had. His rubber-gloved fingers reached into breathing brains, his stimulator touched a point, the voltage was turned up, and something of this patient's past, distant or near, was also turned up. If the electrode was held long at a point, the dream might expand, there might be more of it, more of the story, the electrical field perhaps spreading. And this might happen too if the electrode was moved to different points.

What the surgeon was awakening with the electrode was a hallucination; his patient was seeing something or someone not in the operating room, hearing something or someone no one else was hearing. Sights, sounds, smells arose, and there was nothing in the outside world to account for them. They were all inside the patient's brainmind. Hallucination could be stimulated only from temporal lobes that had trouble—here it was an old scar—not from normal temporal lobes. Illusion could be stimulated also—there was something in the outside world but transformed into something that was not in the outside world.

The temporal lobe is anatomically like a fist with the thumb pointing forward. It extends toward the middle of the brain, where the diencephalon is, and toward the base, where what is called the hippocampus is. The neurosurgeon when he operates is apt to turn back a larger flap than the nonprofessional would think he needed, but it must be remembered that only diseased temporal lobes have so far shown the capacity to hold and to give back the acquired, experiences, dreams; no other part of the brain has shown this. It does have some undeniable crucial capacity for binding past experiences to the present.

Epileptic attacks that originate in the temporal lobe and its environs, which is the commonest place, are apt to be accompanied by déjà vu; the patient feels that everything that is happening now happened before. Hughlings Jackson long ago pointed this out also. Someone suggested it might be owing to a partial shutting off of the blood supply, the nerve cells here not getting enough oxygen, which if proven would of course alter nothing as to brain mind. The attacks here may have other bizarre accompaniments, such as bizarre odors.

For a long time there has been evidence that the temporal lobes contain the highest levels for hearing. They should therefore be closest to the hearing mind. Some spots in some temporal lobes, when touched with an electrode, draw from the person a gross sound or block a sound. That is, if the person is counting out loud (under the drapes on the operating table) and the temporal lobe is stimulated at a point, the counting stops. The patient does not know why or what happened. He just stops counting. (Only somewhat similarly, but at another point of cortex, a person might know that his hand had moved and that he had not moved it, and that if he reached over with his other hand, he could stop the movement.)

There have been many observations on damaged human temporal lobes, and some research on animals. Both lobes have been removed in a monkey. It produced a manifest, frightful mental maiming, called temporal-lobe syndrome. All in the neural fields know about that syndrome. It was well described, nevertheless difficult here also to be sure of just what was being described, but at least the appearance was a shattering of a world. Even so, the removing of one lobe in a human being might be not shattering at all. It depends on which. In a right-handed man the left lobe has been spoken of as noisy with mind life—many regions of the brain are spoken of as silent. In spite of what has just been said of the left temporal lobe in the right-handed, a surgeon can snip two and one-half inches off its forward tip, and no mind will be lost. Apparently, no mind lost. However, let the surgeon not snip off more than that, because "if the scale do turn but in the estimation of a hair" there will result the desperately maimed, reminding us again how close the chasm of mind may come to the edge of brain.

As to the noisy left temporal lobe, someone described it as playing its part whenever the mind was concerned with things seen and heard, and the right, whenever the mind was concerned with space and was getting its data from touch and sight.

Telegraph lines, gross lines that have nothing really to do with the work of the temporal lobes, run low in the neighborhood of each lobe, and when these are ripped across by tumor, there are vision defects, smell defects, taste defects, not much more learned from such tumors than might have been surmised from the neighborhoods damaged.

With some deficits, the person seems to be receiving his sensations normally but is not able to give significance to them; this is called agnosia, and there can be individual agnosias for the individual senses.

One might suppose, if one were supposing, that because the right and left lobes are so profoundly important to the creature, they were set far apart on opposite sides of the head, with the brain stem thumbed up between, and then if one lobe were destroyed by screaming savages come down from the hills in an automobile at 2:00 A.M., the other would have a maximum chance to escape; but one must immediately admit that evolution has never seemed much concerned with you or me, always only with the big us.

MEMORY TRACE:
If Democritus Had Explored for It

What is the impression left in the flesh of our brain by an event? What is the trace? What is altered? And what is the unit of such alteration? Dare one look for that? The ultimate mark? Democritus would probably have looked for it. One does shrink from the idea of a unit of mind and tries to get one's foot on structure.

The unit could be a neuron. It could be a linking of neurons. It could even be the surrounding of neurons. In our day many surely would be tempted to think of something into which they could shove a microelectrode. Many would be tempted to join the chemists and try to find some giant molecule able to alter an atom here or an atom there, then able to hold the alteration.

Whatever the unit, it probably would be the same kind of unit in beetle, fish, frog, man.

There might be an electrical twist. There might follow, or there might just be, an oscillation of sorts. The oscillation would lead to an etching into the flesh of the brain. Which is a sequence developed by a winner, not of one, but of two Nobel Prizes. He may not have had much evidence, but he had plausibility and vitality. The etching would not be visible, not today, but someday might be visible by means of an electron microscope or some future microscope, and it might be that one could not rightly use the visual word *etching*. But that would be the memory trace. That would be the mark. If it does not yet have an identity, it has long had a name—engram.

For some persons with their heads over microscopes the attention has gone to those terminal knobs of the neuron. The knobs are visible with the ordinary light microscope, were described by Cajal. Dendrites → cell body → axon → terminal fibers of the axon knobs. Evidence of changes in those knobs has been searched for. A few workers were convinced that with activity there was an increase in the number. It could not be proven.

For a time it was thought the size of the knobs held the answer. They swelled with use and shrank with disuse. If a knob swelled, it would bring that neuron nearer a neighboring neuron, possibly touch it, a region of touchings, a region sensitized, a region abler to recognize a signal that earlier had passed that way, hence a memory. Unless there is a retaining—this never was in doubt—the creature would have no way of carrying past to present, present to future. And he has. He has the briefly enduring, sustainedly enduring, short memory, long memory, and he has that long long long memory that would be species memory.

American students spoke for a time of reverberating circuits. Reverberate means literally to strike back. A nerve impulse would go round and round in a chain of neurons, reverberate in that sense, a round-and-round that maintained itself, the chemistry of each link, each neuron, supplying the energy for that link. The round-and-round would be the memory, the engram. Everything about this is reasonable. But the search for even a single dependable engram has been like the search for the Holy Grail—no pilgrim pure enough to find it quite.

A circuit might lie still for a time, be nudged, begin again to circuit, dig deeper into the flesh, lie still, be nudged. That seems right for a memory. It is not to be trusted too easily because it seems right, but it is one of the ways a human being memorizes.

How long might a trace last? By common sense, long. By experiment, what dependable experiment there was, short. In one nervous system, a cat's, the trace lasted four one-thousandths of one second. By another technique, eighty seconds. By another, six hours. For six hours a second message would find it easier to get through where a first had gotten through, hence a memory.

Electricity is everywhere in the brain. That cannot be gainsaid. Small voltages sum to large. Voltages shift. Battalions of electrical Lilliputians march north, east, south, west. What a commotion does exist. This is no obsession in a neurotic at MIT. This is one of Pavlov's "facts." The potentials are there. They can be measured. We trust what can be measured. Whether and where and how such a memory is embedded in the commotion, the future may tell, but no need to wait on the future to say that if all the studies of all the laboratories were pieced together, every region of the brain could be proven dynamically connected with every other, as it is structurally connected. Anatomists from their side long have heaped up the evidence for our brain being one world. Our mind to each of us always has been one world. A scientist of the nineteenth century was notorious for having said that mind, therefore memory, was manufactured by the brain as bile was by the liver, and we of the twentieth century smile patronizingly, because we believe we know that mind, therefore memory, is manufactured by reverberating circuits, or the like. Socrates and Plato groped along the lanes of reasoning, whereas we tramp the straight road of laboratory research. A plain citizen needs to pinch himself to keep knowing that what the laboratories have up to now found, besides a complicated chemistry, is complicated electrical patterns, that could be related in more complicated but exceedingly flexible fashion. Memory running alongside them is not an extreme assumption. Memory being the output of oscillation, reverberation, patterns worn among nerve fibers, swellings of synaptic knobs, anything comparable, might be nonsense. But probably not. A wilderness of electrical swayings the brain inescapably is, a *Waldweben*, a wilderness in which the pioneering of man, after so many tries, might be delineating the ultimate order, ultimate precision, ultimate chill mathematics with which Nature has put together this, her prime instrument, if the human brain is her prime instrument.

WORM MEMORY:
Little to Wash

The planarian is a leaf-shaped flatworm, lives in fresh water, has been a convenient dab of material for the use of zoologists and others, and has recently been taken up by the psychologists.

Planarians—worms—were conditioned. It was a modification of Pavlov's technique with his dogs. Pavlov would have been interested, would have paid those worms a visit if they had made the trip to Leningrad. He never cared to travel.

The worms were "trained" to "remember" accurately and to "respond" wisely, and if they did not they were "punished" and if they did they were "rewarded." It was found they did. Their nervous systems could make engrams, whatever they are. The worms had memory.

Next, in some laboratories, the worms were sliced, and now it was claimed that the head end remembered, but the tail end remembered too.

That dropped experimentation to the molecular level, where all would be stable forever. Furthermore, since the mid-fifties and earlier, biologists have been diversifying their investments, have drawn them somewhat out of the proteins and deposited them freely in the nucleic acids. (Tell that to Aristotle if you get there first.) One nucleic-acid molecule, DNA, has the code, and another nucleic acid, RNA, carries the messages from the DNA, which is in the nucleus, to the ribosomes, which are outside the nucleus, aided and abetted by still another RNA, and promptly the building stones of the proteins, the amino acids, get restless, get marshaled, fall into place, form chains, and the chains because of their great number and the rearrangeability of their links can be as different as our lives. They are our memories.

It is toward the better understanding of such and other memories that much of the experimentation of our time has been directed, is directed.

Natural in our mid-century for many scientists to conclude that memory is chemical, is not deposited in some pattern of neurons, as was thought for years, but in those molecules of DNA and RNA, which have the almost infinite possibility of arranging the nucleotides that compose them, hence the almost infinite possibility of the numbers of memories. Soon some researchers reported that it was specifically the RNA molecule in the worm that learned, because if the sliced tail end in the throes of regenerating itself a new head was chemically treated so as to alter its RNA, the worm behaved in a random, erratic, psychotic manner. It had lost its memories or they had been thrown into disarray. The worm had gone off into the who-knows-where.

A rash of laboratory activity ensued. The RNA and the protein manufacture is basally determined by DNA, and DNA is basally inherited, but somewhere there is also the capacity to retain the acquired, which is this-life memory. Neurons have millions of RNA molecules. The neurons are stimulated, the RNA gives orders; there is some newness in the orders, some newness in the proteins, some newness in the memories. So the worms learn.

And some of this learning can be transmitted to others. Psychology can be passed on. Animals can be taught. Life can be taught. Animals are said to have been taught fear, the fear passed on, and it was not vague but specific fear.

Contrary opinions all along the line were voiced. Verification of the planarian experiments in some laboratories was declared impossible. The worms in those laboratories refused to cooperate. For those scientists the worms simply would not remember. Slicing them was therefore silly. But the first group and their protégés stuck to their guns. They were more enthusiastic than ever.

In short, memory and mind at the molecular level were proving as troublesome to agree upon as at any other level, and that conceivably is the moral.

Today, educated planarians seem illusory, but who can tell about tomorrow? Who can tell what may yet bob up?

HUMAN MEMORY:
Never Washes

Whatever once is in our mind, is in. We are assured of that. We think we forget, think it unhappily or nervously, but are assured, on the whole convincingly, that nothing is lost, may be hidden but not lost, except what came in immediately before a concussion, or before a sleeping pill. The last we all know except the few who have the sturdiness and the nobility to pass up this one night's sleep. Whatever truly took the trouble to get in cannot get out unless it is cut out literally (can one believe it?) with the attached flesh by a surgeon, or by an apoplectic stroke, by a neat small hemorrhage, by a neat small blood vessel narrowed until the blood no longer flows through, by an island of brain if not cut out, rubbed flat by the erosion of old age. Nothing is merely erased, particularly when printed in by feeling, or had specific reasons for staying unerased—reasons of guilt, shame, embarrassment, pride. There was a dance way back that lamentably failed. There was a dance with a redhead who let one think one was possessed of the rhythm of the wind. There was a dance that stood still like a summer night. There it is—now. There it is as it was five times ten years ago. A track of rain on a window pane, a figure in the design of a lace curtain as twilight settled toward evening. The psychiatrist seeks his patient's earliest memory. He believes that it was the one that clung so long and could be brought back in such detail because it had good grounds for clinging, and it would be worth his knowing. The years may have reshaped it, but the reshaping at each point had good grounds too, and it would be worth knowing.

Sometimes a memory seems merely an emphasis in endlessly extending mind. A beam of light strikes there, and that emphasis looms out of the dark, at a supper, at an all-night wake. Why? Every human being is as perplexed by the trifles he remembers as by the importances he forgets—that influential citizen's name, the mayor's, which for a hundred civic advantages one ought to remember. "I know your name, Your Honor, as well as I know my own—what is it?"

Any day, any hour, his mind may play such hide-and-go-seek with him. He tries to recall the theme of the second movement of the Ninth Symphony. Has known Beethoven's Ninth from when he was a child, yet all that will sing in his head is "White Christmas." In the restaurant last night they played it over and over, pretended the holiday season was gay, and this morning his brain plays it over and over, enough to drive him mad. The restaurant was full of other noises, but his mind kept filtering them out, only not "White Christmas." Nevertheless, tomorrow morning, Tuesday, while he squeezes an orange, over the earth, pounding and clear, surely will come the theme of the second movement. Must have been there yesterday, else how could it be there tomorrow?

A poor unhinged young man sits in an outpatient clinic, and a doctor plies him with questions. The answer to every question is the same bare phrase. Inadvertently, or so the doctor's skill makes it appear, a shot of color hits the young man's face, and the next instant it is as if a child's crockery savings bank had smashed, and memories spill and roll over the floor till the clinic-hardened doctor wearily shakes his head and the man looks haggard. Another doctor picks at a detail in a woman's story, picks and picks; suddenly tears fill her eyes, and he scoffs at the tears with that occasional studied brutality of his profession, but she remains as noncommittal as the front door of the U.S. Treasury. In desperation he puts her under a light anesthetic, and from her, too, memories spill and roll. They had all been locked behind that noncommittal front door. An eminent psychiatrist relates of a person who had lived on a ranch in Montana until she was three; years later in her dreams she saw the old house, but always the outside. He put her, too, under a light anesthetic—into the house she went. That interior had been in her those many years, and it would be past believing except that each of us has had some experience to match it. One gasps at the weavings from inexhaustible spools, hum of inexhaustible looms, that must have been in the skulls of Dostoevski, Dickens, Hugo, Zola, and in that stubborn Irish skull of James Joyce. We who are not geniuses cannot understand how there could be crammed into a single brain, where it apparently was, the stuff of *Bleak House*, of *Crime and Punishment*. Ours are meagerer savings yet impressive in their way.

A man at a public dinner is fidgeting because he is the speaker. He bites into an olive, grows absentminded, finds himself in his grandmother's house, the pantry, and he turns and tells the toastmaster in the new tuxedo about his grandmother. He always remembers his grandmother pleasantly. The old lady was a spiritist. Nothing wrong in being a spiritist. He has himself never met a spirit, but we have different talents. His tall uncle . . . He loved his tall uncle more than the other uncles, and there were five, and five aunts, and that was only on his mother's side . . . But the toastmaster in the new tuxedo, from Detroit, now is absentminded too, is looking down toward the table, possibly seeing his own uncle, possibly seeing his soup. Tall uncle, tall mostly because his nephew was so short, had reached his bony hand into a crock on the third shelf in the pantry and came out with lard up to his elbow. The old lady had switched crocks because someone was stealing cookies. On and on he went. Why was all that registered in the first place? Why does it speak out now, out of that olive? He has eaten ten thousand olives from that day to this, and it was not the olive, not the whole of it, but the seed. When his teeth bit the seed, that was the trigger, the kind the psychiatrist seeks, the memory that hangs on to the seed, the earliest if possible, on the same principle that if it stayed that long it had good reason for staying, and what was the reason?

To Toscanini the story of Mozart's feat in the Sistine Chapel may have been comprehensible. Toscanini is dead and we cannot ask him, and maybe it was not comprehensible to him either. Mozart on the Wednesday of Holy Week went to the Sistine Chapel to hear the *Miserere* of Allegri, a carefully guarded composition; even the singers were not allowed to take home their parts on pain of excommunication. Mozart listened. The *Miserere* over, he hurried to his boardinghouse, wrote it down, all of it. It was in old church form, a double choir, no strong accents to help a mind remember. Two days later, Good Friday, he went a second time, heard the *Miserere* a second time, had put his score into the bottom of his high hat and while he listened made corrections. His family was frightened. The news did leak. He was summoned. Kapell-meister Christofori demanded the score, examined it, must have been as dazzled at the power of that memory as everybody since, found the score accurate, showered Mozart with praise.

Some memories step boldly to the front of the mind, as that with the olive. Some are timid. Some are methodic: a letter of the alphabet leads to a name, a name leads to last year, last year to eleventh-century Scotland, to Lady Macbeth, to Shakespeare. He said that memory was the warder of the brain.

If a man has an experience like that with the olive, he tells it next day at lunch, and if the story goes well, tells it at dinner, and if that goes well . . . Each time he tells it, it enlarges. He is surprised how he plucks fresh details right out of the air. Were those part of it? One definitely was not. It occurred this morning at Third and Walnut, but it fit so perfectly that anyone would have included it. Thus does a memory become a luxuriant tree. Sends out twigs. Sends out branches. Each twig has its bud. Each bud has its reason for being there. The brain is strange—stranger than any physiology. The mind is strange—stranger than any psychology. Memory may seem airily indifferent to what soil it is planted in, indifferent to what it helps itself to for the sake of its growth, and that is why memory underlies invention, imagination, tomfoolery, and the congenital liar.

FRONTAL LOBE:
Investigation with Gunpowder

Phineas Gage was a competent laborer on the construction of the Rutland and Burlington Railroad, had an excellent mind, acceptable personality, good character.

That was in 1848.

Phineas placed a quantity of explosive and a fuse in a drill hole, was starting work with a tamping iron 3 feet 7 inches long, 1¾ inches in diameter, weighing 13¼ pounds, when something went wrong. The iron plowed into Phineas's face, on through the orbit of his left eye, into his skull, through the frontal lobes of his brain, came out the middle of the top of his head, landed some distance away, was bloody and greasy. Phineas was stunned. He was driven in an oxcart three-quarters of a mile, walked a long flight of stairs to his hotel room, was lucid, pulse sixty, hoped he was "not hurt much," was seen by a doctor, survived twelve years. The tamping iron amputated the forward part of the frontal lobes on both sides, achieved, as proven at postmortem, what in our decades has come to be known the world over as a lobotomy.

It was a prefrontal lobotomy.

Phineas's trouble was in his two frontal lobes. The small girl's had been in her one temporal lobe. Memory is associated with the temporal lobes. What is associated with the frontal lobes is even harder to say. One feels uncomfortable when one begins to speak of any part of the brain in isolation, the parts and functions always proving interconnected, grossly or subtly.

In the history of medicine, every so often a solitary person has been struck down, in this case Phineas, the result some enduring advantage for the rest of us, for the health of our bodies, the richness of our minds.

Phineas's character changed. He grew profane. (Did not inhibit his self-expression.) Started enterprises that each time were abandoned because they failed to lead quickly where he expected. (Did not think ahead.) Was obstinate. Lost his adult point of view. Retained his passions. Something had happened, as stated at the time, in the sexual sphere. "The equilibrium of balance, so to speak, between his intellectual faculties and his animal propensities seemed to have been destroyed." That was a doctor speaking. "A child in his intellectual capacities but with the general passions of a strong man." For a time Phineas was a Barnum exhibit For a time he planned a line of coaches in Valparaiso, stayed in Chile eight years, but no coaches. Went to live in San Francisco. After other vicissitudes of a human life, it got to be May 1861, when one day Phineas was laboring in a field, the next was stricken with convulsions and dying. Tamping iron and skull can be examined by anyone in the Warren Museum, Harvard.

Something was learned from Phineas about the frontal lobes. Something about the brain. Something about mind. Something about brain-mind. More might have been learned but the time was not yet ripe, would not be for another hundred years, when the surgical removal of great numbers of frontal lobes would provide great numbers of facts, not easy to interpret, differing with the person, none to be trusted too far, and seeming more to be distrusted as mind on all sides was studied apart from brain, by psychologists, psychoanalysts, psychiatrists. This studying-apart-from-brain might change now any day, might never.

PSYCHOSURGERY:
Investigation with Knives

Two chimpanzees continue the story. The two had appointments at Johns Hopkins University, were volunteers (so to speak), were to take part in a study of the common cold. The one's name was Lucy, the other's Becky. Having finished their obligations at Hopkins, they went to Yale. At Yale they received intensive testing (Yale was early and always a place for psychological testing) to establish in numbers the power of their minds. They were given a battery of psychologicals; then their frontal lobes were investigated with knives, were surgically extirpated. This was considerable. One of the two before the operation had been "difficult," had a "disposition," let the world know it, flew into a fit of temper over any laboratory problem she could not solve, got worse from day to day, but after the surgery was a saint. The word used was *saint*. Later that year, 1935, an adventuresome neurologist in Lisbon, Egas Moniz, who knew about Lucy and Becky, and knew about some earlier attempts (frowned upon) to cure human insanity with operations on the brain, thought it worth risking the injection of alcohol into the frontal lobes of an asylum patient. The alcohol destroyed where it went. So that tissue was beyond function, good or bad. In later procedures the white fiber tracts joining frontal lobes to thalamus were cut through on both sides of the brain, the lobes in that way varyingly disconnected from the rest of the nervous system.

The operation of prefrontal lobotomy was comparatively simple yet might worry the surgeon because it was to disconnect what long had been thought the highest flesh. A special kind of courage there must have been in Egas Moniz. He later received a Nobel Prize, and later still it may have been regretted that he received a Nobel Prize. He had performed a planetary miracle, but miracles do not as we know remain miracles—this one for only a few years—but men learn by them both during the period of faith and afterward during the period of disillusion. Surely Egas Moniz had by leaps increased the experimental familiarity with the human brain. This pioneering began in November 1935, and by January it was possible for him to report on some twenty asylum patients whose mental illnesses he thought he had helped.

The original procedure was called leucotomy, after the instrument that did the work, a leucotome, a blunt hypodermic needle with a knife hidden in it. The leucotome was plunged into the brain, the knife was released, rotated, and the brain flesh was thus cut through. Later the surgeons, styled now psychosurgeons, largely gave up the leucotome, and there ensued a long series of modifications of technique till the operation got that name that for years was on the tongue of the world, lobotomy. It was less on the tongue after a time. However, the almost vulgar familiarity with the human brain continued. Aspects of traditional psychology were boldly reexamined by surgery, always in the course of operations that were attempting to cure human ills, and often did.

If the frontal lobes were lifted out of the skull, the term was *lobectomy*. Lucy and Becky enjoyed lobectomies.

Following that first gamble of Egas Moniz, the number of disturbed persons operated upon rapidly mounted, so that by 1948 four to five thousand cases could be reported at a congress in Lisbon, a record that was afterward bettered, if bettered was the proper term. The modifications of technique indisputably improved the technique, which became more exact and controlled, the brain extirpations less extensive, even small, the original slashings rarely employed. Then the whole procedure began dying out, and now is dead, by and large.

The clinical successes or failures or any moral meaning attaching to this surgery are not under discussion here. Drastic ills may require drastic measures, the patient's misery driving the physician to steps he otherwise might regard as barbarity. Life sometimes requires barbarity, as sometimes homicide, as often suicide. Whoever visits an asylum finds his judgment about treatment in a fluid state, unless he numbs himself by the frequency of his visits. The human mind is darker in an asylum, but even while thinking that one should not numb oneself, should not get used to other people's dark, one does. A few who early began performing lobotomies remained routinely enthusiastic, others were reserved, others condemned the procedure, others were "intemperate doubters." The general field gradually got comfortable with its name, psychosurgery. Animal experimentation, as for all things in modern medicine, was brought in to test methods and, where that had any common sense, with results.

For the human being the operation was mostly performed through holes bored into the top or sides of the skull, one on the left, one right. With time, less brain was disconnected, all operative steps were more economical, instruments more precise. Only the bottom quadrant of one frontal lobe might be separated off. Whatever the fiber tracks slashed, the cortex might be skillfully helped to retain its connections with the central system. Or slivers of different sizes of the outside surface might be merely undercut. That is, procedures were diversified, might in themselves be good, bad, better, worse. In the earliest days one occasionally had the impression that the surgeon was simply weighing off tissue, thirty grams for schizophrenia, twenty grams for involutional psychosis, fifteen grams for something else. It made one think of Shylock.

The probability is that brain should never have been imagined that hashable.

A thin electric wire, a cautery, was sometimes inserted an exact distance at an exact spot, and after an exact voltage of current was delivered, the spot coagulated. Voltage, time, kind of current, became standardized. The thin electric wire (insulated except at the tip) might instead be directed toward a spot in the thalamus, the spot coagulated, that procedure called thalamotomy. Claims were made for all the methods, always with some balance struck between psychiatric relief and permanent loss of mind.

What may have been the most extraordinary of the operations was carried out, not in an operating room, but in a psychiatrist's office, the patient after a short rest on a couch returning home to enjoy his lunch, the whole no more bothersome, locally, than the extraction of a tooth. For anesthesia he was given an electric shock through the brain; then an instrument with a sharp point, which it is not nice to call an icepick but which is called that, was driven by a mallet through the bone of the upper inner corner of the orbit of the eye, shoved in a measured distance, swept from side to side; a certain amount of frontal lobe was thus separated off, and the operation was accurately named transorbital lobotomy. It has been stated (in print) that an enterprising psychiatrist could do fifteen transorbital lobotomies in a morning, which is not to belittle the operation but only to state what has been stated (in print), and no reason at all why a psychiatrist's office should be a less proper place than a hospital, though in those years one discovered oneself not immediately able to throw off one's outmoded prejudice that a psychiatrist's office was a place for talk, with now and then a suggestion thrown in, now and then a prescription. Some surgeons, questioning the method, thought a surgeon ought always to be seeing where he was working, his eyes in every instant focused on the operative field, especially so because this flesh was agreed to have more than a casual relation to a man's inner life. But did it really matter? Hamlet, striding after the ghost, cried out: "And for my soul, what can it do to that, being a thing immortal as itself?" Which rhetoric would be directed not at lobotomist or psychiatrist but at life and death and salvation and whatever else. Let it be said, furthermore, that lobotomies were conscientiously debated by most doctors recommending them (like heart transplants today), conscientiously carried out, conscientiously examined a month or a year or even years afterward.

Lobotomy did, or could, or should, without question have taken us to one more station from which to view mind. One station cannot be expected to be as attractive as another. We are all of us huddled together, and each of us is also huddled alone—young surgeon, crazy old woman, crazy young man, poor muttering human lump squatting among the pigeons on the curb at Trafalgar Square. On most days we do not think of the huddling.

THE LOBOTOMIZED:
It Is Dimmed

The result of frontal-lobe surgery? To state it bluntly: it produced tens of thousands of Phineases.

They differed. Their illnesses had been different, the surgical proceedings different, the persons different to begin with. But, taking this all into account, the result of the lobotomies upon the minds was rarely satisfyingly described, and not explained. The surgery was neater than that with gunpowder and tamping iron; skilled toilers of the brain had taken up the quest, and they were adventuring, not in dog or cat or chimpanzee, but in the creature with frontal lobes par excellence, the creature best known to man, best able to explain what was going on inside himself. Nevertheless, interpretation was not easy. No such simple formula as: this much frontal lobe cut away, this much mind cut away.

The lobotomized after the surgery had less usable brain substance, might or might not have less mental illness, might or might not have more definable mind. It was difficult to make one's way through the reports.

It might be enthusiastically stated that a patient after lobotomy is no longer a custodial problem, no longer requires hospital care, no longer is an expense to the state, can return to his family, get back his job. All of this is without question a medical result, and it should tell us of the work of the frontal lobes, were one not immediately stepping into quicksand. Even if the patient (or experimental subject, it seems) had had a normal mind willing to cooperate, one might have felt engulfed, but he had an abnormal mind carrying the handicap of a brain underneath that had been surgically cut through. It would take a most rare psychological acumen. That is itself rare. And here it is confronted with apparently so many items.

It might (in a report) be enthusiastically stated that the lobotomized is less destructive. He has not the zest he once had to rip a fixture from a wall. This could mean that the presence of frontal lobes even in a sick person makes him less predictable, less inflexible, less socially acceptable, more interesting.

It might (in a report) be flatly stated that the lobotomized had a self no longer troubled about self. This could mean that the frontal lobes normally inspire an individual to trouble about himself, his private self, to dwell on that, possibly to seek to understand that. This is reaching rather far. Troubled he is, but hardly in any such clearheaded direction as to give himself the problem of trying to understand his private self. He is sick, usually seriously sick, mentally sick.

It might (in a report) be stated that the lobotomized has the aggressiveness to go out and meet a world in which previously he had been timid, and for this reason might even be earning good money. This could mean, freely interpreted, that the frontal lobes normally prevent a person from earning good money, and that could mean, freely interpreted, that the frontal lobes lure the mind from preoccupation with the outside world to preoccupation with the inside world, which surely would make it too bad to lose the frontal lobes. But that is of course more quicksand.

It might (in a report) be stated that the lobotomized is no longer worried about the future, and that could mean that frontal lobes deal with the future. Now they are gone and there is no future, nothing ahead to worry about, nothing ahead to expect.

Many things might be stated.

It should be pointed out, as it has been, that the gains from lobotomy might be as great as those surgeon-psychiatrists said, yet the person, if one imagines him equipped to stand off and look at himself, might not agree that he has gained, rather lost. A psychotic, an insane man, in a lucid interval might very well swing around, curse the surgeon meditating lobotomy, or any comparable brain-damaging therapy, tell him straight: How can one human being dare decide for another that it is better for him to escape his tortured self in order that, for instance, with a dusty cap on his head, a dusty cigarette between his lips, a dusty cloth bag for an apron, with majestic dignity via a nail at the end of the stick, he may pick up paper on the hospital grounds.

The original massive amputations of the frontal lobes were meanwhile less and less frequently performed, and only on the bitterly ill, schizophrenics with ceaseless hallucinations, voices that came from someone who was not in the room, sights that no one else saw, schizophrenics on whom all other therapy had been tried and none worked. Also, on the incurably depressed in whom psychiatric treatment plus medication had completely failed. Also, on those suffering the pain of cancer, death waiting, opium addiction established and no longer overcoming the pain.

When the amputations are reduced in size, as they were, there is reduced blunting of mind, reduced deterioration of personality, reduced scarring, therefore reduced chance of epileptic seizure. The last is a side effect in half the cases, one body-mind misery traded for another. Even when the surgery is no more than an undercutting of a small part of cortex, there may be scarring and epileptic seizure.

Summarized—so far—and drawing from scattered reports, the frontal lobes normally might deal with self, future, imagination, no money in the pocket. None of that has anything like a single meaning, a single direction, does not filter out a frontal-lobe function. That inside a man's noble jutting forehead remains a part of his mystery. A peculiar fact, among hosts of peculiar facts in the mystery, is that the IQ of the lobotomized is unimpaired. That apparently has been a consistent finding. It is even improved in those patients where there had been continuous agitation.

Immediately after a massive amputation, the patient lies in bed with eyes closed. He can hear, can answer, but is profoundly inert. He has lost sense of time. He thinks minutes to be hours, tomorrow to be yesterday.

Gradually he appears to return to the world. When he does return, it is at once evident that something has been lifted off his mind, a Bostonian restraint. There may be an uncalled-for gaiety. Coarse joking. One patient swats a nurse on the rump. And this now might seem to mean that the frontal lobes normally, instead of making a person less socially acceptable, might make him more socially acceptable.

Summarized again—so far—it appears that the human stump left after lobotomy depends upon what the human being was before lobotomy, and upon his illness, and upon the surgery.

Confusion as to the work of the frontal lobes is natural, as must be obvious from any listing of the impressions of these honest-enough experts. Even the brain in this neighborhood, though it may as flesh look like brain flesh elsewhere, may one day prove chemically or otherwise not to be the same. (There are definite chemical differences in parts of the brain.) The frontal lobes are a late neighborhood in the brain's evolution. They are possibly not as eon-frozen. They are possibly not as unvarying. They might accordingly not be as analyzable.

A deep depression sometimes settles permanently over the lobotomized, deeper than that which may have been there the years before, though observers usually emphasize reduced depression. It is manifestly reduced if it had been an agitated depression previous to the lobotomy. Selfabsorption (from all the reports) appears reduced. Anxiety reduced. A flattening of mind, a bleaching, is emphasized. The mind-flattened schizophrenic even in the heyday of lobotomy would accordingly not have been considered a good candidate. After lobotomy the pathologically groping grope no more. Nice not to grope, but not nice not to be able to grope. Most lobotomists claim that the IQ is unaltered, the intelligence unaltered, the capacity to pay attention unaltered; the implication could be that the personality is unaltered, but that is not true. Even the patient may claim the contrary. Members of his family recognize losses.

It is easy to see in all of this that the surgeon-psychiatrist (psychosurgeon, lobotomist) is trying like the rest of us to get an impression of mind and is moving from point to point to see if he can ever once have a complete vision.

Naturally the surgeon-psychiatrist (psychosurgeon, lobotomist) tries to see some good accomplished by his surgery, yet only a foolish surgeon could imagine that he creates mind by cutting away brain. He may cogently argue that the brain left behind after the surgery is thereafter less interfered with, as is argued (and proven) for the hemispherectomies where the removal of the bad half makes the good half work better. Those minds in any case are always altered, maimed, though they were of course maimed before the hemispherectomies.

Today there are fewer descriptions, fewer explanations, and almost no lobotomies; not much psychiatric interest in them; nevertheless the reports of the past voluminous surgery have for purposes of throwing light on brain-mind continued to have interest. Some intellectual observers keep reiterating how in the lobotomized the power for abstraction did dwindle. The possible disposition to abstraction of the observer must always be kept in mind—so impossible to see anyone else and not to some degree see oneself.

With the practical, with the obstructing disadvantages that the lobotomized may seem sometimes to have lost, one does understand what was said earlier about the lobotomized being able to get a job. Some physicians say, too, that though he gets a job, he drops it, does not want it. His ambition is gone. Why work? Some who are not lobotomized have come to that conclusion too.

LOBOTOMIZED PERSON:
A Bicycle Rider Observes It

Single cases were described by a Swedish physician. As might be antic-ipated, to observe a single case imaginatively, to observe it before and after lobotomy, or even only after, might tell something dependable about the frontal lobes, and the observation need not necessarily be by a physiologist or psychiatrist or physician, though these would bring the advantages of their techniques and experience, but observed and described by a mind that had the cunning to enter another mind. To observe such a single case as against testing five hundred with batteries of tests would of course not produce graphs, but might help us some, as against not at all, to understand how the frontal lobes relate to the rest of the brain, and relate to the mind.

This Swedish physician not only described single persons instead of populations but used ordinary terms in describing them. He had undoubted talent. He wrote his own histories. He did not neglect to perform the routine psychological tests, but what was impressive was his common sense and his interest in these persons. Like a good doctor he wanted to keep in touch with them afterward, when they were less ill, in some instances for years. He visited his patient's home, invited him to dinner in a public dining room, went bicycle riding with him to watch how he behaved in traffic, talked to his family, to his friends. All in all the Swedish physician appeared free of the compulsion to push a human mind into a textbook class or to report it in textbook fashion. He never forgot that these were single human beings with single hu-man minds, that each was unique whether sick or well. He reported a wife as saying she had been given a different husband by the loboto-my. This was not the same man. A mother said of a daughter whose frontal lobes had been disconnected that the deep feelings in her, the tendernesses, were gone; she had grown hard somehow. Another who had preferred classical music now preferred dance tunes. A wife said of her husband, "His soul appears to be destroyed." That last remark has been repeated many times by those who treat the mentally ill.

One might conceivably come nearer an understanding of the work of the frontal lobes if one made a generality of that wife's remark: namely, the frontal lobes piece together those delicate shades that some have thought to be a man's soul, the headiest distillate of his mind. The distillate could be simply the unanalyzable. The Swedish physician was definitely not thinking of religion when he quoted that wife. It has been claimed that religious feelings are helped by lobotomy; the person becomes less tense, less depressed, lighter in mood, though lighter may not be the character toward which religious feelings should move.

An English neurologist thought that the frontal lobes were necessary during the period of life when mind was acquiring, and that after that, one could more safely lose them. He drew his conclusions not from persons who had been surgically lobotomized but from persons who lacked frontal lobes from birth. The English neurologist thought that the frontal lobes enabled our minds to acquire, and that the acquired was then transferred to another part of the brain. If true, we would be needing our frontal lobes less after our mind had put away its savings and was living on them. The Englishman warned that many persons got old early, settled into that humdrum, but that others stayed young late and for them lobotomy would be too bad.

Human frontal lobes are huge compared to those of the ape, more so compared to those of cat and dog, and if one looks lower than mammals one cannot be sure that there is even anything corresponding. This of itself says something about the frontal lobes. It has also put the burden of understanding them upon human studies.

SIGNS OF EMOTION:
Masks Its Feeling

What a person is feeling, whether he has a sick or a healthy mind, may or may not be evident in the signs he gives out, though it usually is, and it depends again on the observer, one so often better at this than a hundred others.

A scowl appears to be gathering by the same stages in two persons. A grin appears too. A woman in the clinic each time her husband is mentioned curves her lip upward on the right side, then slowly the sign fades. Sadness has its stereotyped signs. Gaiety has. One often thinks, masks. One thinks, ancient masks, those that covered the body head to toes. Eons fashioned and supplied each species with its kinds.

Throughout his life, the individual remolds his species masks, but only somewhat, and mostly only the human individual remolds, but always it must be remembered that it is the human eye that sees, and it might not see perceptively the signs given out by a fruit fly.

Emotion is always somewhere behind the mask, strong feeling or weak feeling, subjective behind objective, and the scientist no less than the plain citizen may find it difficult to keep the one separate from the other, the *deus ex machina* from the *deus*, the painted laugh from the laugh inside .

As for the scientist's experiments, it is more feasible to perform them on the *machina*. Goltz successfully removed those two cerebral hemispheres from those three dogs. In his warm way of writing, and he could be accurate and still warm, it was sufficiently clear that signs that previous to the surgery had been held in check, after the surgery were released. He did not use the word *released*. That would come later. But signs seemed already to burst forth.

Years later, there was a series of experiments by an American. What he was seeking to do was locate accurately the site of control in the brain of the signs of one emotion, rage. He used cats. He removed more and more brain in successive cats until he approached near the tailward end of the hypothalamus. Here he found a few highly special cubic millimeters. These he must preserve if the cat was to display the signs. These millimeters were the rage center, which still today remains an established landmark.

Such a mutilated cat bursts into the signs on slighter provocation than a normal cat. Something that had been acting as a brake on the signs was lifted off by the surgery. With time the meaning of this display became clearer. The geography of the brain became clearer. If the experimenter's cut went farther toward the tail, even a trifle, there were no signs. But if the cut went farther headward, there would be some modification of the signs, something probably added; it might be difficult to see. Techniques continued to improve. It was possible now for an electrode to be implanted right within those millimeters, and the animal allowed to recover; then when the electricity was snapped on, there was again a bursting forth. Signs of rage could be snapped on or off. Depending on the position of the electrode, the signs might seem like fear, like terror, like aggression, like defense. It was a grossly mutilated animal, and yet every now and then there were the evidences of nuance of control. The implanting of electrodes grew technically by bounds, as did the surgical cunning, the shrewdness of the instruments. Everything suggested increasing delicacy in this experimentally induced behavior, but far from delicacy when contrasted with what we know, or think we know, of our own signs. Everything warned that there must be care in interpretation.

Interesting, or again just what one would expect, those millimeters (the rage center) were by evolution relegated to the same neighborhood in cat, in dog, in man.

When one sees a hissing cat—Willy, the tom, sending greetings to the hound next door—one recognizes that Willy's private life has indeed modified Willy's evolutionary life. One needs to look closely, but then one cannot escape seeing that somewhere inside this respected male there are touches of complex individual signs.

By the old James-Lange theory, signs would come first, and they cause the emotion. A man looks sad and that causes him to feel sad, and unquestionably looks and feeling, whatever the final truth about them, are intimate. The back-and-forth between body and mind is everywhere quick. That the signs and the emotion could keep each other going requires no theory. Othello feels, shows what he feels, feels the show of what he feels, feels more, this accelerating until the Moor can do it, strangle Desdemona, she so lily-white and frightened, he so dusky and unhappy, and in the audience not only the excitable Italian, the mad Irishman, the hot Spaniard, but the cool Englishman apprehends personally what is transpiring behind the footlights. Apprehends, but with that anguish of the theater that is enjoyment. So much in our own personal signs of emotion is layered and matted and not quickly unraveled, but is enjoyment, too.

The laboratory signs of rage were christened *sham-rage*. That was long ago. Nevertheless, better to stay out of the way of that cat's claws, for the claws are not sham. They unsheathe, then scratch, they tear, and they appear always to have in them, even the claws, something besides signs. The more brain left intact, left attached to the rest of the nervous system, if it is the appropriate brain, the more lifelike will be the signs. For perfection of signs of rage, or of any other emotion, any experimental worker would concede that the total nervous system should be intact, all of the controls.

A cat in rage has its hair stand up, its pupils get large, blood leave its skin, its pulse rate rise, blood pressure rise, spit accumulate. It hisses, lashes its enlarged tail, does its nimble footwork, struggles, bites. The normal can be thrilling. The mutilated is an interesting but a sorry sight. A human being in a fit of rage also may be a sorry sight, the observer's own emotions coming in, his disgust, his contempt. He thinks, if he is given to this way of thinking, that what he is witnessing in this enraged human being is maximally the signs, minimally the emotions. The modulating parts of this human being's brain have by his rage been uncoupled from those critical signs-controlling millimeters. The signs are in control. They are free. They are giving a sheer display of themselves.

In other, later experiments it was discovered that when a brain part called amygdala (deep to the temporal lobe) was destroyed, rage also freely flashed forth. Amygdala is small. Apparently neither the tremendous destructions of two entire cerebral hemispheres nor the still considerable destructions in that later series of cats were necessary for the freeing of the signs of rage. It was enough to destroy amygdala. This should be on both sides of the brain.

In monkeys—this now was bewildering—with amygdala destroyed there was no lashing out. Quite the contrary, a killer monkey after the destruction was ready to be petted like a kitten. He could be handled without gloves. He was sociable. He wanted plenty of sex, wanted to investigate everything with his mouth, lick everything, and if he set his teeth into friend or foe, he did it with consideration. In trying to account for the difference between cat and monkey, it is possibly worth recalling that the monkey dines largely on vegetables and goes with colleagues on picnics, whereas the cat eats meat and slinks along alone. Cat is carnivore. Cat possesses the bloody beautiful nuance of signs of carnivore. Monkey and cat are different in their stars, their behavior, their genes.

With amygdala destroyed, a monkey behaved as if he adored the world, a cat as if she detested it, and the adoration and the detestation were both separate and intimate geographically in the brain. The vicious Norway rat followed roughly the behavior of the killer monkey, the dog followed the cat. If the brain above amygdala was carefully removed, so that amygdala was free to play with unimpeded force upon the rage center, the cat was milder, less cat, less quick to defend her slight body.

And the theme applies to us. We might in fact anticipate, simply because of our more-brain, more disarray of signs, especially among sick human beings, or among normal human beings when the times are sick. There is a voluminous literature on tumors and surgery of the brain, but only a few cases show anything resembling rat, cat, dog, monkey. Yet, every now and then, a patient is reported as displaying the periodic bouts of meaningless rage, or meaningless sexual passion. The chronicle in its total does read like a latter-day *Pilgrim's Progress* with Rage and Placidity and Fear and Defense and Amygdala and Hypothalamus written as proper nouns. When the nervous-system localities are left as nature decreed them, the creature, if a human being, shows the signs of one who is watching with appropriate decorum interspersed with appropriate excitement a street fight under his window, a murder trial, a big-league baseball game. Destroy Amygdala, he becomes Placidity, being of the monkey family, and the fight and the trial and the game are as stimulating as a cup of tea to Hercules.

FACE:
Refines Its Mask

In a man's face, especially, his thoughts and his feelings express themselves to anyone who concentrates on the face. We may learn much about that foreigner of whose language we know nothing. His face tells us. Muscle contraction, the tone of those small muscles, the state of the small blood vessels that keep changing the color of the face—all together may speak very clearly of what the man is feeling. During a conversation in one's family, one often does not take the trouble to hear, because one is seeing. The story is there. Who has not tried to read the dream in the face of a sleeping child by some minimum of contraction around the eyes, around the mouth, over the nose.

Animals who have no words to lay over their feelings reach us in the same way.

Some of the muscle fibers of the right side of a human face cross anatomically over to the left, interweave, and that emotion-loaded human creature, wittingly and unwittingly, is pulling his muscles in all directions with his nervous-system strings.

They pull outward the angles of his mouth and a smile remolds into a laugh, but then remolds the other way, into gloom (bad news has come over the telephone), remolds into petulance (he has said what he shouldn't). This is gross statement for what may be infinite refinement of the signs of emotion. Darwin carefully observed facial expression and set everything down in his *The Expression of the Emotions in Man and Animals*. Painters observe facial expressions. Writers do. We all do professionally and nonprofessionally. The thieving merchant does—observes any slightest weakening in our expression.

In the face of a cat the expressions may be more lightly laid down, may not be. In the face of the enraged male canary, expressions surely are more lightly laid down than in us, but in some manner we see or sense what we are seeing. In the female canary the placid thoughts, if we may be permitted to call them thoughts, seem no more than a breeze over a wet surface, versus the tempest that a moment before lifted the features of the male and with his feathering darkened the color of his face.

Your chronically happy colleague, whom you must day after day work next to, has his silly expressions that reveal his silly emotion, that bores you to death. Your TV tenor may be famous for the silly expressions that are coupled with his silly ideas. So it is too with your blustering drill-sergeant. Your football coach. The lamb of a man made temporarily courageous because he has had four martinis, or permanently courageous because he has syphilis of the brain. In all of these the signs of emotions are, and should be, not only more vivid but often more genuine than the emotions. The signs have been of more use, must be at least as ancient, and as rehearsed and exercised by the species. In an old tart, the muscles of her face may appear stripped; she contracts them, not with emotion, but only with the signs of emotion, with her will, which makes it all the plainer that she is only a time-resisting corpse.

Animals have had electrodes implanted at critical points in their brains, and these animals, as newspaper readers know, press keys to stimulate happiness—press, press, press, happiness, happiness, happiness. The interest in those animals is that the mimicry of pleasure can so mechanically be given to flesh—just an electric key.

A pleasure center in the brain has been defined, verified. The white rat will go on pressing the key, press, press, press, happiness,

happiness, happiness; from the pleasure center, muscles will be directed to operate in unison, and the happy animal will press without stopping, forego body needs, may become so fatigued with happiness that it collapses. Meanwhile many microelectrodes have been and are being implanted into many animal brains. Once a researcher gets a notion, he can be as busy with it as a monkey let loose in a roomful of chairs. Of course, he also is a monkey, the highest, the highest primate, and that curiosity could be quite honest, monkey-honest.

An actor manipulates the signs by what to an outsider watching the rehearsal of a play sometimes seems a pressing of keys. He up there on the stage sometimes easily reveals a separation of the signs of emotion from the emotion. The actor orders a grin. Orders a grimace. Orders dignity. Has trained himself to release signs on command. Some of the greatest actors have written on the subject, have claimed they were best when cold, their brain pulling the strings, the puppets performing. The trained brain knows that it can depend on its body parts; the proper muscles will properly perform. Earlier those actors had signs and feelings united, but tonight signs are marching onto the stage alone, the mind having learned how mechanically to cut off or draw in parts of the nervous system. Other actors have claimed that they must feel six nights and two matinees. (How tired they must get.) One does not need to be an actor to recognize the two alternatives. A coed on a date recognized the alternatives: ought she tonight let herself freely express the inside of her, or just again play her expression machinery? The dictionary defines a date as a social appointment with a person of the opposite sex. Dogs have dates too.

EMOTION:
It Feels

Emotion is old in the world and in each of us. A green fern has none, we think. A green snake has much, we think. We have more.

A list of the emotions might begin with the vivid one so much studied for the signs, rage. Far down the list might be a dull one, but an

emotion, determination. Somewhere in between would fall joy, exaltation, love, adoration, heaviness, dejection, misery, gloom, then, to
move upward again, fright. Odd how even each word starts something, smog to sunshine.

Somewhere far up that list would have been loneliness, the missing
of somebody, anybody, everybody. A human being may be ashamed
of loneliness. He may hide it under something else, for instance deportment.

There are the complicated emotions, several layers, that sometimes
are begged out or snared out by a physician or a dramatist or a nov-
elist. There are for instance the chagrins. There is the passion of the
stingy to save every penny, to eat every cookie, to let nothing be taken
away from them, this posing as morality, no-one-will-cheat-me, I-nev-
er-cheat. Then there are the many aspects of to-have-and-to-hold,
sexual excitement, sexual amusement, sexual maladroitness, sexual
ludicrousness, sexual faking, sexual disgust. There are the more plebeian appetites, all tapering off with lessened body vitality, with surfeit,
pampered by idleness, going on mechanically. Hunger then may be
almost noble.

Envy.

Jealousy.

Humility.

Hostility.

Yearning.

Enthusiasm.

Affection. Can be an insufferable woven-in emotion.

Maternal feeling has affectation usually, has pride, has narcissism.
A fine unused word that once was; nevermore. All of woman born
know how genuine maternal feeling can be, nurtured by the mother
while she is carrying inside her the fetus that is getting more and more
acclimated and comfortable while she is nervous, planning her future,
planning its future. Then, finally, comes the Fifth Act, the pushing,
the stretching, the tearing, the cutting with the scissors, she convinced
that now surely she has earned the right of possession, but, as the
pains die down, remembering again, bitterly, that no one can possess
anyone.

Fear.

Fear seems a reasonably pure emotion. Purest? Alongside fear the physiologist might write: overrapid heart, rising blood pressure, temporary breathlessness, pale skin, eyes wide open, muscles set to let the creature flee or freeze it stock-still. There may be trembling. There may be sweating. There may be many kinds of body imbalances.

In and around and at the back of everything, at the back also of happiness, there would be all the anxieties, the normal ones of every Johnny, the abnormal ones of some Johnnys, life-enhancing, life-depressing, continually shifting. The word *anxiety* is overused. Everybody says that it is overused, then uses it. Its crescendo and decrescendo are a running accompaniment with each human being's story. We need it to keep us alert, to prepare us to attack or be attacked. Lacking it entirely, we are ill. It may be a mood, a state of tension more pleasant, less pleasant. Someone has defined it as "concern with future consequences." A few persons may experience it only rarely or slightly. They are the adjusted, the equilibrated, or the righteous, who are able every night to go to sleep with their self-approval, and every morning to rise with it, and with their shallowness. Someone else has defined it as "unresolved tension."

Music, even music, has been described as anxiety. A musical theme becomes a passing from the unresolved toward the resolved, but just before the resolved is reached there is a quick turn to another unresolved, and so on, the listener moving from blissful anxiety to blissful anxiety. Macbeth moves to the murder of Duncan by a succession of anxieties, and after the murder anxiety piles upon anxiety, keeps everybody in Scotland anxious, and everybody in the theater, until at the last, immediately before Macbeth is slain, the long unresolved meaning of the witch's prophecy that "none of woman born shall harm Macbeth" resolves itself in the revelation that "Macduff was from his mother's womb untimely ripp'd."

The uses of emotion?

Without emotion there is nothing that could be called mind. Clarification of thought depends on it. Who has not lowered his voice to cool his thought, got it cool, tried to reason in that cool, failed to reason, put all off till tomorrow, yawned, grew dull, grew annoyed at being dull, blazed into anger, awoke, stayed awake, then toward morning reached an obvious conclusion? What a fool one can appear to oneself toward morning. On a historic occasion the conclusion has been more than obvious. Galileo, facing the Inquisition for a mathematical derivation, that our earth was a rolling ball rolling around the sun, did not make his additions and subtractions without emotion, because the man who grasped that our earth was a rolling ball grasped that fact hard or would not have grasped it. A mere mathematician, no. A mere mathematician would not have faced the Inquisition. Galileo was willing to face prison or death for a fact. Socrates died. Christ died. Those are extreme examples, but the feeblest scientist would admit that emotion, besides readying our muscles for action, intensifies our convictions concerning the action.

Hate may marvelously clarify. Shouting out may. Psychiatrists advise us to shout out. A high love clarifies. Arm in arm for an evening and everything is so clear, for an evening. In the depressed young couple who travel from the hills to the large city and sit on the stone steps, there is no furnace, no clarity, and they sit. In Beethoven there was a furnace of fire, and clarity. In Shakespeare there was a furnace of many temperatures, and clarity. What those two brains were like! What a brain and a mind can be like! Yet a physiologist comes with an electrode and a psychologist comes with a black box, and both come with a white rat.

Emotion prolonged is mood. Emotion getting out of hand is mania. Often one thinks that emotion exists in packets that have compression values, low behind the enjoyment of the sight of a child waving shyly from the back of the automobile just ahead, or the sight of the calamine blue of a January sky, and high behind a murderous plan, behind an erotic anger. He bites at his wife, knows that to bite at his wife is uneconomical but who else is it legal to bite at, and he has trouble remembering what the biting costs, the bad effect on the children, the bad effect on his sleep, therefore on tomorrow's gain, therefore in that equivocal state he begins to work on next year's tax return until the fire in him is ash, not enough even for a tax return, so he slumps before his TV like any simpleton till he decides to go to bed and see what can be done to make up with his wife. Galileo was willing to live for a mathematical derivation, to rise to heaven, to find the derivation wrong, but he and his psyche would have risen together, we hope. To die must be a breathtaking emotion, and no pun meant.

EXPERIMENTAL NEUROSIS:
Svengali

The scientist can imbalance the emotions of an animal. He can push farther and produce neurosis. Persons with experimental talent have been doing that, producing neurosis, for a third of a century, in dogs first, later in a variety of animals—goats, fish, spiders. If a spider is driven neurotic, he weaves neurotic spiderwebs and is claimed to reveal the type of neurosis in the flaw in the web, type of the flaw, place of the flaw.

Pavlov produced experimental neurosis originally by chance, then produced it in dog after dog, soon was doing nothing else, producing, studying, treating neurotic dogs. He cured them too, not entirely. While he was doing that he was steadily reconstructing his scheme for his higher nervous activities, and after a time had put into his scheme the thinker, the artist; and then these sick brains, as he saw it, of the neurotics.

Pavlov had known neurotic dogs in the streets of Leningrad, as we all know them in our own streets, and he found them among the strays lured to his laboratory for conditioned-reflex experimentation. But by 1930 he was manufacturing them, became highly proficient at it, and on the basis of neurosis in dogs laid down the laws, as he saw it, for neurosis in man. Later he established a small psychiatric human clinic. He was old by then. He began reading psychiatry. He never had previously, in general was not a reader, only belles lettres, as he said, on his summer holiday.

He was utterly convinced that the wrecked human psyche that goes to the psychiatrist, and that came to his small clinic, was in mechanism not different from the wrecked dog psyche that he produced. He was upsetting the higher nervous activities of these dogs—he never would have said psyches and still never said minds. With the upset nervous activities he upset the bodies, and he knew how his experimentation and his thinking would extend outward, and he would not have been surprised that widely separated movements, like brainwashing, like hysterical snake cults, would later seek their nervous-system explanation in his dog findings.

The first neurotic dog was another of those laboratory accidents that writes history. Pavlov had presented the dog with the problem of distinguishing between an ellipse and a circle, no meat for the ellipse, meat for the circle. Soon the dog had made that differentiation. It drooled saliva for the circle, none for the ellipse. Pavlov knew, therefore, that that dog had learned.

The ellipse was a long narrow one, a ratio of 9 to 2. Pavlov altered the ratio, 9 to 3, then 9 to 4, and so on, waited each time until the dog had learned, had made the differentiation, and continued this until the ratio was 9 to 8. It may even give the human eye some trouble to differentiate that from a circle, and for the dog it was too great a problem. Pavlov persisted. The problem became an agony. Pavlov persisted. Abruptly the dog's behavior changed. It was peculiar behavior now. This was a mentally sick dog. Pavlov would not have said it so. He would have said no more than what he regarded as fact, that this dog had a brain, which was a delicate instrument, could be overexerted, or wrongly exerted, subjected to stimuli that were too confusing. There was a clash. In a human brain the same confusion would cause the same clash. Months of rest with sedation were required before that dog was able again to do a day's work. (*Work* was Pavlov's term for what a laboratory dog did on a laboratory day.) Even after recovery, there might be relapse. Pavlov called what he had accomplished experimental neurosis. The phrase stuck. For sedation he used bromides.

A flood in Leningrad upset the minds (brains) of a large population of dogs. The River Neva rose in the kennels, some dogs were drowned, others escaped but the water had been up around their necks, and in consequence of that terror the behavior of these dogs was so altered that anyone would have admitted that they were neurotic. Pavlov would also not have used an emotional term like *terror*. But this was life neurosis as the other had been experimental neurosis. Those River Neva dogs also required months of rest before returning to their work. Superficially they might seem cured, but they were not. Pavlov would play Svengali. He would trickle water under the door where a dog was alone in its kennel. That was enough. That would be enough, too, for you and me, make a Poe or a Maupassant tale, water steadily advancing under our door locked from the outside.

In the course of his long study of neurotic dogs, Pavlov became convinced that whether a dog did or did not easily slip into neurosis depended upon its temperament. Dogs were born with their temperaments. Dogs' higher nervous activities were genetically different.

Pavlov decided that any extended study must recognize those differences, and accordingly began classifying dogs. No classification proved
completely satisfactory. The one he settled on was that which the
Greeks used for human beings: choleric, sanguine, phlegmatic, melan-
cholic. He decided he would find those temperaments and then breed
them. Then upon standard dogs he would perform quantitative experiments and arrive at quantitative laws for the deviant mind. It was ambitious for this early point in this type of animal experimentation, and
it was like him. He was well on the way with it when he died. More
and more enthusiastically he waved his wand and altered minds, dog
minds, but whoever saw those dogs found it easy to believe that he was
seeing the characteristics of any neurotic mind. Pavlov's dog hospital
was an old-fashioned state hospital.

In a dog that was by nature an inhibited animal he would intensify inhibition, which was to intensify shyness in a human being born
shy. He would undermine a strong temperament by castration. He
would present an intelligent animal with a problem that was simply
too complex for it, like that of the ellipse and circle. Collision in the
nervous system was Pavlov's statement for what had happened there.
He performed hypnosis as any psychiatrist would, but animal hypnosis. He continued to believe the hypnosis explicable on the lines of his
theory of sleep. For him hypnosis still was partial sleep. It was a brain
with a patch of excitation and all around it inhibition. Pavlov's way of
wording it was: one or more analyzers asleep.

Pavlov would make a dog cataleptic. He would say that he had
produced a plastic condition of its nervous system and on that account was able to mold its body into bizarre postures that the dog
would maintain, like the cataleptics known for generations in lunatic
asylums. He would crack a nervous system by pounding it with stimuli
too powerful for it. Would produce what he called trans-marginal inhibition, which would spread, which would irradiate. He excited what
he called paradoxical and ultraparadoxical states. It was a new language for new mechanisms. No need precisely to understand. No need
fully to believe. Pavlov believed. The states have before and since been
turning up here and there on the human scene. If Pavlov ever thought
of chemical mechanisms for either neurosis or psychosis, he never said
so. Everything was environment versus nervous system. He thought
he had produced paranoia, and it looked like it—a dog with a fine

intellect regarding most things but mad toward one. The inmates of his laboratory were undoubtedly inmates. They worried the observer. Pavlov's interpretation always remained conditioned-reflex interpretation. A psychoanalyst digs back into the history of a human patient to find the destructive experience, and to find it is to treat it; Pavlov supplied the destructive experience, as that problem of ellipse and circle poisoned with meat-powder meaning, made the dog sick, then treated the sick dog. Pavlov often said that if stimuli are strong enough or the nervous system is weak enough, neurosis results. He believed psychosis does too.

A CASE:
Method in Its Madness

This is not the case of a dog.

This is the case of a man not young, but certainly not old, a blond, soft-spoken, and one was always having to prick up one's ears to hear what he was saying, and one wanted to hear. He looked as if he might have a story in him. But everyone has a story in him. On his side, he wanted to be heard, nevertheless regularly slid off into lowering his voice until it was out of hearing. A feminine type but a father. He had sired three children, had gotten them as far as high school, supported them, and they were healthy enough. Three was the right number; families were having three. That is, everything argued this man's more than willingness to meet his social obligations. For a long time he had been having trouble with his foot. It was really bad, deformed. It had chronically gotten worse and looked as if it were going to go on getting worse. A spasm in the muscles of the leg had shortened the leg, had twisted the foot inward, and at the present time it was bound down with contractures. That word *contractures* explains itself. Not only was the leg drawn up on the thigh but the thigh on the abdomen, the whole limb withered. It was really absolutely useless. Painful? Of course. He had been bedridden since March. His back ached like a toothache. Indeed, touch any of a dozen places on his body and he squirmed. An intern checked off gout, arthritis, joint tuberculosis, a list of textbook diseases. (Then the intern went off to his coffee break.) As for living with his illness, as people say, this man did it He was not in the least the complaining type. His life was in no danger, and he did not pretend that it was, but occasionally he did say that that was unfortunate. There was not one of his friends who would not have defended him hotly had someone suggested that he had done this to himself. But he had, in a way.

It began far back. There was this and that and the other thing, slight signs. They had not seemed worth mentioning. He recalled how he had begun to be uncomfortable when he was just sitting thinking. He had the habit of doing that, was philosophic, his friends said, went over matters in his mind. As to his condition, he remembered scrupulously everything he was ever told. Once he was told that he must have had an undiagnosed poliomyelitis. But this was not true; medical opinion was definite, the deformity could not possibly have been left behind from an old poliomyelitis. On and off he wavered about that. How could one be sure? The doctors ought to know, but they do make mistakes. There had been a galaxy of doctors, he listened to them all. He had a genuine interest—objective, he thought. He thought he was the kind of person who could honestly have that kind of interest. Even so, the diversity of medical opinion was bound to do something toward sapping one's courage. Plenty of nonmedical diagnosis had blown his way, of course. It always does. All he had been told, or half-told, some of it entirely apocryphal, and all that he told himself—the complete account amounted to a conspiracy in which he took part. Everything presented itself to his mind in as unfavorable a light as possible, and he helped in that. Everybody helped. Tongues do wag too freely. Everybody whispers, and nobody thinks that anybody hears, but everybody hears. Everybody knows everything about another person's illnesses, thinks he has the right to know. So-and-so has had a colostomy. Now keep that a secret! So-and-so has had a second heart attack! The beasts in the field do have one advantage over us: they do not gossip.

Rheumatic fever had been suggested by one doctor, but rheumatic fever leaves marks on the heart, and another doctor said there were none on his.

Of course, his pain was fact. Pain was all day and all night stamping about his body. He said that, but he made light of it. The contractures themselves would have caused pain enough. The pain was utterly explicable. Some of the pain may not have been as explicable but it did not on that account trouble him any the less. When one of our limbs gets out of order, the other limbs get out of order too, it seems.

He had repeated the history of his case so many times that he was tired of doing it.

In one of those repetitions he ought to have noticed, and noticing might have waked him to something about his mind, that the time of the onset of each new set of symptoms corresponded to some new difficulty dropped on his life. The last was when his dear wife without announcement deserted him for a real man. That word *real* was a neighbor's, but he heard it. From the day of his marriage, the neighbor went on, his wife never had been able to stand having him around. She was quite frank about it. Once in a rage she blurted out: "He makes me puke." That was a strong thing to say. She probably did not say it twice. She possibly did not mean it.

She undisguisedly used him as a male escort. This he recognized and accepted. "She suits me fine," he said with such regularity that a friend of his came to the conclusion that her neglect of him, her cruelty toward him, made up to him for the luck of his having a superior wife. Everybody saw that she was superior. The difference between them was so great. Everybody liked her. She was bright and she was contemptuous, interestingly. He liked her too. "You can't help liking her." He also said that with regularity. He would call her his "little wife" and say she was "pretty enough." In the last three years on a dozen occasions she did not come home all night, and when a neighbor remarked about that, he said overquietly that every woman had a right to some private life, and marriages were just too sticky. It was before his latest relapse, which was the worst of his relapses, that she finally dumped everything on him, the three children, the household, fled. That did upset him. Desertion, humiliation, children, debts—all put on a person equipped to meet only the luckiest of lives.

The line from a disturbed life to a disturbed mind to a disturbed body is seldom as straight as this, and the line here can be trusted, as far as any such line can be.

So, he lies there. He likes a neat bed. Wants to express himself to his physician in good English. Wants to give the least possible trouble to his nurse. It would have surprised him, enraged him, then would have begun to worry him if anyone had implied or said that he was hiding away in that snug bed, hiding from the world, a big-bodied man, let-ting a nurse take care of him.

Actually, the nurse was impressed by how he ignored his deformity. She was just as impressed as everyone else.

One thing was fortunate—his religion. He folded his hands in the old-fashioned way when he prayed. His piety helped him accept. He mashed his toe when he knocked the lamp off the table next to his bed, insisted it was nothing, whereas everyone cringed at the weight of that lamp and the look of that toe. His mother sighed. She said: "All his life he has been that way, met everything that way, cheerfully." An excel-lent woman this mother, the neighbor said, was so considerate of him; when he was having a visitor, she took it upon herself to answer every question directed to him, saved him the exertion. She was outspoken about his trouble—his wife. The neighbor agreed. The wife was the villain. Everything began with that marriage. He had been such an affectionate lad even as a baby.

With so much having been said about him by everybody, so much suggested, so much having happened, anyone would have figured something, but it should be kept razor-sharp that no one who had seen him would have been apt to underestimate the severity of his de-formity. It would have been prima facie ridiculous to get the idea that that part was in his mind. That was in his body. It was beyond repair. A surgeon might help in a palliative way. A wise and compassionate phy-sician might clarify to the man's mind how these things came about, give him a better view of himself, and this might help or harm or prob-ably do nothing. A psychiatrist might do the same more professionally, might also probe back farther into his past, or nudge him to do the probing, especially that far past that it is difficult or impossible for an-yone to probe without assistance, the time before any of this started, before any imagined poliomyelitis, any rheumatic fever, any wife, back there where the hell of infancy is supposed to smolder. If the man sur-vived the inquisition, and the psychiatrist, and the bank account, and if he were not ashamed to look into the face of his mother, or even if he were ashamed, some deep reason for the illness might have been discovered. This would have helped his thoughts. But it would not now have helped his deformity. That was past helping.

If the psychiatrist were skillful, he might even have uncovered the special occurrence, because there must have been a special occurrence, that led to the foot. Why just the foot? How did the foot get into this? Most of the psychiatrist's findings would have been provocative, might not have been clinching. (Better that way.) Any such clarification might ward off some future added trouble, first to the mind, then once again to the body. (Who can tell?) So he lies there. One hates to embarrass him or embarrass oneself. One finds oneself never looking into his face. One finds oneself in one's mind right there in his room reviewing his story, trying to fasten when and where and how he began deceiving himself, became so expert at it, so addicted to it. A stranger entering the room, even a friend who knows so much as anybody, would have his thoughts, at least his eyes, drawn toward the man's foot and away from the man's face and mind. That is what the mind wishes.

The foot is doing its job.

PSYCHOSOMATIC:
The Head Has a Body

Twenty-six centuries after Homer, women got the peptic ulcers, and the next century, ours, men got them.

Since God Almighty fashioned Eve from Adam's rib rather a long time ago, and since anatomy and physiology could not have changed much in the interval, something else must have. It could have been society, which does change enough to nag at the human emotions, and the nagged emotions nag at the human body, and if the nagged body is damaged, that is the psychosomatic.

A generation or so ago an authority writing *On Simple Ulceration of the Stomach* pointed out that after an ulcer had healed, it was apt to recur "from strong mental impressions." Ulcer is frank disease, can be seen with the X ray or during surgery or at postmortem, and something in the mind did gnaw at the flesh of the stomach; the gnawing was related directly or indirectly to the "strong mental impressions."

Freud (Austrian), Pavlov (Russian), and Cannon (American) brought their different talents to work on aspects of the psychosomatic. Freud defined the unconscious, defined conversion, gave them reality, left illustrations, cases where mind then body were disturbed.

Pavlov at about the same time was introducing experimental neurosis, producing animals with severe neurotic problems, mind-sick, body-sick animals that, after he had made them sick, he found ways of treating and curing.

Cannon dwelt on the physiology of pain, hunger, fear, rage, leaned on Darwin's study of the signs of emotions, kept his laboratory experimentation in that direction.

The achievements of those three men, different in value and in force, were largely completed in the first third of the century. Since then, practicing physicians have frequently made incontrovertible the fact that body illness can result from emotional illness. Psychosomatic medicine—the illnesses, the causes, the diagnoses, the treatments—will in the future either have become fatally defined or shown to mean something else.

The illnesses are believed to occur in one of two ways: as symbol, as altered physiology. As symbol: an example would be the hysterical paralysis of an arm, the arm symbolizing the useless or the unwanted, person, thing, event. As altered physiology: an example would be the excessive battering of an organ by an elevated blood pressure.

The damaging blows across the border from mind to body excite hypotheses, observations, experiments, broaden our view of mind, and give us nothing too clear. In conditioning techniques, during the period of delay between the stimulus and the response it is simple to produce disturbing signs in the stomach, in the heart, elsewhere, these far outlasting the stimulus, the heart still pounding and racing after the experiment is over. That might be one way heart damage could occur. In hypertension—high blood pressure, because of, say, an unrelieved silent rage—blood vessel walls are battered, the vessels narrowed, notably those of the kidneys, the heart having thereafter to move blood through narrowed vessels, this costly both to heart and kidneys, the sick man possibly stricken by uremia, possibly dying after three days and nights of Cheyne-Stokes breathing that reverberates through the house. In another man or woman, instead, there may be a holding back of tears, a holding back of rage, a holding back of secrets, and that might go over into a holding back of breath with damage to the bronchial tubes, the diagnosis asthma, though the cause of asthma still often seems a mix of physiological factors and medical misunderstandings. Asthma has been called the "granddaddy of psychosomatic diseases."

There are physicians who insist that even in a disease proven to be caused by bacterium or virus, it is the side effect on mind that is the reality. These physicians believe what they say. Their opponents—the two occasionally are near to enemies—insist that mind never causes disease. Disease to them requires the microscopic, or the otherwise visible, or the countable, or the chemically traceable.

In practice, psychosomatic medicine uses all the tools. It brings in all the disciplines: physiology, chemistry, pharmacology, surgery, internal medicine, psychiatry. There are psychosomatic books, psychosomatic journals, psychosomatic societies. Seminars. Conferences. Lectures. Clinical investigations. Laboratory experiments. Chairs in colleges of medicine. On some days our world seems only ambitious and adolescent.

Human beings have been typed in ways reminding one of Pavlov's typing of dogs. There is the hypertensive type, the ulcer type, the accident-prone type. One investigator argued that an abnormal body-part resulted from an abnormal gene, an abnormal behavior from the abnormal body-part, an abnormal mentality from the abnormal behavior, a back-and-forth between them all, until there is such alteration in the flesh that even a hardheaded county coroner must admit there is pathology, damaged flesh.

The wish to connect mind precisely to the flesh is also a sign of the times. Sherrington's dualism has lost by lengths to Pavlov's monism. Pavlov's fusion of subjective and objective falls within the comprehension of any Simple Simon. Even the matter-and-energy-are-one of the physicist gives support to the mind-and-energy-are-one of the psychosomaticist.

Yonder in the morgue lies a tramp. Yesterday he had a hemorrhage in his brain, collapsed in the street, could not hear, could not see, could not speak, today is dead. The brain cutter looks into his brain. Plain as the man's bulbous nose is the burst artery and the clotted blood at the precise point the intern in the ambulance said. Hypertensive type, said the intern, wrote it into the chart. Dead, said the intern, at 3:05 A.M., wrote it into the chart. The tramp's mind and body lived together, were broken down together, died together.

NEUROSES:
Works Its Way Out

We are more concerned with the neuroses than with any other illnesses excepting that hostile triad, cancer, stroke, heart disease. What in generations past a person thought his private, his personal problem, he now turns over to a specialist, as he turns over his tax affairs to another specialist.

We choose to explain our friends, when they are not like us, as mental aberrations. With the specialist this is something else. With him the terms, when he tries to clarify them, even when he clarifies them wrongly, as a good part of the time he must, help to bring some order, or some further order, into the eternal bewilderment of mind despite his often adding his own bewilderment.

Changing terms with changing times are somewhat the changing knowledge, not change in the bothered minds. These change largely on the surface. These are affected by a society that changes largely on the surface. In the depths even the bothered minds apparently change most slowly, if at all.

Psychodynamics is the mechanism of mental illness. Psychotherapy is its treatment. That essentially has not changed, only the terms.

The overall difference between the neuroses and the psychoses, until there is more complete understanding, is most safely considered a difference at their roots. That essentially also has not changed, and not the terms. Both are different from feeblemindedness, from mechanical or gross chemical damage to the brain, from addiction, kleptomania, criminality, possibly from states listed under character disorder or neurotic character.

In the neuroses the physician looks carefully at the society around the patient; in the psychoses he looks at the mind.

Hippocrates knew hysteria. Hippocrates' remedy was marriage. Marriage may have been the best remedy in the Greek world, as it often is in ours. Hysteria in Greek medical opinion was attached to the womb, *hyster* meaning womb, the womb imagined to get loose from its moorings, skate about the body causing signs and symptoms. In that general neighborhood, the womb, but mostly in its psychological suburbs, we today, too, believe that signs and symptoms arise. The idea as accepted by Hippocrates was accepted for centuries. In the ancient world there would have been no hysteric men.

When late last century psychiatry entered its modern phase, hysteria became for a while the overall title for the neuroses. Today it is one class, still the largest.

In the neuroses, illness of mind frequently becomes illness of body, a turnover called conversion. It has been called that for half a century. A mind in trouble finds itself a body illness. If not a body illness, a behavior, a determination never again to dress like a mouse but to wear a vermilion waist with a midthigh purple skirt and the most bouffant hairdo in town. It may be any quirk, anything that provides a pillow for the mind to lay its head on. The damage may come another way. A man may have an excessive, a hopeless feeling of futility. He may be convinced that he can do no more with his life than he has, so he drinks, keeps that up until he can in fact do no more, must rest his dead-drunk head on a pillow of drink. The neurosis usually spreads around. His wife—disappointed—flirts. One of their older children—disgusted—commits armed robbery, and all society is touched by that, and the police, and the psychiatrists. The social conversion is seldom that clear-cut or widespread. But the inclination to look at everything widespread has grown with the generations, and parallels the increased prying of society. But it began with the man, his need to get relief from the gnawing at his mind if he must stay alive. His neurotic behavior had been—weary word—a defense.

As a result of the growing tolerance, so far as it is tolerance, so far as it is not neurosis in the total society, the mind of the community blames single minds less, which sometimes is fortunate, sometimes unfortunate, both sides of which we all know. Murderers are blamed less. They blame themselves less. One eminent psychiatrist has brought himself so far as to be able to see the punishment as the crime. Anyone having perpetrated anything may cry expectantly from his prison cell: "Have pity on the poor neurotic." We are very much products of our environment, but most of us, when alone, know that our genes, or whatever, include our way of reacting to our environment.

Hysteria, today's hysteria, that one class of today's neuroses, has many guises but probably the same mechanism. A woman suffers a sudden frightening reduction in power, cannot stand on her legs, sinks to the pavement; a man thinks she is having an apoplectic stroke, nervously asks everybody around whether he should go for help, while another man keeps slapping her face as she lies there on the wet cement The reduction in power may be in a single limb, may be anesthetic and not paralytic, but, either way, not explainable by any anatomy of nerves. Anatomy is the same for everybody, but the losses of the hysteric would require a new anatomy. A stocking-foot anesthesia has the sharp upper border of a stocking, and there is no anatomy for that, and a glove-hand anesthesia has the sharp upper border of a glove. Most frequently the hysteric is a woman. She extracts satisfaction from her suffering. She endures in silence. She is apt to be young. Has had life's usual difficulties. May not be too intelligent May have a good idea of what her mind is doing. "I know I am a neurotic." She says that while she puts lipstick on lips that have too much. She must not have too clear an idea of what her mind is doing or she might be discharged from the clinic as a malingerer, and that might make her worse, or better, and even a skilled physician might not be able to predict which. And there would be two points of view also about what malingering is.

Each of us sees the operating of some of the trickery in his own mind. A person feigns a headache to escape the boss's esprit-de-corps dinner, after a while has the headache. The boss's dinner is the illness of mind; the turnover, the conversion, is the headache. That is a normal enough conversion. There may be most abnormal ones. There may be blindness—so as never to have to see the foul earth. May be mutism—so as never to have to speak on the foul earth. Persistent vomiting—to make plain what one thinks of this earth. Immobility—never again to have to move. These interpretations are too pat, but the actions may be also, may keep the observers fooled for a lifetime, the person himself fooled even if he gets as far as suspecting he could be.

Hippocrates knew anxiety. We are all apt sooner or later to run the full scale of normal anxiety. Anxiety can be a symptom of any neurosis. It can itself be an unmistakable neurosis, called then anxiety neurosis, a severe mental illness. The sufferer frequently is a man. He says there is no reason for the way he feels. There is. There is, probably, some hidden conflict—another weary word. He lives in unrest. It is genuine. Contrary to the hysteric, he feels not less body power but more. He too experiences that turnover called conversion. It may go to any organ—heart, stomach, a joint. The symptoms of the anxiety neurotic may be diverse, a pressing in the belly, in the orbit of an eye, in the skull, anywhere. A mind let loose is inventive. When the conversion is to the nervous system, the man may think he is going insane; if he thinks he is, he probably won't.

Depression, some level of it, also is normal. It varies from person to person and from situation to situation. A person who has had misfortune and is reacting to his misfortune is reasonably depressed; the depression is called reactive. Many depressions are not reasonable, are neurotic, may be slight, may be painful, but really not harmful, are spread through the populations of the world and our town, increase disproportionately around the holidays, seem then to come out of their corners. Everybody has experienced degrees of them. Then there is psychotic depression, profoundly different, probably not merely some unresolved conflict in a man's life, probably some difference in the molecules of his brain. Psychotic depression is not quickly over.

The sufferer may slash his wrists, attempt hanging, take an overdose of sleeping pills, some unoriginal method of suicide and usually bungled, but not always bungled. There is that in his brain-mind that would wreak the final havoc upon itself. This psychotic needs to be watched if we continue to think it is our business to keep everyone alive. Whoever attempts suicide is peremptorily classified as neurotic or psychotic, and (this is remarkable) criminal. He may be neurotic, of course. He may be psychotic, of course. A supremely sane suicide never gets credit.

Hippocrates may not have known what we call traumatic psychoneurosis. The person may have had an accident but he may also not have had an accident, a real one. Psychiatrists say that if he had had the accident, he would be less apt to have the neurosis, would have "cathected" the neurosis into the accident. Or it may have been a small accident, not much body damage, nevertheless blamed for a prolonged illness. "Never was right since that day." The man's family and his doctor may keep the neurosis going, though it is likely that the person was a neurotic from far back, merely pushed his mind's burden onto his body. Blames the soulless factory. It may deserve the blame. If the money compensation promises to be sizable, that helps the neurosis. Accident may follow accident, the man may have a talent for accidents, accident-prone, needs accidents for the importance they give him, breaks his bones, and his friends say he is nice but awkward, and his wife screams at him to keep his hands out of the china closet.

The wars showed neurosis latent in many soldiers. Combat or thought of combat was intolerable to their minds, and symptoms appeared. To be sure, bullet-lacerated flesh is real, as is death, but the equilibrated mind is presumed able to meet adverse reality either with successful illusion formation or unflinching intelligence.

Hippocrates probably knew compulsions and obsessions. There are elegant compulsions—ways of eating potato chips, snuffing cigarettes, peeling off nail polish, twirling a dinner ring, counting without seeming to count one's change—little scenes that multiply before anyone's eyes. We all are compulsive, could not get through our normal life without being, or are convinced we could not. But, again, there are the body-destructive, society-destructive compulsions of the abnormal, persons who must shoot persons whom they have not so much as met, persons who must wash their hands two hundred times a day. A night watchman in a college goes from room to room, and before he opens each door, puts on a glove, refuses to turn that dirty knob without putting on a glove. The psychiatrist says believably that that glove covers something else in the man's past, the man's mind, that he shrinks from touching.

The neurotic is living in the same reality that we live in, mostly. The psychotic is cut off from that reality, mostly. It is a long-standing conclusion that he is living in the only world he can endure. He requires his delusions. His profound responsibilities. He requires his voices. He lives with them.

This could all have been said long before Freud plunged in.

FREUD:
Helped to Comprehend Itself

Often Freud has seemed a dark fisherman off in the morning with rod and notebook, stubble on lip and chin and cheek, darker as he got older, fishing for species never yet classified, fishing where others fished but they did not know how. Occasionally what he pulled in was a glimmer, not a fish. Occasionally he seemed to have a fish, but it slid off his hook. Occasionally a haul. At night that fisherman wrote into his notebook the catch of the day, not what he had hoped to bring back, what he brought.

Freud would have allowed this figure of speech, Sherrington been vexed over it, Pavlov vomited upon it, Cajal turned his back on it.

One was Austrian, one English, one Russian, one Spanish—four who in the first half of this century were the pillars under all study of nervous system and mind, and dead now, still are the pillars. Each could express himself, and each was allowed a span of life not long enough to get himself expressed. Freud died at eighty-three. Sherrington at ninety-four. Pavlov at eighty-six. Cajal at eighty-two. Each was born in the short span between 1850 and 1857; the explosive *Origin of Species* with its smash hit for determinism was published in 1858, when they were children. Each was scientifically creative. Each had a tough mind and a tough body, had to have, worked as men work who change some patch, and change some patch each did, worked *comme une bête*, Claude Bernard said and Freud repeated.

That sex begins in infancy was Freud's most revolutionary discovery, and the fact that he could make it sketches the first lines into that mind-lined face. Sex was what he meant, literally, infantile sexuality. In 1900 that was repulsive. Now it is matter-of-fact, or shrugged off. When the idea that sex begins in infancy occurred to him, he thought, as everybody knows in this sophisticated day, that the parents or the uncles were the sex aggressors, attacked the infants; thought he was forced to this extraordinary conclusion by his own and his patients' self-revelations. But impossible—too many wicked parents. What to do then except suspect the infants. They were the incestuous; as they grew older, they fantasied, pushed the incest over onto their parents, made them responsible. That region of the mind where the incest was pushed became increasingly central in Freud's files on the mind, and as the years passed, he filed there more and more, everything that had been or could be repressed. The term *repressed*, and what is barnacled to it, we of this generation have known all our lives, first had difficulty with it, then became amateur analysts, like everybody in the Western world, to the disgust of the analysts. Sometimes we were annoyed at the roadblocks we thought Freud placed with terms, though we granted he needed to introduce them because his subject was fresh, improbable, then probable, then improbable, and so on. Frequently we were openmouthed at his insight. Frequently we wondered whether this was not all enlightened charade.

For the nonspecialist, what Freud first in the world presented to the human mind was the human being with the dubious privilege of crossing the threshold of sexuality not once but twice in his life—as infant, as adult.

An idea like that required, besides observation, fancy, and Freud's was a fancy everlastingly checking itself, cruel to itself. Undoubtedly, Freud was the creative writer as well as the scientist who writes, should be read with that in mind so that his interpretations do not get a literalness they probably do not always have.

To the Victorian, spoken-out sex was obscene, and the thought that every move in life on the surface or underneath was saturated with it seemed evil; more evil, and preposterous, the thought that birth should have begun the long melee, that incest should have been part of our innocent infant mind. Indeed, the thought of this lifelong life of sex could weary anyone on some rainy day. Who before Freud would have conceived the sexual as from the start built into the upper end of the digestive tract, lips for play? The newborn mouths everything, a rubber nipple, a rag doll, slobbers over the dirty roll as twenty years later over the beloved. The daydreaming of the adult could therefore be construed as a playing with imaginary dolls long after those of Christmas past had been stored with their broken legs in the attic. Freud said that every nursemaid knew the facts. Where is that remarkable nursemaid? Freud remembered his own nursemaid, dead, repeatedly fished her, one felt, from the dim back of his mind into the forward part. It was not surprising that Freud's mind, his utterly unique mind, should have gone further and found symbols for the sexual in the simplest matters, in the activities of the nursery, the activities of the garden, of the kitchen, of course in all church steeples whatsoever, and that his thinking should now overtly and covertly reach into the street life of the world, into the hippies, into the writers, their novels, their movies, as well as into the treatises of philosophers and churchmen, even into the writing of one pope.

At four or five the child forgot its sinful infancy, buried it, resurrected it when society and biology thought proper, in the glow of puberty, sex with nature's cool intent. From that long sojourn of it in us came so much that motivated our mind, gave it its shades of color, unexpectednesses appearing in unexpected places at unexpected moments in unexpected forms, making psychology even more scene-within-scene than it had been before, filling the stage with Hemingway beards, geisha eyebrows, bouffant hairdos (for a time), miniskirts (for a time), belladonna eyes, the submerged powdering the nose of the manifest before the lady herself opened the door and with every gamin in the street staring went down the front steps into the moonlit plaza, the candlelit church, the chandelier-lit ballroom.

Freud would have allowed that figure of speech also.

In his interpretation of the mind, and every interpretation interprets the interpreter, Freud granted hypothesis, as how could he not, though he talked often as if he stood on bedrock, and often did. More persons realize this today than in 1937, when he died, and many more than in 1900, when *The Interpretation of Dreams* was published and did not seem an important book.

There at the turn of the century this stubbled fisherman fished day and night, with cryptic bait, with conviction that fish were just waiting to be caught, as it seemed they were. He lured the single neurotic mind, was even better at this than at theorizing, also lured the mind of the century (the mind of many professions beginning therefore immediately to deny him) to look at facts like infant jealousy, infant love, infant hate, emotions that we today watch expressing themselves in our infants playing on the lawn, are able to watch with recognition and intelligence and often amusement because of the way Sigmund Freud and a few others watched mind in infant, in child, in the whole human family. Society may be watching too much, watching for more than is there, watching here as everywhere too factually, but that would only be society's overpay for Freud's originality. It would be too late now, if we foolishly wished it, to escape an observation and an imagination like his. He is somewhere about us, seen and unseen, day and night, interprets our waking state, interprets our sleeping state, partly creates, no doubt, our present comprehension of both states.

Freud was willing to share credit for his crucial insight into the oedipal with Sophocles, that Greek, that first Freudian. It would have been a temptation to eavesdrop on Freud and Sophocles in the halls of Hell or on the other bank of the Styx, modern Vienna meeting ancient Athens, Freud assuring Sophocles that he need not have been thinking oedipal for the oedipal to have been the motivation of his great drama, *Oedipus Rex*. His depths had come up and had done the thinking for him, better thinking on that account. Freud would have gone on talking, was always the teacher, would earnestly have defined the oedipal to Sophocles, who would have been listening with some credulity, some incredulity, squinting his mind's eyes, his experience having been that poetry might lurk in the dawn of any new idea, and one must miss no spot of poetry.

Having discovered the oedipal in *Oedipus Rex*, and the kernel of a new psychiatry, and having come to regard as universal the impulses of that infant boy who became king by killing his father and seducing his mother, Freud sought to plumb literature everywhere, urged others to. He tried to solve what was underneath that Scottish monster and pathetic human being Macbeth, Ibsen's Rebecca West, other fictive characters, particularly of the theater, was convinced one could find why they thought as they thought, behaved as they behaved, why the plays were written. Freud believed that the greater the dramatist the more of his own life would have been introduced into his drama, not obviously, but concealed, and the dramatist not know it, was alerted as all of us to the surface but not to what was below the surface, the persons of his play being fusions of their minds with his mind, the writer's.

And what scientific justification that might seem to the literary and psychiatric critics whose ambition anyway was to peek everywhere beneath the covers of the writing to discover what could be of the writer, drag him out where the reader might see him naked, and the critic have the satisfaction of being seen too.

The name William Shakespeare was to Freud a pseudonym for Edward de Vere, Earl of Oxford, who wrote the plays, and Edward de Vere loved his father, who died, and hated his mother, who too quickly married someone else. Edward de Vere *must* not have been able to kill his uncle, Claudius, because that would have been to kill himself, who in his depths had been covetous of his father's bed, not his throne. Imagine Queen Victoria, let alone Shakespeare's Elizabeth, sitting comfortably in the theater with that notion in every head in the seats around. Times change. In 1964, the Shakespearean year, if the notion did not come through the Hamlets as they were played, its absence was called attention to, even by the people in the audience, who are all Freudians, or not.

It is possible that Freud never believed that Shakespeare in life was the Earl of Oxford, that he simply snatched at that chance because it fitted his theory of the neuroses, and because it was startling. Writers snatch at the startling.

Curious to reflect that while scholars and pedants with more or less literary competence, and latterly with more or less psychiatric competence, have vied to explain Hamlet, the prince has continued to stand there timeless and still enigmatic. Of generations of explanations, Freud's was the most unexpected. He thought Shakespeare had put at the heart of the play a conflict that had existed in readers and theatergoers for the last three hundred and sixty years and all the two thousand years before. The oedipal of *Oedipus Rex* was direct, because Oedipus did kill his father and did fecundate his mother; and the oedipal of *Hamlet* was hidden, because Hamlet merely did not recognize that he wanted to kill his father and wanted to fecundate his mother. The two thousand years had driven the oedipal underground.

It was the fire from below. It was the harassment under the play, and there must be harassment, some kind, under all theater. Who is not ready to pay Broadway $15.00 if he can be sure of two hours of harassment? Up to Harold Pinter's last play and beginning with the late great Eugene O'Neill, the American theater would have been thin diet had Freud not left it his psychological conceptions.

Freud was a neurotic, said so, noted that he had gotten more normal than he had been four or five years before he started on his psychoanalytic self-cure. A dark neurotic fisherman, who even looked so as with the years he weathered and grayed, said we were all neurotic. We are, no doubt, but surely most neurotics who reach the psychiatrist are suffering a different neurosis than that he was saying we all suffer. Freud was able to study his own neurosis, did with persistence and a tough objectivity; one always forgets he was a small-bodied man. His neurotic torments were the common ones, but uncommon for what he extracted from them, how he wrote about them, with economy, vividness, without technicality, straight at the reader, simple, sometimes almost garrulous. He was a voluminous correspondent in spite of his pinched time, for letters enabled him to battle his way to conclusions, to clear his thoughts; most of us must achieve our clarifications by expressing ourselves to someone. He was also a diarist. It is reported that he burned one batch, all of his papers, in his twenties, again in his fifties. To say that a man was a diarist, even if not dictionary-true, then to say that he burned all of his papers, might cause anyone to nod his head, and it does sketch lines into the portrait. We see in another way that he was small-bodied. We think with sympathy of a troubled person.

A bad period followed his father's death, "an intellectual paralysis such as I have never imagined," and then it was that with determination he began to interpret his dreams. This—everybody also knows—was the womb of psychoanalysis. The first psychoanalysis in the world was by Freud upon Freud. He could trust himself to start with himself, expose his mind to his mind (what that came to was in a new way to expose all of our minds to our minds), peer through the fog into the water below, fish for whatever swam or floated. Every scrap had meaning, he was convinced, and was to go far to convince the rest of us, every word from our mouths, choice of phrase, verbal slip. Freud's self-analysis remains singular. No one who reads the history of it could regard it as literary fancy, though it was that too. He kept at it, a half hour a day apparently, for the rest of his life.

In the course of it and of the analyses of many patients and of a ruminating that was different from Sherrington's or Pavlov's or Cajal's or anybody's, he put together his idea of the structure of mind, gave it the three levels, not proving that three was necessarily rigidly the truth, because Freud had also the human brain that goes on helplessly dividing, but, important to him, he could thus effectively hammer into us his theories that were fresh, difficult, more so at the time, especially at the time. Freud said that the higher animals could be assumed similar to man; their minds would have the three levels too. The three levels began increasingly to be spoken of well beyond the psychoanalytic world and its university rim. The levels are, everybody knows, id, ego, and superego, words worth defining, not for the definitions, there being definitions of them enough, and in matters of mind all definitions are too clear, Freud's also, but mostly to etch more lines into his portrait.

The id was the primal, wilder in one and not so wild in another, shaded from the daylight of our life. It was the at-birth, the before-birth, the ancient, the racial. It began far back and in places far off. It was the compressed source of the mind's energy. It was the spring of the wound clock that through our lives unwinds. Saint and sinner drew from the id. The closest translation from Freud's German, as translators simplify, was *instinct*. To this id, year on year Freud added, and if the added was not always lucid, it could mean that our depths never are. Freud never thought they were. Our depths would have, then, both what was shoved under by us and shoved under before us—family, history, species, much doubtless that the intellect cannot ever resolve, though Freud believed everything down there finally resolvable, what appear warring forces, wrestling opposites that clash and thrash.

The word *id*, like any word, like any fact, gathered layers. For Freud, these were especially the masquerades of the fecund—sexual rage, sexual glee, sexual deprivation, sexual anything—that whole dry firewood always ready to burn down the house. It, the id, seemed always the infinite roguery of the reproductive urge plus a fraction of self-defense. Over it Freud threw the dusk of his often romantic prose—that presumed realist. He separated the love instinct from the death instinct, the latter brought to him with force, it was said, by file death of his daughter. The death instinct clove apart, the love instinct pulled together, love and death to Freud and to most of us being both fact and symbol, introducing into life flower and weed, art and insanity. Freud spoke of the Kingdom of Illogic. He grasped at phrases. He had talent and power for descriptions of the mists of the mind, descriptions that his followers often froze into laws. Sometimes Freud's descriptions leave over the reader a Wagnerian music-drama rinse, which washes off, however, as soon as he chooses to tell of the analysis of some single disturbed mind. It is there that time will have deserted him least, where he recognized the id's part in a man's or woman's life, which drove the life, which was always a stirred past inside the present, and another past inside that past, and another, as the analyst finds when he pages back and tracks the subsequent in the previous. Because of Freud, at every hour in every country some worried mind is encouraged to follow paths back in the direction of the id and forward to where those paths are leading, thus revealing to the mind why the present is what it is, this revelation clearing the emotions and leaving the person more than a hope that the future may hereafter be less destructively servile to the past.

As for the ego, it was sculptured from the stone of the hills of the id by the chisel and hammer of reality, always under the watchful eye of the superego, the boss. The sculpturing proceeded wherever and whenever the id went forth, early and late, on its *Wanderjahre*. Freud's went, ours goes, off in the morning, on through the day, never stops, cannot, the restless thing, around the nursery, around the earth, around a city stupid or not so stupid, Vienna or some other place, a few streets, up the one and down the next, a parlor, the bedroom, the butcher shop, the church, the synagogue, the slushy rush of New Year's Day, sweltering July, an excursion to Schönbrunn on an August Sunday. In the course of this the ego acquires ego strength or sinks into ego weakness, but always remains tied to that infant ego that was shaped when everything was fresh, had its uneroded power, cannot therefore be reshaped easily. This we know because of Freud and those who followed him, though we have the illusion, perhaps not illusion, that we knew without them, that everybody always knew, which is one kind of historic illusion. The infant ego had under it the id's hottest fire, but with time the fire cooled, was ashed over, the heat lessened, but never lessened so far as not to have left the energy to squabble at home, to squabble abroad, squabble, squabble, to fight it out with oneself, and with anyone, to the last cackle.

Finally—the superego. No need to be exact about that either. Freud was not. He frequently dropped such a phrase as "more complicated than we have here described" or "my latest formulation." Freud was the best of the Freudians. The superego was the authoritative, could be people, could be objects, could be ideas, could be ancestral precepts; it accumulated somewhat in a lifetime but mostly was the past. Superego and reality confronted id—as is said every day. Superego was the eternal parent—as is said every day. Stern parent or not so stern, and not the one or the two but the entire corseted crowd, tradition, history, the hierarchy that in our time seems usually to be leading toward the sociologist or the psychologist. But it could be the secretary-general of the longshoremen's union, the chairman of the board of the stock exchange, the president of the United States or the university, Socrates or Billy Graham, all past or present who loom over us, advise us, get us grants, and, if we do not keep one eye on them, choke us. Superego differed from culture to culture but principally in the form—the tilt of the hat, length of the skirt, diamond in the tiara, whether or not on a holy day we went into the temple wearing straw sandals. It was the orthodox. To Ibsen it was the *ideal*, a trap. To Freud it included what was left of the conscience. Capitalize it, and to the existentialist it became the embodiment of what to less ironic minds would have been Jehovah or Buddha or Christ.

To comprehend all or any of this we must, in the opinion of some analysts, wait till we have ourselves been analyzed, which would enrich us, surely, but most of us will have to get along with our poverty. Each of us has a mind and an idea of mind and must agree to be satisfied, must also insist on being allowed to do his thinking about mind with what mind he has, no matter how that annoys anyone. Sherrington gave succinct statement to this inevitability. "Science, nobly, declines as proof anything but complete proof, but common sense, pressed for time, accepts and acts on acceptance." It would be best if that common sense had no indoctrination—not religious, not psychiatric, not scientific.

The brain as brain, the 1,350 grams of near-to-far, did not enter Freud's scheme, nor did Cajal's microscope, nor Pavlov's higher nervous activities, nor Sherrington's reflexes. Freud sidestepped the work of those other three giants of his day, may not have known much about any of their work, would have sidestepped it if he had, sidestepped also today's computer mind, macromolecule mind, information mind, anthropologist mind, though he did himself dabble in anthropology, and psychiatrists ever since have dabbled in it. Of course he would have sidestepped today's brain experimentation, which appears sometimes to expect to latch the small brains that compose our big brains to the small behaviors that compose our lives. Mind sans brain was the territory of Freud's explorations. He never doubted brain, his determinism including everything or trying to and often having the possible misconception that it did. Early in his career he attempted to fit mind squarely on brain, drafted a manuscript, postulated clever nerve cells, not as Cajal had shown them to be but as it would have favored Freud's theory if they were. He never published the manuscript. It was a naive performance on the part of a genius who was many things but rarely naive, though naiveté is part of the portrait too. His conclusion? We were living in too early an age to fit mind on brain, so he let it be mind for mind's sake, psychology pure and simple, his own. He summarized: "I have to conduct myself as if I had only the psychological before me." It was the psychological detached from the flesh of the brain, but grossly or subtly attached to the sex flesh of every Adam and Eve. The riddle of mind's inner nature he did not pretend to solve. That naive he was not. His words were "defies all explanation or description." This fisherman fished through layers of fog, through depths of water, made his hauls, then noted that the greatest haul of all eluded and might always elude the hook and net of the intellect. Near the end of his life he spoke of "the still-shrouded secret of what is mental."

That last must be recognized about Freud's view of the mind, insofar as he placed it or was able to place it where we could recognize it. This was closer to Sherrington's view than to Pavlov's, not as impoverished as Sherrington's, not as imperious as Pavlov's.

In Vienna, he occupied that famous house up the steep side-street, 19 Bergasse, where he had three small rooms—a waiting room, a room for interviews, a study. There he conducted his daily sessions, nine hours, thirteen hours, guided neurotic after neurotic to pick up layer after layer. His reading and writing, important to him, he put off till night, typically offered a lady, in whom he was interested as colleague, and as lady, an hour of discussion if it could be after ten o'clock at night. Vienna medicine lay below him. He said he hated Vienna, where he had spent most of his life, toyed with the idea of quitting it, going to England, Australia, never went till a crazy world drove him out. Possibly he loved Vienna. No one knew better than he how love and hate cohabit. America he hated; Rome, the ancient city, he loved. Vienna gave him isolation. He considered the isolation, not given to him, but forced upon him, he left with it, but probably preferred it, certainly profited from it. To the people in that street his days must have had a monotonous order. His clan was small. He broke with one group of it, held to another. A group might desert him en masse, as did the Jung, Swiss group, and though this may have saddened him, he was dry about it, anticipated it, understood it. He was outspoken, enjoyed his humor, coarse often, in controversy was logical, ruthless, truthful. Social opinion in the world was all of the time changing. He spurred the change. With bluntness, argument, hypothesis, fancy, fact, he did his part to wipe out the old moral order that he believed had produced the neurotics he treated. Unwittingly he may have done his part to wipe out all moral order, in the respect that, at the moment, this sometimes seems to have occurred, the ethical replaced by the psychological, the old world loosened before the pillars under the new are firm. The sick came to him from many places. Psychoanalysis spread through many countries. It sprouted branches that forgot their trunk. Sometimes it was as if a flame were being fanned in desert grass, and as if curiosity about sex were the wind. The righteous recognized that they too had the curiosity. Events preceding World War II drove him as a Jew to England, where he died a loitering death of cancer that gnawed its way through his jaw and cheek and into the orbit of his eye. Some canny and evasive border, it is not possible yet to say how broad or solid, was by Sigmund Freud added to the edges of the human mind's comprehension of itself. His body was cremated, the ash put into one of his Grecian urns; his approach to the mind lived on, not so much what he thought mind was, but his technique for thinking of it.

DREAM I:
Freudian Wish

Freud's book *The Interpretation of Dreams* (*Die Traumdeutung*) was his greatest, most think, he thought. From the time he published it, more than seventy years ago, up to the time of his closing illness, he revised it year after year. He had youth when he wrote it. He wrote rapidly. The book runs. It may be somewhat slow toward the end, may be too long, may be or feel unfinished, but Freud was expecting the subject to go on developing from where he left it, knew no doubt that it ought to feel unfinished, ought to be incomplete, allowed open ends to remain open ends, had the competence to do that, recognized that there was surmise, that there must be error. Besides textual revisions, he made revisions also in the overall ideas. He knew this must go on. Of the foundations he felt confident, thought they were sound, and two-thirds of a century later they stand in a way that would make anyone think they might a long time, long after their superstructure has crumbled, and it will crumble, has crumbled, but of course is not the rubble on the ground that one might think if one listened to some, these often the ones who have dwelt most inside the edifice. That is somewhat as with Pavlov.

Freud's book is dream cover to cover. Psychoanalysis rose from it. Psychiatry and indeed the life of our time have been profoundly touched by it. People read it who read nothing else of Freud. Not a difficult book to read. Freud knew and said that with it he had struck his great blow, would not strike another such, no one having the power for two. "Insight such as this falls to one's lot but once in a lifetime."

The essential conclusion of the book was that if the meaning of a dream were searched for and found, it would invariably reveal a severe logic running through its ostensible illogic. Its insanity was sanity. It was as determined as everything else in body and head.

The dream laboratory and dream research support purpose in dream. We require it. When we are deprived of it, we are ill. There are many experiments on dream deprivation. Nevertheless, the intent in the content of the dream the researcher does not much concern himself with. Freud did.

Freud was the first to see economy in the content of dream. In sleep everybody has always seen economy in the content of dream. In sleep everybody has always seen economy, for body and mind. And Freud thought that dream existed to guarantee sleep. Dream enabled us to go on sleeping. The body must get rest, though Freud was not concerned with that, but the mind must get its kind of rest, and dream helped it to that. Again, today's dream researchers have come to a similar conclusion. Dream is physiological. It is part of the normal machinery. For this the dream researcher brings laboratory evidence in an era when laboratory evidence is sacrosanct. It is safe to say that the researcher never has had Freud's insight, keeps off from where Freud spent his life, in the hidden, sometimes thinks that he would not care to have Freud's insight, sometimes seems unaware that dream has content. Freud not only brought attention to the intricacy of the content and to what he believed led from that to aspects of our human life that have and that have not been revealed. Freud's genius gave him an earned indifference to the anatomy of the brain, and he might have been indifferent too (for himself) to that apparatus that would be invented to study dream. His study needed no grants, needed three meals and himself.

He tells dryly how *The Interpretation of Dreams* was received by book buyers. Two years after publication the sales were two hundred and twenty-eight copies, three hundred and fifty-one after six years. (Should be comfort to authors.) This did not make Freud doubt himself, but deepened his natural irony, probably depressed him on and off.

The book had its origin in his own dreams. He recounts a number, chills his prose, states that he left gaps wherever the substance would too much expose him, drew in a friend to advise him. He wanted to present the necessary facts for the interpretation, but was no Jean Jacques Rousseau, kept his private life as far as possible to himself, but autobiography like biography is nearly impossible, the biographer often getting himself in all the more when he is trying to keep himself out. Freud did try to be frank with his reader; trying will not always manage it.

The first of his general dreams to be publicly analyzed was his famous Irma dream, and when one read it, so many years ago, one's astonishment was that all of that could be in a dream, or could be put into a dream. One was apt not really to believe any of it.

He continued analyzing dreams. He says that by the time he was writing his book, which may have been halfway through his career, he had analyzed more than a thousand. Analyzed the dreams reported by others. Repeatedly he referred to his famous Dora dream, described elsewhere in his writing. Often he dissected a dream episode for episode, phrase for phrase, word for word. He explained that symbolic interpretation of the totality of a dream, saying what the whole of it means, which had been the common way previously, was not his way. He must break the dream into pieces, examine each piece, afterward fit them together, make a history.

Out of this type of study he reached his conclusion, not entirely original with him, that the dream was the fulfillment of a wish. It remained his theory. It was not all he thought of dream but it was overall. The dreamer was given something he wished, in the simplest case something he would have to wake to get, and so, because he had gotten it, in his dream, he could go on sleeping. The dream had had purpose. It defended his sleep. This Freudian interpretation became with time the classical psychoanalytic interpretation, spread through the psychoanalytic world, the psychiatric world, medical world, literary world, world. Dream researchers would sometimes think it invalid, or think worse, would come up with a precise contradictory example, and Freud himself in successive editions of his book, as he had foreseen, modified his interpretation, but the remarkable fact is how little he did modify it, and how it branched.

Every analyst recognizes *The Interpretation of Dreams* as the beginning of his field. So that field began in dream, and not figuratively.

Early in the book Freud recounts other theories of dream. The ancient ones were not theories, not hypotheses to be tested; the dream was simply a direct message from the divine. Freud put the theories in their historic order, commented on them, picked out what led to his, was meticulously careful about crediting, and one might say that he could be because his own theory was by then so secure. He related what he learned. He developed his conception of the dream's structure, mechanism, purpose, revelatory power of the character of a mind, especially neurotic mind. Dreams of patients appear in his voluminous writing. He started a new language, made a new vocabulary.

Freud's original reason for studying dreams was to study the neurotic mind, help in its cure. Disturbed though these human beings were, he nevertheless dragged from them what one might think must disturb them further. He scissored out fragments, sought to relate each fragment to something in the symptoms, brought together the fragments, laid the resulting picture, as completed as he could get it, before the neurotic, who, seeing his mind from earliest infancy illumined by his dreams, was able to contemplate his whole panorama, the whence, might get a grip on the whither, be helped.

Freud believed that via dream he was making each neurotic more intelligible, neurosis more intelligible, also psychosis. He believed the psychotic to live a life of dream, to be always or mostly in dream, that only by staying in dream could he maintain some balanced relation to the world, and it remain bearable, his suffering being unbearable as soon as he stepped into so-called reality. Here too dream, this life-dream, was serving as wish fulfillment, a long-sustained fulfillment instead of getting something to enable one to stay in bed for a night.

Not only did the psychotic require his dream, and the neurotic his dream, we all did. Dreaming helped, in its way, to maintain the level of the universe, a level demanded for galaxies and stars and planets and atoms, and of each of our infinitely small selves with this halo of mind around them.

Somewhere along the line every dream researcher and every psychiatrist is apt to pay lip service to Freud, die father. One well-known researcher into sleep regularly paid that lip service but also regularly added a touch of sarcasm about the father, possibly eased himself for regularly finding his own thoughts so much less significant. The same researcher added his touches of sarcasm about Pavlov too. Jealous world. We are all part of it. Even in our sleep, in our dreams, we do not escape it.

In spite of our familiarity with Freud's wish-fulfillment conclusion, it still remains startling that he could think we dream to preserve our sleep, that dream had that as its function. Formerly we might have said that it had no function, which now from all sides we know it has. Freud thought it satisfied our night worries, night wishes, let us sleep on oblivious of the clock.

Freud's plebeian illustrations are often quoted. A man is hungry between midnight and morning, so dreams himself a banquet, therefore does not need to wake, put on his shirt, walk to the corner, buy a sandwich, and because of that agitation not be able to fall asleep the rest of the night. An intern knows he should be awake, visiting a patient in the women's ward, therefore dreams himself into the ward, not as intern but as patient, double-locks his sleep, is in his bed where he wants to be, and compelled to stay there because patients are forbidden to get out of bed. Freud himself had a recurrent dream in which he drank a glass of water, induced the dream by eating anchovies, sated his thirst with dream water, did not so much as have to lift his glass. (One of those informal experiments that would not even be remembered except that it came off the pen of an accepted master.) Most of the time, wish fulfillment was less revealed. It took on disguises. The interpreter became an expert at seeing behind disguises, even when they were superimposed, his adroitness at this sharpened by his daily work and by his knowledge of the symbols that dream uses, his conviction that every word in the crossword can be fitted into its proper place if one keeps at it.

SLEEP:
Off It Goes

The following was written in 1606. Everybody remembers it, because everybody reads *Macbeth*, and everybody sleeps or tries to.

> Sleep that knits up the ravell'd sleeve of care,
> The death of each day's life, sore labour's bath,
> Balm of hurt minds, great nature's second course,
> Chief nourisher in life's feast. . . .

That could not have been written today, because there is no one to write it, but that almost affection for sleep may come to anyone deprived of it, as Macbeth was.

A child leaps into the day, in a few minutes is as bright as dawn, is dawn, is the whole world, is shy, is asleep, is awake. "I had a nap." The adult does not do that well. The adult during his waking hours lets his mind wander, here, there, forces it to stop, forces it to concentrate, forces himself to explore, to act, is carefree one moment, is anxious the next, suddenly has grown so tired, the eyelids just close, the retinas have the curtains fall in front of them, some touch of light may still get through, some signal travel to parts of the brain, but not enough, the neuron pools stay quiet, and there is that balm of hurt minds, sleep.

During sleep the living machine does not stop, would be dead if it did, but continues at some minimum, stays in one place.

A hedgehog, one species, sleeps twenty-two hours of the twenty-four. The human infant sleeps most of the day and night. Then it is a child and sleeps less. Then it is grown up and sleeps mostly at night and on Sunday afternoons, and even for that requires bourbon or coffee. Everybody knows somebody who can sleep only after coffee. There are the persons who half-sleep all day and solidly sleep at night.

What are the body signs of sleep?

The neurologist's reflexes are diminished, may be absent. The knee jerk, that reflex, may be depressed beyond arousal. Those remarkable reflexes that hold the body up-right falter. Our muscles rest. Single fibers, or great masses of fibers, may not only rest but rhythmically rest. Blood flow lessens, and one could say that in this respect the whole circulation rests. Between each two beats one can think that the heart, the contracting organ, rests. Between each two nerve impulses, thousandths of a second, the nerve fiber, the transmitting machine, rests. A pleasant word, *rest*. Most adults spend a third of the twenty-four hours in one or another sleep posture, change posture, may think they do not but laboratory observation says they do. One American celebrity was partly celebrated for saying that if he slept a small fraction of every hour, four hours in the total, he had had enough. According to the newspapers a student stayed awake for eleven days and recovered from the loss by merely sleeping seventeen consecutive hours. An Arab policeman spent his third in the vertical stance leaning on his gun at a busy intersection in Cairo directing traffic, asleep. At least, his eyes were shut and his body swayed.

Difficult to look accurately at one's own sleep or one's own waking. Sleep, human sleep, has perplexed philosophers and plain citizens from long before Freud and Pavlov and Darwin and Homer. An alive brain lies under it. We know that. The bewildering electrical states of an alive brain lie under it. Sleep is far from fleeting death, though writers have called it that.

The sleeping mind frequently has been considered the reality, the waking mind the unreality, by sentimentalists and mythologists, but by nervous-system scientists too.

Some of us are miserably self-conscious toward our sleep. It does not help us sleep. One gentleman, on the contrary, said that he just told himself to think of nothing, and he fell asleep. The nothing may of course have been something, and exceedingly positive in its effect on brain and mind; the gentleman did not have the kind of detachable intellect that might have known. Another, a bad sleeper, stated that his day mind was always waiting to step out from his night mind. Another, hearing this, stated that his own night mind stepped into clarity and went back into obscurity many times a night, and that he envied a neighbor whose didn't step into clarity until the alarm went off at 6:00 A.M., when there was just clarity enough to reset the clock, to step out into final clarity at 7:00 A.M. That wonderful extra hour! This man drank himself dead with vodka night after night. A man's sleep may be as light as a leaf or as leaden as lead, and all depths are experienced sooner or later by every mind-observing mind.

Gross body fluctuations occur with sleep. The pulse rate drops, it may be twenty beats. The blood pressure drops, it may be twenty millimeters. The kidneys put out less urine, the morning urine then more concentrated. The sweat glands put out more sweat. Body temperature lowers though that of the skin may rise. Carbon dioxide accumulates. Breathing may change in character, become irregular, a gasp, a bump, a snore, when a man's wife wakes him and the breathing becomes regular again. Eyeballs roll upward and outward or in any direction, may oscillate slowly once a second, then rapidly many times a second. Pupils constrict. There are numerous facts clear or unclear. One-half, three-quarters, ninety-nine one-hundredths of us goes off somewhere, the night mind coming into a glory that the day mind recognizes or suspects but cannot ever quite apprehend. The day mind meanwhile has been given surcease from the unutterable, if we think as Eugene O'Neill, or surcease from all the pleasant little duties, if we think as Isaac Watts.

EARLY THEORY:
How It Goes Nobody Knows

There is evidence that the brain never sleeps.

Aside from that, how do we get sleep? How can the top end be thrown out of gear, or into another gear, the engine left idling, as has been said, for three to eight or twelve hours?

Eighty sleep theories have come and gone, most of the early ones obscure enough to be buried in a bibliography, but easily illustrated.

After supper more blood goes to grandmother's stomach and bowel and skin, some of it diverted from her brain, which is deprived of fuel and oxygen, and grandmother loses consciousness, sleeps. By this old theory the brain during sleep is partially suffocated. It never was a probable theory, and blood-flow measurements during sleep apparently prove the brain to have not less blood but the same or more. We could not understand why there should be more blood, except if the mind is intensely active during sleep, the brain might need more blood. At least, there is no evidence of anemia. It is always possible that a small controlling area of brain is getting less blood, and quick shifts of blood from one part of the brain to another are suspected and recognized. No one nowadays believes that insufficient blood or insufficient oxygen gives us this regular, this peculiar, this partial disappearance.

Another early theory had the work of the day pile up acid, lactic acid, and that brought on sleep. But the brain uses lactic acid normally for energy. Also, we sleep when body and brain have not been hard-working, may sleep well after a lazy day. Furthermore, we do not confidently know what a hard-working brain is. As to a hard-working mind, we think we know better, and probably do.

Another early theory had the brain pile up not a by-product but a poison. There was a "hypnotoxic" substance, hence sleep, and when sleep had reduced this, we returned to the waking state. To investigate the substance, spinal fluid was drawn from a sleeping dog and injected into a waking one, which fell asleep. However, a researcher found that when spinal fluid was drawn from a waking dog and injected into a waking dog, that one also fell asleep.

All day our muscles with their nerves are tense and working, or not working but nevertheless tense; then evening comes, they are relaxed, fewer stimuli travel from muscles to brain, the brain is therefore less beat upon, less roused, and we sleep. A light sleeper might insist that his body often was supine and collapsed, he not fidgeting, lying there, 3:00 A.M., wide awake, just did not take the dip into sleep. We all know the dip. It seems to occur unexpectedly, despite the fact that it has occurred night after night after night. We are here and we are gone. That is the kind of occurrence that makes our mind always a stranger to us, the two of us looking at each other. We know our mind better than anything in the world, yet are shy toward it, vaguely uncomfortable with it. Anyway, a bad sleeper who has been a bad sleeper all his life and rather accepts it, cannot easily be convinced that he should look to his unrelaxed muscles and not to his unrelaxed thoughts for his trouble.

All eighty theories, discarded for one reason or another, have had their historic or their literary interest, have also thrown some light on sleep, some light on mind.

Even the way a person approaches his bed has an interest. It may be bizarre. A small girl regularly dives in headfirst, and nothing seems right about that, ought to wake her for a week, but her brain switch is new and young and freshly oiled, and off she goes in half a minute. Her affairs are not as scrambled as they will be. Possibly, too, she does not go off as instantly as she seems to, possibly loiters somewhere in the stages A, B, C, D, or 1, 2, 3, 4. The evidence seems to be that in the first part of sleep we plunge into the deepest depths, pass quickly through stages 1, 2, 3, and 4, stay much in the last stage, less as the night advances, generally go back and forth.

A prominent citizen approached his bed methodically, allowed himself time to unbend from the strains of the day, nagged at his wife, patted his dog. Sprawled in a chair. A second chair. A third chair. Gradually he got himself to the room he knows so well. Removed his pinching shoes, his squeezing collar, slipped into his gaudy pajamas, slipped into his bed. Decidedly, he did not dive. Pulled the bedding up to his neck, expensive linen. Folded back the top sheet, neat as an envelope. Unkinked his joints to the accompaniment of self-conscious yawns. Had shut the windows against noise. Turned on the air conditioner, not to cool the air but to produce the steady hum that would screen out other noises. Previously he had drawn shut the tailor-made curtains against the morning light. And now he drew shut his eyelids. (This hero never takes a sleeping pill.) Following two hours of deliberately shutting off his thinking, he decided on a warm bath. Immersed his skin in the monotony of water not too hot not too cold, eased himself into it, stretched himself out in it; if only he dared enjoy it. (This hero never takes a half-pint of vodka.) He came to a resolve. He would make adjustments in his calendar. As it was, he was spending a month in Florida in winter, a month in Europe in spring, two months in the White Mountains in summer—but he would take off more. Also, he would play more golf. Had he been born in the low categories, he would now have dried, sat in a chair, begun reading the fine print of the stock quotations, which would cause no agitation to his brain and would tire his eyes, those lights turned down, turned out.

RECENT THEORY:
Even Now Nobody Knows

The inside of skulls has increasingly been the researcher's laboratory. Sleep has been investigated with gross surgery done upon the animal brain, and in the human brain, fine electrodes have been applied to the surface, or pushed distances into it, even implanted, left for periods of time.

A cat's thalamus has had an electrode implanted, time allowed for healing, the thalamus shocked, and the cat has fallen asleep. Before she did she curled up as cats do, and it was natural sleep, because if the researcher approached with a package of sausage, she sniffed and woke. The experiment appeared to favor the existence of an active sleep center, a place that when shocked induced sleep, real sleep, not just depressed wakefulness. Other experiments favored a wakefulness center. Experiments can be clear-cut. Sleep still is difficult to define, has relations to the brain that are difficult to limit.

Sleep. Arousal. Wakefulness. Attention.

The four are linked in our everyday thinking, linked also in the psychiatrist's probing of the psyche, and the physiologist's study of the working of the brain.

Sleep inside us does often feel like an on-off event. Surgeons know an area of brain where they go carefully, which might be considered an on-off area, life and death uncomfortably close, sleep and waking close. Sleeping sickness has shown at postmortem nerve cells destroyed there. Latterly the pharmacologist has seemed to prove that the blessed-cursed sleeping tablets act there.

When the brain stem is cut across high up, a profound sleep settles on the animal. That experimental surgery was performed long ago. At the time, the thought was already developing that the low stem kept the high stem aroused, that the high kept the cortex of the total brain aroused, and so we were awake. Numerous experiments made it almost unarguable that the stem forever intervened between a sense-stimulating world and the cortex, to rouse it.

When that cut brain stem of the profoundly sleeping animal was on the raw surface toward the top of the brain, electrically shocked, the animal waked. An electrode woke it. On the contrary, if the core of the brain stem was surgically scooped out, a sleep descended on the animal from which it never woke. If there was no strict sleep center, that neighborhood nevertheless plainly had a primordial concern with sleep and wakefulness. Loosely speaking, a switch could be thrown there and good night till tomorrow. The body's thermostat was not far away, and the daily cycles of body temperature were sometimes described as paralleling and even somehow determining those other cycles, sleep and wakefulness.

It was fifteen or so years ago that sleep experimentation began to concentrate on the brain-stem core, on the reticular formation in there, that intricate weaving and crisscrossing of nerve cells and nerve fibers which runs the length of the core. Sense paths (pain, vision, hearing, smell), on the way to their specialist areas of the cortex, sent side branches into the core. That is, sense messages on the way to the cortex telegraphed collaterally into the stem. Thus that sequence: core stirred, cortex stirred, creature stirred, in a state now to pay attention to the messages that were being delivered. Messages also traveled down into the spine, where there were the final controls for neck, arms, legs, the parts that perform anything the body is called on to perform. So, as has often been said, the mind is alerted above, the body alerted below, the creature able to think and to move.

To repeat: sense messages, pleasurable or dangerous, cold toddy or hot lady, stimulate the brain stem, it stimulates the cortex, sleep departs, our mind is aroused, and we attend, act, think.

The cortex also stimulates itself; at least, each of us seems all day to be finding that his nervous system at its top is able to stir itself. Or, to speak with unphysiological abandon, the mind is able to wake itself, to know itself to be awake, though only God knows just how it knows.

The part played by fatigue? That again is not as demonstrable as common sense suggests. Gradually vast numbers of the nerve-to-nerve linkages do probably in some way become depressed, are harder to reach, their thresholds raised, a block thus placed in the way of the stimuli of life, the creature getting drowsy, settling into a light sleep, a heavy sleep, passing through the stages, in and out, up and down, till tomorrow morning the thresholds are dropped to day levels again, the networks reachable again, the mind attentive again, enslaved again.

Nothing has been said of sleep's periodicity. Some of us prowl through the night and some prowl through the day, but most sleep at night. On a day-night planet resulting in a society with day-night habits, learning could have set the periodicity. In Bergen, up near the Arctic Circle, the streets were crowded in what ought to have been the sleep period of a twenty-four-hour June day. Citizens claimed they needed little or no sleep during that period, which may have been partly the traveler's too-literal acceptance of whatever he was told. People anywhere who claim they never sleep usually lie or are mistaken. Each of us does, however, find that he can swing around his twenty-four-hour schedule, change the one-two ratio of sleep to wakefulness, also find that he is sleeping too much or sleeping too little. On the other side of this is the fact that in numerous arctic expeditions, where observations were made not casually but exactly, men did maintain the one-two ratio despite the all-dark or the all-light of the planet.

Frederic Bremer, eminent and imaginative physiologist, was largely responsible for our knowing that the brain stem wakes the cortex.
In his picture he brought stem and cortex together. The cortex he
conceived as made tired by the affairs of the day, the stem as watchman on guard over it, its master. Fatigue spread in the cortex. The
master nodded. The watchman watched. At a point the watchman
decided, now. He threw the switch, and that was good night till tomorrow. Bremer did two unforgettable experiments. He used cats. In
the first cat he cut between brain and spinal cord, produced a brain
separated from trunk and extremities, a brain with no nerve lines
coming to it from below the head, but with those from the parts of
the head undamaged. This cat slept and waked as whole cats do, in
rhythm. Its eyes during wakefulness followed persons as they moved
around the laboratory. In the second experiment Bremer cut higher,
between cortex and high brain stem, produced a cortex separated
not only from trunk and extremities but from the parts of the head
and face, especially from the labyrinth, those parts of an unmutilated
animal that batter the cortex day and night. This cat slept a deep
sleep, and as in any sleeping cats, its eyes rolled downward and there
they stayed. Its eyes did not follow persons as they moved around the
laboratory. It never waked. Never.

PAVLOV'S THEORY:
He Thought He Knew

Pavlov's favorite word all his laboratory life was *facts*. He wanted nothing but facts. His theory of sleep he did not consider a theory. It was
one of the facts. Sleep was the spread in the cortex of *internal inhibition*,
as he called it. He had discovered it during his conditioned-reflex experiments. Internal inhibition spread and the creature slept. Hypnosis was partial spread. Sleep and hypnosis and internal inhibition he
considered essentially the same process. Much that is known today
about inhibition, the chemistry, was not known in Pavlov's day. Many
physiologists not only doubt Pavlov's internal inhibition, but his theory
makes some of them scathing toward him as a person. His fame meanwhile persists and grows, in recent years appears to outgrow Freud's.

Pavlov is one of the figures of the past most present to the present. He turns up everywhere. He was many-sided as an experimental worker, and complicated.

Internal inhibition was to him as real as the sun and snow of the north. The inhibition was part of conditioned-reflex action, and with the conditioned reflex Pavlov was prepared to account for inanimate and animate, for Theodore Roosevelt and William Jennings Bryan, and the crowds that listened to both, and, later, Mussolini saluting, giving a conditioned response with his chin to the conditioned stimulus of the crowd at the Rome International Physiology Meeting, where Pavlov pointed it out; was also prepared to account for the behavior of kittens and cats and dogs and neurotics and psychotics, for everything that the sleeping and the awake creatures of earth do.

This special inhibition, sleep, as Pavlov understood it, was protective, prevented exhaustion of the chemistry of the delicate brain cells, particularly of the cortex. A brain was active day and night. Inhibition got it its rest, moment's rest, prolonged rest.

Fail to give the dog its dab of meat when the bell rings or the buzzer sounds or the light flashes, inhibition begins. Repeat the fail-to-give and soon the dog sleeps. The man in the street might say that we dull when things we expect do not happen. Pavlov would have agreed. Give a difficult problem to a dog, it sleeps. Lengthen the delay between stimulus and response, between bell and meat, the dog sleeps. In a hundred laboratory situations, corresponding to a hundred life situations, the dog sleeps. Inhibition begins at a point in the cortex, as Pavlov back in the first third of the century understood it, or the inhibition begins at many points, the points coalesce, inhibition spreads, moves down into the stem, where by many lines of evidence there is a final shutoff for the waking mind.

Instead of inhibition having moved down in the brain, it might have moved up or left or right or in or out or in all directions, only not into the vital controls for circulation and respiration; they must never sleep. Ostensibly we would die if they did. Ostensibly parts of the brain were always accepted as always awake. Yet, in late years, what is suspected is that no part of the brain is ever fully not awake.

Pavlov studied sleep originally because his laboratory dogs fell asleep. This disturbed his experiments. It made him irritable. So he decided to study the situation, to find with conditioned-reflex experimentation what sleep meant.

In one well-known experiment he selected twenty notes of a musical scale, made inhibitory stimuli (no meat) of nineteen of them, an excitatory stimulus (meat) of the twentieth. With the dog brain thus prepared, the experimenter related that he played the musical scale. First there was a succession of inhibitory notes and the dog fell asleep, stayed asleep until there was the one excitatory note, when it waked, ate, fell asleep again when the inhibitory notes of the rest of the scale once more depressed its waking mind.

Hypnosis supplied facts. Pavlov thought hypnosis was an inhibition in perhaps a single analyzer, as Pavlov conceived analyzer; this asleep, the hypnotist was able to perform Machiavellian maneuvers with the unresisting parts of it. A dog's limb could be lifted by the experimenter where no dog would lift it, the limb remain there, as a poor woman's arm may for hours in an insane asylum.

It should be illuminating to those who watch human behavior that an occasional scientist has seemed to scorn Pavlov exactly because of his imagination, his fancy, his vitality. He was light both in body and mind. All that he did, and he did many relevant experiments, and no doubt thought that he did many more, has with scrupulous avidity been combed by those who do not believe as he believed.

Had anyone ever asked Pavlov whether dog and man slept by the same mechanisms, he would have bristled. There could be no question here. Brain was brain. He would have said peremptorily that he could produce sleep, vary sleep, measure sleep, in a dog, and, if the circumstances were right, in a man, who was an evolutionarily later dog.

DREAM II:
Laboratory Steps In

In 1952 a young scientist observed that the eyeballs of sleeping infants moved under their eyelids. The movement was back and forth and rapid, fraction of a second, and the eyeballs were moving together. Later, this was labeled rapid-eye-movement sleep, REM sleep. It came in bursts, spurts. Still later, the time between the bursts of REMs, when there were no REMs but the person was asleep, was labeled non-rapid-eye-movement sleep, NREM sleep. This brought the study of dream into the laboratory, because older persons were found also to have the REMs, and when they were waked, they said they had dreamed. The movement went with dreaming.

In the ensuing years, dream laboratories sprang up in place after place, United States, France, elsewhere, and before many more years the National Institute of Health was offering grants of more than a million dollars in support of dream research. There was a society of dream researchers. There was an international meeting—a Frenchman, a German, a Russian, a South American, talking different languages around one subject.

The dream laboratory?

It is practical to take the moment when the researcher enters for his all-night job. His subject, a man, is paid. He is lying on a couch in a cubicle. The researcher is not in the cubicle, has his own cubicle, or just a place where he can, without disturbing the sleeper, watch him and make his recordings. The watching is apt to be through one-way glass.

The subject has fallen asleep. He has passed from the awake stage through the various sleep stages. These are well established, though with some arbitrariness. An electroencephalograph is recording the stages.

Electrodes of the electroencephalograph are attached to the subject's skull. An electrode for recording eye movement is taped near his eye. From under his chin runs a wire for recording body muscle activity. From inside his nose, one for recording respiratory activity. From the left chest, one for recording the action of the heart. Nevertheless he sleeps. First the sleep was NREM sleep. After a while it passed from NREM to REM, the dreaming sleep. His eyeballs plainly are oscillating under the eyelids.

A hypodermic needle has been stuck into his vein and taped down so as not to break if samples of blood are drawn—usual in surgery but not usual in dream research. A stomach tube may be in his stomach to record what goes on there—still less usual. A special kind of plethysmograph may be registering changes in the size of his penis—not usual at all. Psychiatrists have made everybody, and themselves, and researchers so aware of sex behind every door that the sex organ itself gets into some of the experiments. A rubber tube, hose, is belted around the man's chest to record in the old-fashioned way, also, respiratory activity. Other apparatus. Yet despite all apparatus the subject continues to sleep. Pavlov told us over and over that the animal's brain was built tilted toward sleep, and that a spreading inhibition was sleep. What Pavlov did not know was the degree of excitement there could be in a brain while arms and legs and trunk are limp in sleep, which is one of the facts that dream researchers have found and taught. If the dream subject has trouble falling asleep on the first night, that is called first-night effect, and the subject gets over it, or if he does not, if he is a healthy insomniac, he is discarded.

In the researcher's cubicle the paper of the electroencephalograph keeps rolling.

It is ninety minutes since the first REM sleep for this night, and according to previous research there should now be another spell of REMs, and there is. The electrode for eye movement is recording. There has been a reduction of general body activity, also recorded. The subject has passed through the various sleep stages, the electroencephalograph recording them. There are the periodic spindles of sleep, the "bursts" of fourteen cycles per second, which is stage 2, then a sprinkling of so-called delta, slow and high voltage, which is stage 3, then all delta, stage 4, and now back to stage 1, which is light sleep, dreaming sleep, with REMs. There will be three to four or five more spells of REMs before the night is over.

When the sleeping subject passes from NREM sleep to REM sleep and the eyeballs report, that is the cue. The researcher rouses the subject, questions him. Eighty percent of persons so questioned say they had a dream, relate it, or part of it, or hesitate, are blocked. Such blocking of the content of dream, as other somewhat unusual characteristics of it, has always offered to the psychiatrist or the psychologist the possibility of uncovering something with meaning for that mind. But the dream researcher is more concerned with the mechanics of the occurrence of the dream than with the dream.

Dream research is a comparatively new research. The first high enthusiasm may have dampened some, which would be the history of any new research, but undoubtedly it has excited speculation, has caused a reexamination of old beliefs, old superstitions.

Hard muscular labor decreases dreams. Hard mental labor increases them. The decrease and the increase can now be definitely established, or so it seems. Attention on dreaming increases it. Anxiety does. Illness. Sexual preoccupation. High REM activity is reported to accompany the dreaming of the sexual act. The night brain as brain plainly is busier than was formerly believed.

DREAM III:
Dream of Infants

It was in infants that the rapid-eye-movement sleep was first observed. That could mean that they already are dreaming. However, the chance is that the infant is not dreaming. Confronted with REMs in the newborn, the researcher is apt to relax his original conception of "dreaming" sleep, look for some additional explanation.

The newborn sleeps much of the twenty-four hours, and may be in the "dreaming" sleep eight hours of the twenty-four. He also, when he is being studied, has the electrodes wired near his eye, to the top of his soft skull, and thus wired, he blessedly sleeps. Does the day-and-night hammering of the world in the newborn, as also somehow the REMs, stimulate the developing brain to develop? That has been a question. We do of course not know what mind in a dream essentially is, any more than what mind in the waking state essentially is, but, slurring over that, does the dream of infants assist their nervous systems to become completed nervous systems, and their minds to become minds?

Why do we dream?

From the analyst's point of view, there has not for half a century been much doubt. Different analysts would say it differently, emphasize differently, but the dream is some kind of mental clearing, mental riddance, allows for an escape into disorder from the tightly ordered strictures of the day.

From the researcher's point of view, there has also of late been less doubt. He suspects purpose, not in the dreams of the newborn, which he hesitates to believe in, but in the REMs, which are fact. He thinks he has some beginnings of an understanding. In the adult,

REMs concern the mind. It dreams. By dreaming it rids itself of what would in some way clog it, not to dream being destructive for the mind. In the fetus and newborn, the REMs concern only the brain. Lacking them, the brain fails to get something it needs for its maturing. Dream in some way stimulates the brain to become a completed, a connected, a working organ, whatever the consequence of that for the mind. The infant brain is not yet sufficiently nudged from the outside, so it must be from the inside, and by that nudging the brain and then the mind get to be what in our lives we seem to know them to be.

The newborn may have needed to dream in the womb. He may have needed to dream in order to climb out of the womb into daylight. He may have dreamed his way from another life into this life. Reincarnation may be fact. Mind may go on from mind to mind to mind. A present life may be a literal continuation of previous lives. Ancestral memory may be dream. In that case the researcher may be making the incomprehensible comprehensible, may be carrying us back to Wordsworth, farther back, to the Middle Ages, farther back.

DREAM IV:
Creativity

"Kubla Khan" is the most famous instance of a dream doing direct work.

Coleridge had taken laudanum, very likely. He had toothache, very likely. "In consequence of a slight indisposition, an anodyne had been prescribed, from the effects of which he fell asleep in his chair." Those are Coleridge's words. He is speaking of himself. The "slight indisposition" does for some reason, one cannot put one's finger on it, start a feeling of invention, which may not have been intended, and the words are ordinary, but whether or not, out of the ensuing dream rose a vision that for its kind may not be surpassed in any language, except for the visions of Dante.

> In Xanadu did Kubla Khan
> A stately pleasure-dome decree:

> Where Alph, the sacred river, ran
> Through caverns measureless to man
> Down to a sunless sea. . . .

Of Coleridge's taking an anodyne we know bitterly. He took laudanum until it destroyed all of his capacity for literary production, alienated his family, his friends, including Wordsworth, finally left him only Charles Lamb and some indefinite persons who seemed to roam around him and in whose house he lay year after year. The laudanum did, however, not destroy the strange Coleridge quality, dream quality, that still was there when the emaciated head was sinking dying between two pillows. For this we have the testimony of Charles Lamb, who visited him every day, was Charles-Lamb honest, a Charles-Lamb reporter, which is to say watchful and compassionate.

In the explanatory note printed with "Kubla Khan" Coleridge tells us that he was writing down the dream when a visitor interrupted, and the rest was lost. Literary analysts accept that account, though one cannot escape remembering that Coleridge's "Christabel," where there was no question of dream, at least none was raised, also for some reason, some block, some something, stopped before its end, and the second half of even *The Rime of the Ancient Mariner* may not be as finished as the first half.

The introducing of the "slight indisposition" may have had concealed in it the hope to cover somewhat the fact of his addiction. No one likes everyone to know he takes drugs. De Quincey at the beginning of *The Confessions of an English Opium Eater* states carefully that his addiction was under the, to use his words, "coercion of pain the severest." Coleridge does not particularize toothache or laudanum in the formal note, only the indisposition, the dream, the writing, the visitor, the unfinished "Kubla Khan." The poem is the most internally finished Coleridge ever wrote. It is short, swift, visual like a dream, breaks off as do dreams. The hook-and-eye connecting of images (hook-and-eye is his phrase) and the glossed language may both have been facilitated by the laudanum, but warp and woof the fabric was Coleridge fabric, as De Quincey's was De Quincey, as the genuine inside of any lesser creativity is, opium or no opium.

Essentially, whether Coleridge did or did not have the dream is irrelevant. He almost certainly did. We all dream so-and-so many dreams per night, as the dream researchers have documented for us, and so-and-so many daydreams per day, meaning three billion dreaming heads and their dreams rolling with the planet every twenty-four hours; and one "Kubla Khan."

More satisfying for the literary analyst or any other analyzing analyst, one would think, to reflect on the instance of Edgar Allan Poe, that mind that tramped in nightmare beyond dream. Through his rapid uphill fame, Poe everywhere exposed his disorderly downhill life, but the nightmare in that mind was always ordered, always clear.

He had his actress mother, who died when he was three. He had his father, who disappeared when he was two. And a father is a father, but also is the husband of a boy's mother. So, child Poe is child Oedipus, if anyone likes, would in his unconscious kill his father, wed his mother, and this child also is exposed on a wild mountainside, the emotionally wild urban milieu of Poe's childhood. He grows to be a man, takes as wife his child cousin, thirteen years old, and like many another artist, like Dante Gabriel Rossetti notably, merges the lines of those two faces that he knew best, the corpse face of his tuberculous mother, the always dying face of his tuberculous wife, and out of them creates the faces of the women of his tales—Morella, Berenice, Ligeia, Eleonora—also copies off his mind their deathly pallor, their deathly allure. Add that Poe remained childless. Add the tale "Loss of Breath," and give it the overinterpretation it has been given, namely, that loss of breath is Poe's own loss of potency. Poe accordingly becomes impotent. Maybe he was. Impotence itself, by a similar overinterpretation, becomes death and has the hues of death. Add a quick summary of several of his death-in-life stories, and what an irresistible invitation has been extended to a first-year resident in psychiatry to diagnose necrophilia, corpse love. To dis-diagnose necrophilia would require at least a second-year resident. Add an unrelated fact—because it is called attention to, though one does not see why, and it could hardly have much to do with creativity—that he had trouble borrowing four dollars out in Fordham. Add that Poe, like the late Brendan Behan, appeared to drink himself into who-knows-what dreams, but highly creative dreams surely, alternated brandy with opium, they say, the dazed unhappy man carrying his hemorrhaging child-wife from chair to bed, eventually placing her coffin on his writing desk. Dream and reality seem now a frightful mix.

But—he did write "The Raven." One's own temperament may not incline one toward a poem like "The Raven," but that does not prevent one from reflecting that the whole host of drunkards or near drunkards and addicts or near addicts of America and the whole host of its psychology-bitten awake-and-sing critics could in the genre neither produce its equal nor explain its peculiar force. "A limited genre." "Quite so." Poe was a dreamer and a poet, his kind, a singular kind, and let anyone account for the dream in any poet if anyone can. His life seemed rifted, cloven by a too lucid creativity, while the person, the human body merely, was transported through Richmond, West Point, the Bronx, Philadelphia, Baltimore, in which last place he died at forty, presumably of delirium tremens, and his inscrutable restless genius died with him. The scrutable is his life and, like much in anyone's life, is irrelevant. In the foreword to Marie Bonaparte's overdone *Life and Works of Poe*, which foreword she asked Freud, her teacher, to write, he says: "Investigations such as this do not claim to explain creative genius, but they do reveal the factors which awaken it and the sort of subject matter it is destined to choose." Marie Bonaparte may not have been that modest. Freud was modest. Again and again he shows himself the best of the Freudians.

Now move off to the Orient and consider for a moment the way the creativity of one Oriental, a Japanese, can strike one Occidental. He is the greatest of Japanese artists, probably, or one of two or three. Human faces in Hokusai's line drawings often appear as reality exaggerated, reality seen through the illusion gauze that every dreamer knows. Hokusai's sketches are of the hard Japanese daily life, but always with that gauze dropped down in front. Even in his paintings, real and unreal touch the lines, black lines frequently done with black Japanese ink, a bristle of a woman's hair, streak of eyebrow, hollow of cheek, jawbone, the violent thrust of a man's extremity. A face to terrify a child rips through a Japanese lantern, gets stuck there, as one could think, lantern and face immobilized, the lantern being the *mise en scène*, the face being the actor.

Steeped in nation Hokusai was, helplessly Japanese, helplessly himself, helplessly creative, hundreds and hundreds of paintings and drawings. Of the man we know only a few facts, not one with the melodrama of the whole of Poe. Hokusai endured, if you like, this difficult planet until he was ninety. Was poor. Stayed in the same district in Yedo, today's Tokyo, that is on the large island of Honshu; he was always vacating one chilly room and moving into another, sleeping on straw-matted floors inside rice-paper walls, the houses always somewhere close to the Yoshiwara; painted the women of the Yoshiwara but made no specialty of it as did Utamaro. Painted anything living—dog, bamboo, fish, a samurai, a bridge. A bridge can live, of course. Worked always at lightning speed, never blotted out, it was said, apocryphal no doubt. Had trouble with a nephew, like Beethoven. Had trouble with the police, like Beethoven. All of us have troubles, a new trouble little or big every day, and all of us daydream and nightdream; and one Hokusai. Probably he accepted, as did everyone around him, that divinity spoke through dreams, the tutelary divinity, the ancestor who pushed his way across the generations and appeared in his mind, and he made a straight copy of what was in his mind, as Coleridge claimed he did. Hokusai bequeathed us his own face, too, a landscape of wrinkles, harshly believable and unbelievable to Occidental eyes, no doubt to Oriental eyes too: at least what is there was seen through the illusion gauze. One fancies Hokusai in lantern light, squatting by the urori warming his not very warm rice over not very warm charcoal. With chopsticks he shoveled the rice into his mouth from a green bowl, dozed, dreamed, peered curiously, strainedly at something, waked, worked, sketched. Under the ink-brush a robber swung his sword, slayed, was slain, two bloody heads lifted, died. What he peered at was something reflected and refracted in his special mind; born with that.

And now Shakespeare. "To sleep—perchance to dream: ay, there's the rub!"

From those words Shakespeare moved into his great soliloquy. How quiet immediately when one comes on Shakespeare. How far immediately from nightmare. It is not dream either, but nearer. We have had that soliloquy inside us so long that we think it is part of us, think: "I could do it." Try. Paraphrase. Let Hamlet speak your paraphrase. Listen. You will be saddened, because you will have had demonstrated anew your deep lack of the depths of poetry.

This has this much to do with creativity: *One mind can.*

There are times when we wish that Shakespeare had left a journal, then recall what analysts did with Beethoven because he left a journal, or left so-and-so many scribbled sheets of paper. It makes no difference to Beethoven how any analyst analyzes, nor to the idea, Beethoven. The *Eroica* is there where it was. But it could make some difference to us, to what our banal thinking has made of us. A sheaf of Shakespearean letters would be a temptation, nevertheless, even one good-length letter. We might not learn anything about the nature of creativity, or about Shakespeare's dreams, whether he was an insomniac who batted about in his bed, whether he nightly as a preliminary doused himself with ale, but most definitely we would learn that he was not one to write himself out interminably in notes as did poor Coleridge, display himself naked as did poor Poe. Yet a journal, even a short journal, must on some page give us a flash of that towering mind's view of itself.

He wrote the plays we do not know where, corrected them probably on the stage, because he was an actor, played the ghost in *Hamlet*. For this there are the extant playbills. There are the documented evidences of several kinds. There are the copies of the actors' parts, this and that printed play, a few versions exhumed by poor scriveners scribbling under the moon to buy themselves bread and cheese. There is the great fire that burned the Globe Theatre to the ground with the playbooks of the playwrights. Then somehow there are the Quartos and, finally, seven years after his death, the great Folio of 1623. How subject to vicissitude all of it! Suddenly one realizes that Shakespeare might not have survived. No Shakespeare in the world! But God, let alone Shakespeare, is not dead, and Quartos and Folios were saved.

Shakespeare, that man, lived. Shakespeare was born: that mind was born.

Whatever being born may one day mean either to chemistry or philosophy, Shakespeare's mind arrived on earth mostly complete. It never in one lifetime could have accomplished anything beyond a polishing off of some of its roughnesses. Fundamentally, we do not know what mind is, though any man's wife without sending for any analyst may know or think she knows the twists and turns his particular mind has taken under the pressure of birth, infancy, adolescence, mother-in-law, old age, and how that mind has made use of life to modify what it came with. Shakespeare's seems to have been able to make use of everything, and this, the ability to make use of everything, does not characterize creativity either, characterizes dreams only somewhat, but worth mentioning, though it directs our attention more to the substance of the created work than to that special energy, if it can be called energy, that causes substance to live. Substance is approachable. Imagination, the ease of the poetry in Shakespeare, the labor of the poetry in, for instance, Michelangelo, are hardly or not at all approachable, understandable.

Shakespeare may have napped in the afternoon as did the older Hamlet. If he napped he dreamed. We had to wait three and one-half centuries for dream researcher and psychoanalyst to make us utterly sure he dreamed. He had to dream. We are scientifically convinced he had to dream. He probably always snatched a dream when he snatched a nap. And of a night he dreamed so-and-so many dreams. He REM-slept, rapid-eye-movement-slept, and NREM-slept, non-rapid-eye-movement-slept. Then Shakespeare's mind went the Freudian unconscious-preconscious-conscious back-and-forth and the id-ego-superego back-and-forth and whatever other backs-and-forths. Also, we can be sure he free-associated. For Shakespeare we might go scientifically on, no doubt, find a word sequence that fingerprinted his creativity as Ernest Newman found three ascending notes to fingerprint Beethoven. It was an empty finding. Shakespeare and Beethoven, when they slept, staggered their dreams in eighty-to-ninety-minute periods, unless the physiology of sleep was different in the seventeenth and nineteenth centuries than in ours. The eighty-to-ninety makes the planetary dreaming like the grains of sand, everybody, everywhere. Creativity is not like that.

The word *dream* comes into the plays, but it is always the poet using a word, a thin veil of the illusion gauze dropped even in front of the word, which is not at all the word the physiologist uses. Concocted dreams occur here and there in the plays. Then Shakespeare questions whether we may not all always be dreaming, our life a dream, as he says, surrounded by a sleep. Hardheaded Pavlov said again and again that the brain was built tilted toward sleep, that to be awake was perpetual arousal, and he also said that if he were given two hundred years he might fit dream into his scheme. The two hundred indicates that canny Pavlov was aware of difficulties.

Did Shakespeare daydream? Daydreaming unquestionably is not the same. He could not have done much of it, not in the usual sense, not idly, since even with his genius he would have needed to keep disciplined thoughts one following the other if he was to achieve not only the unparalleled quality but the unparalleled quantity of the masterpieces. He finished them first to last in something like sixteen years. Began late—his early thirties. Stopped early—late forties. Whatever dreaming is, whatever creativity is, his creativity did not loiter, his mind did not relent from that balanced imbalance that every self-observing mind knows is necessary if it is to remain itself, not to speak of producing play after play. As for the blank verse that buoyed the plays along, Shakespeare must have been able to produce it as his heart pumped blood, ordered his breakfast porridge in blank verse, blew out his midnight candle. One begins to think that blank verse was simply his prose, and then one comes on a prose passage so limpid, so rare that—well, he could write prose too.

The daydreaming, the not-quite-awake, the half-somnambulistic state, Shakespeare's sharpness would dispose one against believing he used it directly or exactly or at all. That half-somnambulistic state has been considered the productive state. Tchaikovsky considered it so, Schubert did. We do not know how they induced it, brought it out of the who-knows-where, hard walk on a dark night, a plunge of their heads into cold water, vodka for Tchaikovsky, Rhine wine for Schubert. Nietzsche spoke of the productive state as *Rausch*, should have understood it, in some state like *Rausch* wrote the whole of *Also Sprach Zarathustra* in eleven days, or so he said. But Shakespeare? He appears to sit cold sober while his unconscious sprawls over the universe, a loose unconscious in the grip of a tight conscious. One always imagines him calm, inwardly, amidst the hurly-burly of Blackfriars, or any London theater, or any brawling tavern. Marlowe, fellow playwright, was killed in a brawl. Those were hot times. John Aubrey judges that Ben Jonson was jealous of Shakespeare. How could he help but be? But Ben Jonson did stress Shakespeare's gentleness. That was stressed also by Heminge and Condell, fellow actors, in the introductions to the Folios. That is stressed also by us in the twentieth century. Gentleness flows to us from every play, sometimes from every page. We feel, without too much examining the idea, that a supreme and dreaming gentleness might be near to the highest creativity, assuredly part of it. Gentleness helps a mind to stand off from itself. Shakespeare's seems to do that, seems as aloof and nobly dignified as that face, those eyes, that tall forehead, even if each feature is considerably sentimentalized in the Droeshout portrait of him.

Finally comes *The Tempest*. Finally comes the moment when Ariel is dispatched to do his master's last biddings on the island, and then, with those biddings done, to get his promised release, airy Ariel to work no more for Prospero, maybe for no master, free to sail like any old-fashioned spirit or elf or even witch wherever it pleases him, couch in the cowslip and sing:

> Merrily, merrily shall I live now
> Under the blossom that hangs on the bough.

If Ariel was created by Shakespeare to represent creativity, his, it might tell us something of what he himself thought of that process of the mind, his experience with it, how separable from the rest of the mind, how in some respects a foreigner, a servant called in, sent off. "Where the bee sucks, there suck I."

The Tempest and whatever parts of *Henry the Eighth* completed, Shakespeare would have been bouncing in a stagecoach from London to Stratford, there to count his moneys, adjust his affairs, make over his will, loll in the soothing neighborhood of his daughter, Susanna, reflect on this and that, eat, drink, doze, sleep, dream, maintain an amused expectancy, wait for April 23, 1616.

The chance is that with *The Tempest* he was bidding us all good-bye, and how charming if he was. He was making his stage bow before the theater of the world, and by way of ornamenting the bow did once more toss into the stage lights the colored balls of his grace, excite himself with words, Ariel not free at all, not able to be free, Shakespeare not troubling to understand dreams or creativity, just dreaming, just creating. Then there would have been that other indifferent day, that day of amused or musing expectancy, when with the familiar words he would have helped his body to stretch out under that flat stone.

> Good friend, for Jesus' sake forbeare
> To dig the dust enclosed heare:
> Bleste be the man that spares these stones,
> And curst be he that moves my bones.

Desecration of graves was common in the seventeenth century, and Shakespeare had ordered his to be dug seventeen feet deep, presumably ordered.

DRUG:
It Is Bored

De Quincey took opium. Coleridge took laudanum—opium—and wrote "Kubla Khan." The Chinese smoke the opium in pipes, at least did before they became responsible communists. Japanese drink saki. Soviets drink vodka. Collegians drink Coca-Cola, or sniff glue, or swallow a cube of sugar with a drop or two of something colorless, odorless, tasteless. *Brave New World* Aldous Huxley swallowed four-tenths of a gram of mescaline, watched what happened to him, wrote what happened to him, wrote interestingly, and we do not know just what he accomplished toward a fashion in our always-ready-to-be-fashionable world. An estimated eight billion stimulating ampheta-mine tablets are produced in the United States per year. Women in the Tennessee mountains, instead of glue, sniff snuff. Hashish, marijuana, cocaine—curls of smoke from Hong Kong meet curls of smoke from Delhi meet curls of smoke idling up from Cuernavaca or idling down from a cheap high furnished room in an off-Broadway hotel. Much of the American twentieth century puts itself to sleep with sleeping pills, or bourbon, or bourbon plus pills, then next morning looks for its mind somewhere around the bed, cannot find it, then finds it with the help of Dexedrine or just four cups of moral black coffee.

Which could be said another way. A mind pours a drug upon its brain, chases the drug with another drug. Which also could be said another way. A brain pours a drug upon itself. A chemistry pours a chemistry upon a chemistry to alter a chemistry. Why? Why would a brain want to do that?

Some of the reasons are hidden where no chemist or psychoanalyst or sociologist or addiction research center will find them. Some are as plain as a prizefighter's ears. That brain is simply hoping for an hour to escape the pain of belonging to the human family. For an hour to escape itself. Escape its restlessness. Escape its fear of death. Fear of life. For an hour to avoid the minds that come via the telephone, via a letter, via a knock at the door. For an hour to accelerate dullness into gaiety, gaiety into anxiety, anxiety back into dullness, that circle whose diameter gets shorter at each circling. For an hour to be in the swim. For an hour—if an old man—to wander backward into his youth, which they keep telling him was happy all day long. For an hour to dissolve his wife. Or himself. In vodka. In whiskey. In sentimental nobility. (Does the body distill a drug that is sentimental nobility?) In blind hard work. (A drug?) In sixteenth- or twentieth-century Puritanism. (A drug?)

The human species is drugged. Always has been. Has thought it needed to be. Or did need to be. Or wanted to be. The cutting edge of life had gotten too sharp for the human type of mind.

In an experimental mood, a first human brain has administered to a second human brain what would throw the second into a temporary insanity, psycho-phrenia, the second having volunteered for this out of vanity, stupidity, twenty-five dollars—the reasons we all volunteer. Lysergic acid explosively revealed its curious capacities years ago. Lysergic acid became, besides a drug, an idea in the mind. Lysergic acid became a symbol of the right of any human being to do with his life exactly what he pleased. Lysergic acid claimed it produced more than a temporary insanity, produced revelation, final clarification, final expanded consciousness, apparently an understood state by some psychologists. *Psychedelic* became a newspaper word, meant consciousness-altering, applied to any drug that did it. Without a doubt lysergic acid changed John Henry. He changed while he was watched and while he watched himself, got uneasy, got suspicious, got a flush, felt nausea, talked too much, talked of hearing colors, seeing sounds, touching smells. May have begun to wish that he had not let himself in for this damned experiment. Wondered whether he would be the one who stayed changed forever. But—they say—the experience might of itself repeat itself, without the drug, diabolic uncertainty. Some psychiatrists definitively do not regard what occurred to be insanity. Some have administered these popular drugs in the course of therapy, administered them to lift depression, to tone down elation, to achieve at one and the same time depression and elation, loosened someone who was not responding to Freudian interviewing, to bring him where interviewing was possible.

Lysergic acid came to the attention of science because of a laboratory mistake. A chemist got it into himself, afterward took it deliberately. This was in Basel, Switzerland. He wrote in his notebook that he felt dizzy, lay down, had a delirium not all unpleasant, was both unclear and clear, a sense of standing outside himself, viewing himself.

He had poured something via his stomach onto his brain, and somewhere in the neighborhood was his mind. He got on his bicycle, rode to his home, which was close but which in that mind (where all reality is) was miles and miles, time sense warped, his space sense warped, his report no doubt warped. Everybody seems to know more than one version of that famous occurrence.

Pavlov, the Russian, induced neurosis with two metronomes ticking at different rates, conflicting conditioned stimuli, named what he produced experimental neurosis. The name sticks. What happened in the body, the face, the actions of the dog victims did look like neurosis, looked like insanity. Were those conflicting stimuli producing an altered chemistry that left in the body a foreign chemistry that rationally might be called a drug, that acted on the brain, that produced a behavior? Freud long and perhaps always believed that most of his neurotics, those victims that were not dogs but human beings, had resulted from some thwarting of their infantile sexuality. Did that produce a chemical? Today psychiatrists emphasize guilt, shame, aggression, defense, dependence, inturned or out-turned hostility, in-turned or out-turned anything. Chemicals? Drugs? Life produces its neuroses in known and unknown ways, which sometimes appear all reducible to a mind examining itself too much, too destructively. Our outrageous brain is the cause. We use it too much and we have too much to use. In Caesar's words, we think too much—because we cannot avoid thinking too much.

For a time it seemed that new chemicals to produce neurosis or psychosis were being discovered every afternoon. Hallucinogens produce hallucinations, mis-stimulate a brain so that it mis-sees a world.

Adrenaline is part of the jargon of our day. "He had a shot of adrenaline." Normally, adrenaline is broken down rapidly in the body, the breakdown occurring in stages, its manufacture also in stages, and some of the stages being among the potent mind-twisters; so one could imagine—let us imagine—a chemistry stalled at a stage, and sending the invitation that caused our dear departed grandmother to visit in our bedroom without turning the key. Chemists have delved into tyrosine, phenylalanine, tryptophan, amino acids with whose chemistry the body must normally work, and might work abnormally. Chemists have delved into the chemicals that body and brain employ in their operations, that may be present too much or not enough, an unnatural chemistry producing an unnatural mind, hereabouts one proven cause of inborn idiocy.

Tranquilizers, old ones, metamorphose into tranquilizers, new ones. Tranquilizers are part of the twentieth century's chemical faith. Molecules of a tranquilizer, with some peculiar arrangement of atoms, sidle up to molecules of a brain, with some peculiar arrangement of atoms, and a mind changes. Psychopharmacologicals there are of many kinds for many kinds of cures or hopes of cure, some discarded, some arriving, millions and billions of dollars in the turnover.

Mental hospitals discharge the sick en masse as a result of these types of medication (drug therapy), which may be praised as extravagantly as yesterday was lobotomy (surgical therapy) and the day before yesterday shock (electrical therapy). Mental hospitals in consequence are quieter places.

Which all could be said still another way. A host goes to his cellar, accompanies a bottle to his dining room, agitates the bottle as little as possible, proffers its contents in thin glasses. A family doctor proffers a capsule. A psychiatrist, fifty minutes. A nurse, a suppository. A living body—the oozing of an endocrine or combination of endocrines or fractionation of an endocrine or failure of an endocrine or genetic absence of an endocrine. Result? A mind floats, or divides, or quits for the night, executes any of the thousand tricks that the head end of this creature performs to amuse itself, to numb itself, to treat itself, to inspire itself, to understand itself.

This product of earth and air and water plays upon its highest regions with new configurations of earth and air and water. A lofty one studies the configurations, changes a benzene ring, a hydrogen, a carbon, a methyl group, makes a crafty new compound. (That one is the chemist.) Another dispenses the compound. (The pharmacist.) Another experiments with it. (The pharmacologist.) Another prescribes it. (Physician or dentist.) Another forces it to help in a mind diagnosis. (Psychiatrist.) Another informs the younger generation of its curse. (Narcotics administrator, schoolteacher, preacher, rabbi, priest.) Another informs that younger generation how it opens the spirit. (Psychology professor.) Molecules insinuated into a brain built of molecules alter them, alter a mind, alter a person. But do they? Is he inside there altered? Is the observer of himself ever able to be honest enough, detached enough, shrewd enough to state accurately that he is altered, let alone how? What is altered could be facade.

COMMUNICATION:
It Is Lonely

A tomato does not communicate with a tomato, we believe. We could be wrong. The rule possible is that those of the living communicate who move over the earth on their own power: feet, wings, belly muscles of some crawling thing.

A female sparrow plunges into the air, and a male sparrow flutters after. She has told him. One small mind (and brain) has communicated its indignation to another small mind, and with such clarity that it reaches across to our faraway species with its big mind, biggest presumably. A honeybee by a navy-signal technique communicates to a swarm of bees the direction and distance of the nectar, does it in the patterns of a dance, fast if the nectar is near, slower if farther, startling facts dulled only because we have heard them so often. With more reflection, the facts seem not as startling as does that other fact, that a human mind should have figured them out. A red ant zigzagging in a crack in the pavement nudges another red ant, a slight nudge, so slight that we looking down could not hope to detect the communication, if it is communication. Impossible for an ant with its small nervous system to tell more than a short story, but those dolphins that click and whistle, male and female, while a researcher electronically listens or millions over TV look and listen, ought to be able to write a novel. Any of the huge-brained ought to be able to. A dolphin novel. An elephant novel. Tiny nervous systems surprise us, insects'. Foreign forever, insects. Their taste for sugar—the dilution they can detect—is like ours. But what else? They wear their skeletons outside, soft parts inside, and this ought to have made them suspect from the start, that they would show themselves in reverse through and through. Yet a cricket voice to the human ear can have a warm quality, should be even warmer to a cricket ear. In some countries insect voices have inspired the poets. In Japan, Lafcadio Hearn collected a good-sized volume of Japanese poems where insects' rhythms, timbres, pitches are transcribed to human rhythms, timbres, pitches.

Last Saturday night along Stony Brook in Princeton, New Jersey, insect voices filled the air. They got into Paddy's ear. Paddy is an Irish terrier. She scratched her ear, then woke. No, it was not the insects. She emitted a carefully architectured growl, her vote in the night townmeeting, loud enough to indicate how she voted, and having got rid of that was able to fall asleep again. What waked her was a voice evolved much later on earth, a fisherman's. The fool was preparing for dawn somewhere down the river.

When the communication is by posture or movement, that's gesture. Animals use gesture even if the term is not in their dictionaries. A bird will feint with its wing to threaten a trespasser just as a prizefighter will with his arm to threaten the other fighter. The prizefighter will at the same time gesture with jaw, mouth, eyes, or an expressionless expression. Many animals communicate with movements in their faces; monkeys, gorillas. A bird's tiny eye watches a man's huge eye to anticipate what that monster will do next. A macaw creates extraordinary drama with an eye, the pupil of it largely.

Pantomime goes farther. Pantomime is the carrying out of a piece of dumb show, as when a cat cats up to any empty milk saucer and looks down, or a dog marches from parlor to kitchen and stands himself in front of the refrigerator and looks up, or a pigeon pigeontoes to a water bottle that has dried and, if the owner of the coop is too stupid to understand, sticks its bill several times into the bottle's trough, then waits with pigeon decorum while the bottle is refilled.

Vocalization goes farthest. Vocalization must have begun long ago, mind far on its way, the primeval forest full of sounds, and ears evolved to hear the sounds; with a miraculous nuance ears do. This sunlit, moonlit, starlit planet has been and is noisy with communication. Meaning? Explicit meaning, part of the time. A warning. An invitation. A rejection. Less explicit part of the time—the wish with the raising of voice to shove away that age-old emotion, loneliness. The creature has been too much alone, never liked it, wants to put something into the air even if only self-heard in order to make this hour less heavy, not knowing precisely what it is doing, as most of us do not know precisely what we are doing. Therefore, it communicates with just anybody, everybody, nobody. The maid humming in the kitchen is communicating with nobody that anybody could see.

At some place, at some time, in some manner, late in earth life there was evolved a brain that let a mind communicate with symbols. Secondary signals Pavlov called them, secondary conditioned stimuli that excited primary conditioned stimuli that excited conditioned reflexes; the ambitious Russian thus thought that he had brought communication too within his ken. Perhaps he did. Whether or not, symbolization must have appeared and disappeared through hundreds of thousands of years of nervous-system development, long before those colored paintings on the walls of the earliest caves. An understanding of symbols—not of the order of Maeterlinck or Freud—has been proven possessed by the modern dog and chimpanzee. If the symbol was some juxtaposition of human sounds, say the two syllables "table," a dog mind (brain, 65 grams) has been known to attach the syllables not only to the one kitchen table it was good sense for this dog to recognize but to any and all objects with four legs and a top. If the symbol was a coin, one kind among other kinds, a chimpanzee mind (brain, 350 grams) has been known to instruct its fingers to collect coins indiscriminately from a coin machine, separate them into two groups, valuable, valueless, and, after a period of happy hoarding, to choose one of the valuable, insert it into a different machine, buy a banana. In studies of chimpanzees in the wild there has apparently been some evidence that the voicing and hearing of symbols could go over also into the beginning of a capacity for abstraction.

Finally, the human mind (brain, 1,350 grams) collects, hoards, chooses, arranges colors, lines, shadows, employs the twenty-six letters of the alphabet, or the fifty thousand ideographs if the country is

China, and with the combinations of these conveys its thoughts, never quite satisfyingly, to countless other human minds.

PHONATION:
It Is Noisy

If the question is just put bluntly: "What produces a great voice?" anyone answers: "Genius." As part of the genius, there is sure to be an instrument able to operate flawlessly, though not necessarily flawless.

In man this instrument includes windpipe, lungs, the breathing apparatus generally, walls of abdomen and chest, diaphragm, throat, mouth cavity, nasal cavity, so-called articulators, tongue, soft palate, lips, teeth, perhaps gums, then the various muscles that go with these and their blood vessels and nerves. Using this total, man ejects the spoken and the sung, using much of his body as foundation. A voice may shake a foundation.

Phonation is the production of voice. Speech sounds require voice. And speech sounds are at the periphery of what at the center is the puzzle of language.

This instrument has fascinated singers, singing teachers, speech teachers, physiologists, physicists, otolaryngologists, also the architects of concert halls. High-speed cameras have made more understandable the top of the windpipe—the larynx. A man may invest everything he is in his larynx. If a hail-fellow-well-met temporarily loses the use of it because of laryngitis, he might as well be dead. Often it has been described as a wind instrument. It has flexible reeds, vocal cords, two thin folds of muscle and membrane that meet at an acute angle, leave a chink between, and when air from the lungs blasts up at them, they vibrate, which sets the air around that body vibrating, which sets eardrums vibrating, and a voice is heard. Fleshier false vocal cords assist. In a vibrato the singer lets the vibration be heard as vibration. The dimension of the chink, the force of the blast, the length of the cords, their tension—all contribute to produce the initial tone, and each can be changed by muscles inside and outside, cartilages used as levers. When the cords are further loosened or tightened, separated or drawn together by the professional, Joan

Sutherland or Leontyne Price has started hurling a note into the Met, and any person on a Wyoming mountain hugging his transistor in the dark knows where and when and who; and has understood something about the mind of each singer.

The singer herself listens to her voice as it comes through the air but also as it comes through the bones of her skull. Two sources. Two listenings. These correct each other, and that is the way timbre, volume, pitch are achieved, one isolated human being's timbre, volume, pitch. If either of those two sources is blocked, as by a cold in the head, the voice may be so altered that we cannot make out what is being sung or, if recitative, said.

All of this is phonation, and is going in the direction of mind, is mind, but the mind end of the definition is by no means as clear as the rest.

A larynx can lengthen, can shorten, the basal voice accordingly become high-pitched or low-pitched or anything in between. That then is reshaped in pharynx, nose, mouth, these chambers having varyingly sensitive walls, which by fractions of a millimeter are pushed out here, pulled in there. The mouth is subject to vast training. As the voice advances through these passages, it gains in complexity. A voice can be delayed. Can be whispered or shouted. Can be made to explode forth or glide glissando through a tunnel of flesh. Can be articulated until it reveals perhaps better than any part of our body our capacity to individualize. Mind is behind all of this. Resonance of voice is increased or decreased by the caverns of chest and head. Man's indescribable wish to express, and his success at expression, require this only somewhat describable machinery. A consequence of this wish and of this machinery is that on the portico of the country club at night we hear a coo, recognize whose, possibly why, or a shush, an ah, something bogus, and who would not reflect on the vocal confusion in the Ark when the pairs were watching the rain?

A soprano sings a high tone in perfect pitch, seeks expression for the purest in her spirit, and while one listens to her, and when her mind and the instrument are obedient to that spirit, one can imagine oneself in some high tower, high above the earth.

How like some strained sculpture of Michelangelo, besides how vocally knowledgeable, may be the contortions of a great singer's mouth when one looks at her head-on. One sees lips and teeth, around them face and neck, sees the leaping shadows cast by machine parts operating farther back. The grimness shocks one. Animal shows through. Species shows through. Race shows through. A larnygologist sent puffs of air through the larynx of a dead dog, produced the bark. That surprised him. Leonardo did the same for a goose, produced the honk. That did not surprise him. An intellect like Leonardo's would no doubt have distinguished quietly the qualities given to a tone from below the larynx, from above the larynx, and he very likely reflected on how the differences in voice anatomy of man, wolf, lion contributed to their power to spread an emotion, a thing of mind, fright, through the world.

Having arrived at the vocal cords, that blast from the lungs was halted, much, little, infinitely little. The cords vibrated in their lengths for the fundamental, in fractions of their lengths for the overtones. Then, above the cords, eddyings were introduced among the quivering molecules, shades among the tones, till there emerged the voice of a German world, differentiated from an Arabic world, from a Balinese, five-month-old Ann from seventy-year-old Ann, a natural baritone from a fake. If the voice came mostly from the chest, singers said chest register, if from the head, head register, if from between, middle register.

Vowels are somewhat prolonged tones and have varying complication, mind making use of the nuances. Consonants are tones checked in throat or mouth. With *t*, the checking is clean. A nasal tone like *ng* twangs because the soft palate has not altogether shut off the nose, and sound leaks through it. In paralysis of the soft palate, all tones are nasal. The rolling of an *r* occurs when the blast of air strikes the tongue as it rests against the palate, the tongue now the reed.

If mind keeps pouring into these products of phonation the molds of words, phrases, sentences—that is not all of it—there results the speech that marks a nation, a southerner of the Deep South versus a graduate of Groton in New England, East Side New York versus West Side, the guttural, the simpering, the adorable, the affected, the genuine, the warm.

Aeons built these structures that produce the inherited, and besides the inherited are able to produce the results of individual learning, which began the first time the infant ear filtered his mother's voice from the voices around. A sensitive French observer states that meaningful attention has appeared on the second day of life. The infant would first have generalized the sounds, then more and more refinedly differentiated as it more refinedly imitated. One would expect this to take time. It did. Month after month an infant legato was carefully modulated. A year-and-a-half-old male will have traveled a great distance in this world of speech sounds. He produced inflections, hurried tones, slowed tones, changed volume, plugged in a consonant, and while practicing his vocal gymnastics might have had the back of his hand on his hip as he saw his mother with the back of her hand on her hip when she gave orders to the grocer. Everything of his body, if we could see, was joining to express that mind. In due course consonants were not accidentally but deliberately planted. An audible question mark fell at the end of a sequence that asked a question. At last, a great day, the family had the right to be sure that Aunt Susan had been named Hooi. Much of Hooi was stored in that male's brainmind before this monotonous "Hooi, Hooi, Hooi." The name was commissioned to do things. The chief himself had a name, knew it. Names were splattered over objects, persons, actions. The incomprehensible was cracked by the comprehensible. We thought we heard that. Then we knew we heard it. Explicit articulations were attached to explicit situations. Life and voice were increasingly integrated into a total. (So easy, and too glib, to say it that way.) This continued, with diminishing freshness, through kindergarten, school, a trip to Cairo, to Jerusalem, to Istanbul, until from the winds sweeping the earth, from the taxis and the crunching buses, from TV blaring out of every summer window, from the fundamentals and overtones of men and their animals, there rose a mighty surge, reminiscent, vulgar, dull, a splash of brightness, to the sensitive ears of an aging lady of musical talent sitting in her room in the dark of the seventeenth floor of Tudor Tower in New York City.

Americans—this is not on another subject—have frequently been assured that their contribution to the earth's noises is uncultivated, their phonation puerile and unfortunate. A British actress made this clear. She explained them to themselves. "Americans today yell because in the early days of their country they had to yell to their cattle." Yelling had hung on as their mode of communication. She said that

that had been unfair to the cattle, their ears. However, the raw country, she repeated sympathetically, did require the yelling, and Americans never lost it, not in their loud theories either, not in their total minds. Each time the British actress explained, and she loved to, she made a horn of her two hands and bellowed through the horn, but what she bellowed was marvelously British.

APHASIA:
It Is Maimed

Disaster in the brain may wipe out speech, all or some.

The victim will lie there, perhaps the whole day and half the night, stare at TV, a blessed use for TV.

One proud woman had a stroke. Her paralysis was on the right side. She was able to get herself from room to room. Wept too easily, but had an imaginative doctor and an affectionate husband who always found a way to help her. Her speech loss was near to complete. Abruptly a tortured effort showed in her face, not a disabled face. She was wishing to expel a word. The wish was like a light switched on, but the light dimmed, the barrenness returned, then she laughed, could hold back laughter no better than weeping. Began again searching for the word. The word was *strawberry*, but what at last she burst forth with was *peach*. Like an opera singer needing to fill an opera house, her mouth, her face, her neck, her chest, came down on that one-syllabled *peach*.

Aphasia, loss of speech, has taught us much, possibly most, about that no-man's land between brain flesh and mind.

A brisk eighty-two-year-old gentleman paralyzed on his right side sat month after month in a hospital bed, passed his time relishing irony, read his newspaper as often upside down as upside up, and with the military bearing of a reader in a public library who is there to get warm in winter. Once he broke the quiet (a quiet will suddenly fall over a hospital ward) with the single letter *u*, the next moment blurted *universe*, and when his newspaper was taken from him muttered monotonously *university, university, university*. It was as if he were trying to latch that word tightly among his few possessions—latch life is what one thought. One thought too of how there must be some border of feeling around speech, how without that border speech is not speech. Another gentleman in another hospital in another town repeated till it paled out, *"Don't remember, don't remember. . . . "* He was like someone in a nightmare. A single naked phrase was his bank balance after the crash. Another made a ward daft by echoing whatever anyone in any bed spoke, and one realized how correct but literal is the neurologist's term for this, *echo-lalia.* Another was asked to touch his left ear with his right thumb, appeared to take that request duly under consideration, then duly touched his left ear with his *left* thumb, and when he had achieved complete agility with this, elaborately offered that gesture as an answer to any question addressed to him.

The intonation that goes with speech may be lost, bleached out, the skeleton syllables remain. It is reported that a musician lost music, no other sounds. A psychiatrist might be suspicious of that—the musician's underneath could be manipulating for him, so that he never again would have to give a music lesson. A mathematician lost numbers. A psychiatrist might be suspicious of that too. Psychiatrists are suspicious.

Wrecks like these have been studied for generations by the neurologist, later by the brain cutter, wrecks of war, automobile, tumor, hemorrhage, clot, narrowed blood vessels. The details, flesh and speech, difficult to disentangle often, have added to the slow but accumulating analysis of brain-to-language. The overall intent of the neurologist has been to assemble a pieced-together, brain-related speech; of the brain cutter, a pieced-together, speech-related brain, both always working somewhere in the area of general mind. The results often have not been satisfying, though now and then they have been, somewhat, and dramatic, the explanations nevertheless arousing caustic differences of professional opinion. Where there is dissatisfaction, this has been partly that any attempt to pin language to flesh revolts us. Mind seems demoted. Brain seems demoted. We ourselves seem demonstrated to be only more fatally earth, as we probably are, though it is unrealistic for such a reason to allow a depressed mood to settle on us, there being evidence enough, or ignorance enough, to let flicker the hope that we may in the end not be proven entirely masterpieces of dung.

About ninety years ago the French anthropologist and surgeon Paul Broca, having reviewed published cases of speech loss and his own cases, and the reports on the postmortems of the brains, thought he had staked out the territory of spoken speech as the posterior portion of the third convolution of the frontal lobe of the left cerebral hemisphere in the right-handed. (No special merit in knowing this geography; enough persons know it.) Broca concluded that this territory—Broca's area—must not be destroyed if the patient was ever again to talk. By hindsight we today think it is not surprising that this locale should have been in the neighborhood of brain that controls tongue and lips and whatever body parts are necessary to chip from raw sounds:

> . . . the morn in russet mantle clad
> Walks o'er the dew of yon high eastern hill.

Similarly for written speech, the territory was that of the muscles of the hand. Disaster there might not cause a paralysis of the hand, yet its power to write words, the mind's power to imagine the written, was maimed or wiped out. Similarly for reading. The spoken, the written, the read are mind-soaked and interwoven in the brain, and, again, not surprising that this late evolutionary capacity of speech should have had intimately geared into it the machinery of vision, hearing, movement, therefore a crippling of it in some major or minor degree possible at numerous points. In minor degree the crippling might go unnoticed except by an alert neurologist. When the damage was in the association area for all three senses—parietal, temporal, occipital lobes—as happens, the crippling might be most unhappy.

A by-product of the study of aphasia has over the years been speech maps, speech atlases, and there have been speech cartographers, the last sometimes derisively called diagram makers.

Early cartographers were succeeded by late cartographers, by psychologists, by psychiatrists, speech students, speech therapists, linguists, philosophers. An erudite member of any of these groups may have added knowledge. Anatomists have added. Anthropologists have added. Engineers have added analogies from telephone, telegraph, television. Even that geologist or paleontologist may directly or indirectly have sought among fossils for the earliest origins of language in the earliest origins of man.

Years ago a famous neurologist-physiologist set up a famous classification. Verbal aphasia: loss of the capacity to understand or use words. Nominal aphasia: loss of the capacity to understand or use nouns. Syntactic: loss of the capacity to keep a sentence in order. Semantic: loss of the power to extract intelligence from words. The famous neurologist had a tidy mind, an excellent mind, but one does nevertheless see a professor walking among the stars with a grammar in his hand.

Everyone suffers transient aphasia, at the close of a hard day, between midnight and morning if the telephone has waked him, after enough cocktails, enough drug, or when he gets older. Pavlov complained as though a trick had been played personally on him—he was forgetting names! On the other hand, a person's rememberings, as we know, may be so bizarre that he is as fascinated by them as by what slipped away. A distinguished American woman-scientist returned after years to a distant country and was astonished at how she could snatch back names of members of families, attach them correctly to the wrinkled faces, but in a rice country, which this was, she could not recall the word for rice.

Electrical stimulation of a brain during surgery may light up a scene for the patient, a scene that was dark till then, and he may speak out words appropriate to that scene. Everybody in the operating room hears the words. The whole maneuver remains more macabre than any whodunit, and it is a generation since it was news.

The fitting of an aphasic's history to his postmortem runs into difficulties. One man will die quickly after he has had a stroke, no opportunity in the minutes or hours or days he groped in his fog to study the detail of his speech, so no use to study the detail of his brain. Another apoplectic will live for years after a slight stroke, he and everybody forget just what did happen at the time, so again not much use to study his brain. Another suffers a cerebral accident (beloved phrase), is not able immediately to speak, as apparently with President Eisenhower, but in a week he has recovered sentences, in a year everything. He does not suffer a second stroke. Survives twenty years. Moves to another city. Dies there. Is a member of a family with an idiosyncrasy against postmortems. Back there twenty years ago he had told his story to a neurologist, and the neurologist is still alive, but he was always an absentminded fellow. As for the man, the patient, he naturally felt no obligation to contribute to the next generation's understanding of brain and language, so he left no diary. After his death, in spite of his widow's idiosyncrasy and his civic prominence, his brain did get into the hands of a brain cutter but—the brain cutter was incompetent. Remarkable, when the pit-falls are reckoned in, that as much has been learned about brain-to-speech as has been. Remarkable, too, the disillusioned modern investigator with neither religion nor philosophy, the pathetic cold-in-the-head toiler with the ambiguities of his own mind and those of earth and sky and sea, that he should unravel as much as he has.

An occasional psychiatrist has found it possible to turn his back on the ruins of nervous systems. He has found himself able to ignore the sometimes clear-cut damage to flesh associated with suggestive damage to mind, able to detach the intricacies of aphasia from intricacies of brain, call loss of speech loss of intelligence. Why not? It avoids the geographic haggling of generations of neurologists, thrusts language into general mind, and disregards brain. Freud would not have gone that far. Sherrington might have. Pavlov went the whole other way, like a carpenter he nailed mind to brain, nailed in language with conditioned-reflex hypotheses, and did this so efficiently that it might be difficult to pull out the nails if one thought it worth the trouble.

A colored woman sat in a chair in the clinic, had had a stroke, had made an excellent recovery, as doctors may phrase it, but her speech was gone. Her face on the right side did not move, had the smoothness of a basalt sculpture, the paralysis itself contributing some odd loveliness. She was touching. The corner of her mouth on that side drooped, moved with strong feeling though she could not move it of her own will, which is common. No fretfulness was in her. No confusion. "I know, but I can't talk." She was able to say that, repeated it, but it did not sound monotonous. This colored woman's problem was not that she could not see the shapes of words, or that her speech machinery could not produce the shapes, but her hearing could not make meaning of what was fed well enough into her auditory machinery. Testing bore this out. The comprehension of the heard was impossible. The function was gone. Never for the rest of her life would this woman be annoyed by the petulance of verbal wrangling. Spoken speech was henceforth mere tone. She was shown a photograph intended to ask a question. Vivid understanding sprang into her body. She looked as if she were also going to spring into an answer. Something blocked the road. "I know, but I can't talk." She repeated the six words. What was arresting, however, was that she could maintain such poise. She could sit in the midst of this trouble in her brain and in the midst of those doctors and be calm. Far inside her there was a clear mind. One might be wrong but one felt this. Conceivably she was even clearer—in there somewhere—than before the stroke. She always had been clear, exceedingly, her friends said. Except for her speech, they said, she was as she always was. Speech in this woman did seem an overlay on mind. Speech could be rubbed off and mind remain. A human mind was not a collection of language items. It was more enduring. It was more resistant to disaster. It hovered there beneath that basalt surface. She sat in the center of a clinic with thirty doctors and her loss and remained *that* woman.

HAND:
It Is Sly

When the first prehumans rose to their hind feet, wobbled for millennia, finally stood upright, achieved the erect posture, it must have been a sorry day, but a touch of grandeur was on it. The top end was free. The forefeet could help in the development of what was now to be a human body, especially in the development of that body's human hand, which would prove the most flexible tool on earth, invent lever and wheel, subsequently wristwatch and stethoscope and jet plane and missile to the moon and futuristic and surrealistic painting, and while getting that work done would take part also in the production of this most intricate of all types of communication, speech. Henceforth there would be a creature quicker than any to understand, and misunderstand, and—what is paramount—he would spread mind over the planet.

To what extent and in what manner the hand contributed to this we do not know, only that it did. The rest of the body was not idle, but we have more than an intuition that the hand led the way. Aristotle had said, "The opinion of Anaxagoras is that the possession of hands is the cause of man being of all animals the most intelligent." Aristotle did not think so. He thought hands were effect, not cause.

The five-fingered reptile has a plan of hand in general like ours, anatomists tell us; therefore the original sketch was made long ago. Muscles, fascias, tendons, blood vessels, bones, and the nerves that run to the centers and back from the centers, none is drastically different in us. There are differences in what, because of its gross mechanics, our hand can do, but the enormous difference surely is our enormous brain, its special character, the emphasis given to its parts, even the sheer space allotted to the hand in its cortex, as much as to the face, which includes tongue, cheeks, lips, all that is employed in speaking. Over the great span from reptile to us the fingers gained in the capacity to fold over an object, to grasp it, and the thumb in its capacity to touch, to oppose, any one of the four fingers; meantime the hand-brain was expanding into the huge communication center we know it to be.

A good hand plays many roles, strikes one sometimes as having mind tucked right in it, and no one would be surprised to discover how easily it is made self-conscious. In an Easter parade, in photographs of brides and grooms, on that platform at a commencement, abruptly one sees nothing but the hands. That flushed graduate would prefer we did not stare so at his hands, and we are not intending to, but we do, and we may suddenly, and apologetically, feel that we are intruding on some secret we have no right to know. Two hands stay stubbornly in a stubborn man's lap or, the other way, gesticulate overmuch, disclose his nervousness. Pockets were designed, partly, to hide hands, as were suede gloves, but where hands all day hide themselves is in occupation.

A London shop takes advantage of that biological situation, wraps a package in a gay paper, attaches a red string to the package, a polished wooden ring to the free end of the string, and a lady slips her index finger through the ring, walks along the Strand with her package on ahead of her, one self-conscious hand given something to do, forgetting itself, its visibility diminished, her visibility increased, she decorated by a package. Along the Strand also there is always the lady whose package runs and pulls at a leash, again one self-conscious hand given something to do. A cigarette is a package, neat and small. A smart-looking bearded young man lights one, puffs at it twice, rests it, taps off the ash, for a quarter of an hour keeps that up. Canes and umbrellas communicate, in London. A gentleman in black cravat and striped trousers, skipping down the steps of Saint Paul's at three o'clock in the afternoon, swinging his cane to the horizontal, then twirling it through 360 degrees, is bragging to the city that he has just kissed the bride and has no responsibility for her.

Every freshman medical student is taught that the left side of the brain in the majority of human beings is language-dominant, more particularly one part of that left side, the person spoken of as left-brained. He is apt to be right-handed, right-eyed, right-footed, may lean to the right. Not all of these characteristics are in one person usually. Three facts: first, the language dominance of the right-handed may not be left-brained; second, the left-handed may be even less often right-brained; third, preference for left or right hand, called handedness, seems inherited but not fatally. The majority of cats and parrots have been reported left-pawed and left-clawed, the majority of rats and monkeys the reverse, so, were those reports to prevail, human dominances would not be unique. Cats in a more recent report were stated to have either left paw preference or right paw preference, and they stuck to the side chosen; difficult however to be sure that any of this can be established.

Formerly there was the conviction that if a child was left-handed, we should not try to make it right-handed, because bad things would happen to it, stammering and neurosis. There was the other conviction, if a left-handed child were forced to fight its way through a right-handed world, that would produce neuroses. And still the other conviction, that the left-handed were all neurotics to begin with. Some neurologists at the same time were offering their opinion, that right-handedness was a matter of education, emphasizing that we live in a right-handed world, but not always emphasizing what is equally fact, that we live in an education-ridden world.

Man's brain does seem cocked to one side, right or left, most frequently left, and the cocked character of his thoughts, their instability, their inventiveness, may be related to this. His brain is readier to be tipped, and he readier to meet the topsy-turvy of life, able also to perceive and to reflect upon the restless vividness of life.

Hands in countless ways speak directly. There are the hands of the Japanese wood-carver saying things to the wood. There are the hands of the French gardener saying things to the soil and the sprouts of green beans, advising them whether they should or should not come out in this changeable weather. There are the hands of the American jockeys Arcaro, Shoemaker and the long line before them and the line to follow, the hands of the jockey reaching to the mind of the horse, and the horse reaching back to the jockey, each correcting the other, the race won. Actors know the force of hands, understate with them, being professionals. Portrait painters sometimes cut them harshly off the canvas, or else paint them with conspicuous care, make them divide with the face the responsibility of communicating what this person has to communicate. For his $15,000 portraits Salvador Dali asked $25,000 if he included the hands, knew they were troublesome to paint. In photographs, weak hands (Churchill's) may go with melodramatic faces. Then there are the strong hands of the master painters. The left hand of Mary in Leonardo's *Virgin of the Rocks* stretches straight toward us, canopies that holy group, Jesus, John, the Angel, is canopying the world, an unearthly light shining from above, edging the fingers, throwing the palm into shadow, making us know that this is not only the most believable hand a painter ever painted but as gentle a thought as a human being ever thought. Blake in his drawings often used hands to convey his visions. He wrote:

> Tools were made, and born with hands,
> Every farmer understands.

In the infant the hand already shows the cunning of the species, soon starts on the acquisition of its personal cunning, life never letting it alone, it never letting life alone, becoming more and more the communicating instrument until one day in the heat of an argument a woman lifts her thumb, merely that *Homo sapiens* opposable thumb, and her husband stabs her to death. Make a jury understand that. The years were not wasted on that hand. It said something exceedingly nasty.

THE PERSON:
It Is Flaming

Under the cold stars dwells that speaking speck, the single human being. His story, or hers, began long before that night he became a member of the human race. That night we of the family thought he was the spit and image of his mother though the shape of his chin was his father's and he worked his mouth like his toothless Aunt Julia. The truth was, if we had had the common sense to face it, he did not resemble any of us, not anyone anywhere, not even physically. He was unique. He was born unique. Two legs he had—that sort of thing. After a while he would resemble us more. He would begin to imitate us more, lose himself. Too bad. But inescapable. Only, why this fuss about this inexperienced performer who forgot his lines and was pushed without his makeup onto the stage? He is too vague. He is like the great vague out of which half an hour ago he came wiggling. It could be—perhaps—the fuss was that we thought we already saw there in the crib the single person. We would care about that. We would want that to survive, would want the single person proven again and again, might be concerned about saving even this infant's singularity, this piece of our flesh and blood and vanity, to keep up the precedent.

In the highest biological echelon the single person, that personality, is king. Christ, King of Kings, was a person, which is his poignance and his reality. Moses was. Buddha was. Herod. In the highest biological echelon the distinguishing of one from one is a chief concern, chief annoyance, chief delight, this distinguishing of mind from mind. Not tallness, not smallness, not a wart on a nose, not classic Athenian handsomeness, but mind—timid, bold, agonized, confused, assertive, assured, double-dealing mind—is the core. Body parts make a difference, an ugly face, a short neck, a thyroid Throat, a swayback back, a clubfoot, but these too hold us sustainedly only because of what happens to the mind inside them.

Picture a congregation of Jews coming from an Israeli synagogue, Italians jammed in the square before Saint Peter's on Christmas Eve when in his white the pope steps forth at the lighted window, Chinese blocking the narrow streets of Shanghai, New Yorkers seventy-five stories down swarming from buildings at the end of a workday—

even in those piazzas or ghettos the imagination reaches toward the solitary one, toward what one thinks is special in the mind of the tramp, the mayor, the pimp, the merchant, and especially special in the poet. Our human eye is that kind of eye. It goes moist over the personal. Goethe spoke of this. *"Grösstest Gluck der Menschenkinder ist immer noch die Persönlichkeit."* ("Greatest joy of the man-children is ever still the personality.") Notwithstanding, in some silly social-minded hour one wonders whether one ought to pay the attention to it that one pays, everybody or just one staring at everybody or just one, always trying to stare under the surface. Might we not see more if we merely glanced for an instant out of the side of our eyes? Glanced, and not analytically? And if that were not true, ought we then not, for our mind's sake, and all minds' sake, be willing to see less? Should we always be thinking: What is he thinking? What is he doing? Why is he doing it? What is his background for doing it? Should we be so earnestly trying to take apart what God or the Devil or just home-spun evolution, or the three joined have put together? Should we be so humorless now that we know there are the astronomer's and the physicist's light-years? For example, what would happen to a patch in the fabric of a person, say that woman's charm as communicated to that gentleman in black tie, if his intellect began to divide the charm into its component parts, did this historically, physiologically, chemically? Worked out the chemical structure of her charm, found it to be a three-dimensional aggregate of polymeric macromolecules but of course with this person's personal atomic configuration? Would she ever again be able to use her charm on him? Communicate to him what he is hoping to communicate to her, truly communicate, not pass along on telegraph tape or as arithmetic from a computer? Should there not remain something not shrunken by calculation, by penetration, by interpretation? Something unexplained? Something inexplicable? Not measured and weighed? Not quantified? Must everything be flattened into carbohydrate, fat, protein, nucleic acid, and their descendants? Broken up into the billions of particles that build the millions of cells of brains? Must the sociologist dig into the city directory to find the precinct of the Bronx we came from and derive us from that, from our deprivation? Or our possession? Must the priest or rabbi flatten everything into earthly and mystic? Must the psychiatrist flatten everything into hostilities, insights, aggressions, the phallic, the oedipal, the rest? Must even short-lived man of the streets choke what might otherwise burst through from his long-lived past? The personal appears so often to wish to stay with a man or woman or boy or girl, and these appear so ready to destroy it by denying it the free impulsive character with which it arrived on the earth.

How does this uniqueness come about?

To the mathematician it is only mathematics. Considering the number of units in a human creature, the mammoth number of ways they can combine, each simply must be different, by the mathematics, no further explanation necessary. Considering the number of genes on the chromosomes of that original nucleus, the fabulous numbers of possible arrangements of the atoms, and the still more fabulous numbers when the two nuclei relate after that night a male talked tenderly to his female, each must be different. Modify that resistant inherited core in whatever ways are possible by those dialects in the Tower of Babel, or the Ginza in Tokyo, the phrases, idioms, inflections, uses, abuses, each to that slight degree also different. Considering the number of environmental influences that hang around each, the influences each has been taught that he must give heed to, the persons each has to meet and somewhat flatter, the detested female chairman of the women's club, the male chairman of the board of directors, the mother-in-law with hemorrhoids who would like to move to southern California to live, the father-in-law who would like not to move to southern California but play golf right here in this county that is better than any in the whole United States, plus the clock in Tristram Shandy's kitchen, plus the classmates with all their sizes of hips and sizes of hopes in the college dormitory, each must be different. Each is a mosaic, ancient in most of the inlaid pieces of its colored stone, modern in only a few of them, the pieces ten million, or ten million times ten times ten million. Naked number could explain the multiplicity of that daydreaming nightdreaming single human head.

To many a public-spirited citizen today, the person is all interpersonal relations—the bad in the bad boy is blown up so big by the bad around him that the boy in the bad boy is contracted to nothing at all. But start anywhere. Construct a model—to keep up with the times. List the parameters—to keep up with the times. Right parameters.

Economics parameters. Education parameters. Birth parameters. Infancy parameters. Immunological parameters. Metabolic parameters.
Acute lack of oxygen does one thing. Chronic lack another. Lack of
sugar. Lack of vitamin. So the removal of the daily threat of a cold
in the head by a winter in the sun of Arizona. So good news. So the
eating of too much bread.

What stands without argument, meanwhile, what asks no accounting, what still can have no accounting, the person down in,
the person large or small, is to each his faithful flickering light, so
long as it flickers, so long as he lets it flicker. It is he who is assured
how different from all others he is, assured most when he has had
an exalted evening, given a solo performance, but assured still next
morning when he drags his body from his bed, when he sits by himself on New Year's Day, or sat the midnight before in a glum gloom
studying the wine stain at the bottom of his glass and thought his
green black mood was because Cynthia in that dying year had returned him his ring.

What stands without argument, furthermore, that person, however
he may eventually be explained, is streaked with genius rarely, his little
is smothered into conformity usually, altered up or down only slowly
in a lifetime, a happy fact when we admire it, a blooming bore when
we do not. That little may have been the littlest little to begin with and
this farther worn down in a Spartan society that never suspects it is
Spartan, but that little still is headstrong. In its skin it waits. On what
does it wait? A shell of fat insulates it, a shell of adrenal cortex, a shell
of progressively failing heart, failing speech, failing locomotion, failing
eyesight, failing friends, failing mind, failing dreams. Dreams get thinner after childhood and in an old man may be thin indeed or absent,
like his hair. A face bulges so one scarcely sees the eyes, but there at
the middle puffs Harry Senior, who fifty years ago had the desk next
to the window in the third grade at school. We come, we go, we are
the same, nine-tenths the same, or ninety-nine one-hundredths, or
nine hundred and ninety-nine one-thousandths. We come cast for our
parts, if born professional enough to have recognizable parts, Puck
or Othello or Cleopatra, born in our costumes, touched mostly in ostensible ways by religion, epoch, nation, family, the days of the years
of a life. The essence—that miracle in one opinion and militantly not

miracle in much opinion—wanders not far from where it was when it arrived through that archway of the womb. We chisel and chisel, but the stone is hard. The separateness of each—and separate it is for anyone who takes the trouble to look past the obvious—is so old-as-the-hills that what happens in one lifetime, one birth-to-death, could be irrelevant. Yet the sometimes seeming great change that on closer inspection turns out almost to be no change at all, might be a reason, if we need a reason, for our coming, and a reason, if we need a reason, for our going. A pounding in his deaf ears may have modified a composition of Beethoven's, as even his rage at his hearing loss, as even the silence that was always extending around him, as surely the remembered noises of his beloved place in the country, of the stream running through, but Beethoven was Beethoven and not the stream, obviously. El Greco may have made use of his inborn visual obliquity, but El Greco was El Greco, evolved over the long road from masers to us. Coleridge was Coleridge, and those words of his were Coleridge, not opium. Lucrezia Borgia was Lucrezia, and she was, besides a hysteric, a speaking, dreaming, largely inexplicable, glittering person, and she was dead at thirty-nine; all that which would rise like a lavender cloud in and around the human mind for centuries was dead and gone at thirty-nine.

THE WORD:
It Is Noble

Rarely, but it does happen, man states his case to God or Jehovah or, for anyone who feels those terms inappropriate in a universe of galaxies and masers, states his case into the hollow of time. Is that satisfactory?

The voice of my beloved! behold, he cometh leaping
upon the mountains, skipping upon the hills.

My beloved is like a roe or a young hart: behold, he
standeth behind our wall, he looketh forth at the
windows. . ..

My beloved spake, and said unto me, Rise up, my love,
my fair one, and come away.

For lo, the winter is past, the rain is over and gone;
The flowers appear on the earth; the time of the singing
of birds is come, and the voice of the turtle is
heard in our land.

Words may be the highest that the mind of man has achieved. In spite of the late flurry against words, the Song of Solomon is still a good song, and is both grossly and subtly communicating. No other animal has the same possibility of perplexity in its reaching out to other animals. Yet man's words are poor things. They are always getting him into trouble. They never fall off his lips as he meant them. They excite the linguist, excite the traditionalist, infuriate both, are used by both as levers for their argument, notwithstanding the words say starkly what they are. That can be weak, can be powerful. The voice of my beloved, the feelings that rose in him he never was able to express, lost them in words. An entire mind may be lost in words. The winter is past, we shed our overcoats and tall boots but the words, those tricky traitors, still are not free. Or too free. What liars! What a delight! Through them the mind looketh forth even more than at the windows of the eyes, showing itself through the lattice. Words reveal us. Conceal us. Teach us. Bewilder us. "Words, words, words," says Hamlet of what he is reading but he is not indifferent to them, as everything else he says proves. They are vulgar. They are wise. They are nothing. Everything. Statesmen screaming across frontiers, egging each other on to human slaughter, administrators in parley with administrators to keep up the prestige of administrators, scholars caviling with scholars to keep exclusive the club of scholars, fishwives shrilling at their husbands, rough voices at the ball park, a drunkard howling into the ear of his bar sister, or into the ear of his satin–bodiced debutante of the year 1920. Words are the bubble and they are the soap. Within our brains they give the widest association for often the slightest stimulus. Give lightning sharpness to a cloudy recollection. Evoke tears. Wipe them away. Damp laughter. Cover an ugly act with a pretty excuse. Rip open flesh, then lay balm on the wound. Make more truthful. Make more false.

Make more tolerable. Touch with reality. With unreality. Occasionally, occasionally, they ennoble. Occasionally, they reach beyond man. Occasionally, man speaks with self-respect to that God.

In the face of a dog we sometimes see the straining to understand. If only we could explain to the dog—in words—that we are only going around the corner to the movie and will be back in two hours. Does He up there glancing down (we are still so gravitation-minded) and seeing into our hearts, as people used to say, note our own straining to understand? We try with everything besides words, with music, painting, sculpture, haberdashery, test tubes, trigonometric formulae, new translations from the Greek or Hebrew. This body with the head on top seeking to find the meaning of itself, is that the cause of the frown moving across the brow of Jove, or is that brow only the brink of our old hill and the frown only the windy leaf-wrinkled October sky without any communication at all, without rage, without pain, without intent? A mind peering at a mind to find what a mind is, standing before its own glass anxiously to fathom its destiny, should one weep? And if not seeking to fathom should one still more pityingly weep?

In a migraine attack, whereas two minutes earlier it was subject, verb, object, all in syntactic order, now a noun hangs in the air and will not come down to join its verb, and a sentence already on the way cannot remember where it began or foresee where it will end. Nevertheless, behind that blurred outside a mind sits (as that colored woman in the clinic behind her blocked speech), knows how things are, knows that the purpose of words now temporarily negated is to help a mind maintain its orderly relations to this earth, to those other heads that stray around it. Words help us foster the illusion that we know what we are about, bolster our effrontery to push forward on this journey, to discover if we can that it is not all illusion.

Off to the side bolstering us are Shakespeare and Aeschylus and Pavlov and Sherrington and the wordy Saint Thomas Aquinas and Sean O'Casey and Eugene O'Neill. We see their facades, mountainous or neat, and something of what lay behind them.

Another time the words are merely continuing the antics, bells and tassels. The clown goes limp with drink or dreams, opens his mouth, the wind blows in, marvelous sounds come back out, words. Dear clown, when they are best he may not recognize them as his own. May think they are making someone else's music. Whose? The wind's? His own words amaze his own ears.

There arrives that morning, or evening, when each of us lets fall his final words, those his friends repeat to show they remember him. Those, too, can have a rightness that makes them the realest fact about that event. The man may speak them quietly, as Keats, or with ironic gaiety, as Voltaire, or with dignity, as Henry James, may be pretending to himself that he feels no fright, though previously he did always avoid looking straight at this day, his last on the funny earth, questioning how it would be, feared but also hoped that he might have a clear mind, so as to be present at his own death. Now he watches, to the degree he is able, squanders nothing. What he speaks is pure self, however trivial, however reeking of his lifelong greed, lifelong sentimentality, lifelong perplexity, or lifelong tender foresight. Some years ago in a picture magazine there were three pages of death masks so trenchant that one woke in the night seeing them again. Yet even those masks, it was a satisfaction to know, could not rival the words that had issued from those heads. Under one, Jonathan Swift's, was written Jonathan Swift's own epitaph. Here it is. Here is what Jonathan Swift wanted said. *Ubi Saeva Indignatio Ulterius Cor Lacerare Nequit.* He wanted that cut into the gravestone. *Where savage indignation will gnaw at my heart no more.* His mask did have an inadequacy. It could not speak.

Index

If you enjoyed this book,

You might like another Prelude non-fiction classic . . .

Rats, Lice and History
Hans Zinsser
The classic account of infectious disease and
human history; 200,000 copies sold

Loving Each Other
Leo Buscaglia
The classic guide on how to build loving
relationships; 4 million copies sold